Diffractional Optics of Millimetre Waves

Series in Optics and Optoelectronics

Series Editors: **R G W Brown**, University of Nottingham, UK
E R Pike, Kings College, London, UK

Other titles in the series

Applications of Silicon–Germanium Heterostructure Devices
C K Maiti and G A Armstrong

Optical Fibre Devices
J-P Goure and I Verrier

Laser-Induced Damage of Optical Materials
R M Wood

Optical Applications of Liquid Crystals
L Vicari (ed)

Stimulated Brillouin Scattering
M Damzen, V I Vlad, V Babin and A Mocofanescu

Handbook of Moiré Measurement
C A Walker (ed)

Handbook of Electroluminescent Materials
D R Vij (ed)

Forthcoming titles in the series

High Speed Photonic Devices
N Dagli (ed)

Diode Lasers
D Sands

Transparent Conductive Coatings
C I Bright

Photonic Crystals
M Charlton and G Parker (eds)

Other titles of interest

Thin-Film Optical Filters (third edition)
H Angus Macleod

Series on Optics and Optoelectronics

Diffractional Optics of Millimetre Waves

O V Minin and I V Minin
Novosibirsk State Technical University
Novosibirsk, Russia

CRC Press
Taylor & Francis Group
Boca Raton London New York

CRC Press is an imprint of the
Taylor & Francis Group, an **informa** business

First published 2004 by IOP Publishing Ltd

Published 2019 by CRC Press
Taylor & Francis Group
6000 Broken Sound Parkway NW, Suite 300
Boca Raton, FL 33487-2742

First issued in paperback 2019

No claim to original U.S. Government works

ISBN 13: 978-0-367-45432-6 (pbk)
ISBN 13: 978-0-7503-0907-3 (hbk)

Visit the Taylor & Francis Web site at
http://www.taylorandfrancis.com

and the CRC Press Web site at
http://www.crcpress.com

British Library Cataloguing-in-Publication Data

A catalogue record for this book is available from the British Library.

Library of Congress Cataloging-in-Publication Data are available

Cover Design: Victoria Le Billon

Typeset by Academic + Technical, Bristol

Series on Optics and Optoelectronics

Diffractional Optics of Millimetre Waves

O V Minin and I V Minin
Novosibirsk State Technical University
Novosibirsk, Russia

CRC Press
Taylor & Francis Group
Boca Raton London New York

CRC Press is an imprint of the
Taylor & Francis Group, an **informa** business

First published 2004 by IOP Publishing Ltd

Published 2019 by CRC Press
Taylor & Francis Group
6000 Broken Sound Parkway NW, Suite 300
Boca Raton, FL 33487-2742

First issued in paperback 2019

No claim to original U.S. Government works

ISBN 13: 978-0-367-45432-6 (pbk)
ISBN 13: 978-0-7503-0907-3 (hbk)

Visit the Taylor & Francis Web site at
http://www.taylorandfrancis.com

and the CRC Press Web site at
http://www.crcpress.com

British Library Cataloguing-in-Publication Data

A catalogue record for this book is available from the British Library.

Library of Congress Cataloging-in-Publication Data are available

Cover Design: Victoria Le Billon

Typeset by Academic + Technical, Bristol

Dedicated to our father, friend and colleague, all in one, and equally to our mother. Without their help and support, this book would never have been written.

Contents

Foreword   xiii

Acknowledgments   xv

Introduction   xvi

1 Theory of diffraction: brief exposition   1
    1.1   Diffraction   1
    1.2   The Huygens–Fresnel principle 2
    1.3   Methods of computation of the Fresnel–Kirchhoff
         diffraction integral   4
    1.4   Diffraction of electromagnetic waves 10
    1.5   Asymptotic behaviour of the diffraction integral 11
    1.6   Computation of the diffraction integral of the Fraunhofer
         zone   14
    1.7   Geometric modelling of diffraction elements 20
    1.8   On optimization of parameters of diffraction elements 22
    1.9   A complex of CAD programs for designing diffraction
         antennas   23
    Bibliography   25

**2 Lenses based on high-aperture Fresnel zone plate: information
properties**   29
    2.1   Focusing and frequency properties of the Fresnel and Soret
         zone plates   29
    2.2   Brief classification of zone plates 32
    2.3   Fundamental properties of zone plates in the millimetre
         band   37
    2.4   Computation of the phase-inversion profile 39
    2.5   Experimental setup   44

2.6   Field intensity distribution in the focal zone of the zone
      plate   47
2.7   Transverse field intensity distribution in the focal area
      of the zone plate   50
2.8   Information properties of zone plates   50
2.9   Single pointlike source   51
2.10 A system of pointlike coherent sources of radiation   58
2.11 The effective spectrum range of a zone plate   59
2.12 The Q factor of diffractive objective lenses   63
2.13 Two-component diffractive lens   64
2.14 Efficiency of diffractive objective lenses   68
2.15 Phase of the wave: behaviour in the focal region   75
2.16 Off-axis zone plate   76
Bibliography   77

3 **Principles of construction of elements of diffractive optics**   **82**
3.1   Methods of synthesis of diffractive elements   82
3.2   Selection of harmonics of coherent radiation   92
3.3   Methods of controlling frequency properties of flat
      diffractive elements   97
3.4   Elements of diffractive optics fabricated on surfaces of
      revolution of second order   100
3.5   Single-component 'parabolic' radio lens   105
3.6   Two-component 'parabolic' diffractive objective lens   107
3.7   Invariant properties of diffractive optical elements   108
3.8   Conical diffractive element   110
3.9   Correction of aberrations of a given order by choosing a
      surface profile for the diffractive element   118
3.10 Polarization diffractive elements   121
3.11 Diffractive elements without axial symmetry   123
3.12 Square zone plate   128
3.13 Field intensity distribution function along the optical axis 132
3.14 Field intensity distribution in the focal area of zone plates
      with low number of zones   135
3.15 Zone plates with dynamically variable focal area   137
3.16 Diffractive elements in off-design modes   138
3.17 The method of constructing diffractive elements for focusing
      radiation on to an arbitrary focal curve   140
3.18 Principle of design of diffractive elements for spatial focusing
      of radiation   142
3.19 Focusing on to a linear segment   145
3.20 Focusing on to a circular segment (tube)   146
3.21 Focusing on to a cone   147
3.22 Focusing to a disk   148

3.23 Frequency and formatting properties of MWDO elements   148
3.24 Numerical modelling of radiation focusing to a conical area   150
3.25 Zone screening by diffractive elements on curvilinear
      surfaces   151
3.26 Interrelation between zone plate parameters on curvilinear
      surfaces   153
Bibliography   154

4 Alternative methods of synthesizing diffractive elements   159
4.1   Synthesis of diffractive elements that create a flat focal line
      with prescribed intensity distribution along it   159
4.2   Design of diffractive elements using the diffraction method
      of computations   162
4.3   Physical restrictions on the possibilities of focusators and
      their relation to other diffractive elements   163
4.4   Diffractive elements for analysing transversal radiation
      modes   166
4.5   Logarithmic axicons   166
      4.5.1 Axicons of the first type   167
      4.5.2 Axicons of the second type   168
      4.5.3 Longitudinal distribution   170
4.6   Two-orders diffractive elements   172
4.7   Diffractive elements for focusing radiation on to flat zones   176
4.8   Optimization algorithm for diffraction antennas   179
4.9   Optimization of parameters of the 'classical' zone plate   181
      4.9.1 'Compression' of zone plate   181
      4.9.2 Effect of distance to the screen   184
4.10 Optimal zone plate in the Fresnel approximation   185
4.11 The general principle of designing optimal diffraction
      lenses and antennas   186
      4.11.1 Zone-plate-based conformal antenna   188
4.12 Application of holographic principles of diffractive antenna
      design   192
4.13 Microwave antenna with flat diffractive reflector   193
4.14 The possibility of dispersion distortion correction of
      femtosecond pulses by choosing surface shape of diffractive
      optic element   195
Bibliography   200

5 Microwave-range diffractive antennas   203
5.1   Flat antennas   207
5.2   Reflector antenna on flat surface   208
5.3   Lens-type diffractive antennas   216
5.4   Diffractive radome antenna on a parabolic surface   218

5.5   Diffractive reflector antennas on 'non-flat' surfaces   223
5.6   Diffractive antenna systems in the millimetre wavelength
       range, based on the effect of conversion of surface waves
       to bulk waves   230
       5.6.1   Specifics in calculating the beampattern of a
                scanning antenna in the horizontal plane   234
       5.6.2   Electrodynamic characteristics of an experimental
                scanning antenna   238
5.7   Effect of the form of diffractive element's surface on the
       structure of Fresnel zones   240
5.8   Aerodynamic aspects of radome antennas of supersonic
       aircraft   244
5.9   Radome antenna fabricated on a surface optimized for
       scanning   245
5.10   Evaluation of the effect of aerodynamic needle on focusing
       properties of DRA   248
5.11   Creation of DRA on an optimal aerodynamic surface   251
5.12   Application of diffractive radome antennas in automobiles
       and satellite antennas   252
5.13   Integration of diffractive element and aperture diaphragm   257
5.14   Selected problems facing the creation of automotive radar   262
5.15   Omnidirectional zone-plate-based antennas   263
5.16   Omnidirectional antennas on cylindrical surface   268
5.17   Omnidirectional antennas on arbitrary surface of revolution   270
5.18   Omnidirectional antenna with flat facets   272
5.19   Dielectric reflector zone plate   274
Bibliography   275
Selected bibliography   279

6   **Applications of diffractive optical elements**   **282**
    6.1   X-ray diffractive optics   282
    6.2   Diffractive elements for visible and infrared bands   288
    6.3   Diffractive elements in optical computers   294
    6.4   Elements of diffractive optics in systems of millimetre band
           radiovision   302
           6.4.1   Generation of radio images of objects in the
                    millimetre band   303
    6.5   Formation of images using partially coherent radiation   307
           6.5.1   Method of isotropic construction of radio images of
                    three-dimensional objects   308
    6.6   Diffractive objectives in systems of holographic radiovision   317
           6.6.1   Diffractive optics in microwave object radars   319
    6.7   Microwave-band diffractive antennas   320
    6.8   Antennas for satellite communications   321

6.9   Security fence protection   328
6.10   Application of millimetre-wavelength-band diffractive
       optics in scientific research   330
6.11   Semiconductor zone plate   334
6.12   Diffractive optics in acoustics   336
6.13   Zone plates for focusing shockwaves   337
       6.13.1 Formulation of the problem   337
       6.13.2 Synthesis of diffractive element   338
       6.13.3 Results of the numerical experiment   339
6.14   Suppression of shockwaves using diffraction gratings   350
6.15   Zone plates for use in deep space   352
6.16   Segmented apertures   355
Bibliography   357

7   **Diffractive elements using man-made dielectrics**   **369**
7.1   Optical constants of materials in sub-millimetre and
       millimetre spectral bands   369
7.2   Man-made dielectrics   373
7.3   Perforated artificial dielectrics   376
7.4   Metal–dielectrics   382
Bibliography   386
Additional bibliography   388

Appendix   **390**

Index   **395**

Foreword

This book is an important contribution to the field of diffraction optics, including zoned lenses and reflector antennas. Drs Igor and Oleg Minin have brought together much information, both analytical and experimental, that has largely been unavailable to the non-Russian speaking portion of the world. Their two earlier books and numerous journal publications (in Russian) have not been readily available to researchers in the western world, although the authors have personally made significant contributions in this area. In addition, they have included the results of many other Russian investigators. The Drs Minin have probably made more millimetre-wave measurements on Fresnel zone plate antennas than anyone else, and made rigorous comparisons with the theory. They treat the curved versions of the zoned lens as well as the more frequently used flat zone plate, and include such cases as multifocal and multifrequency zoned lenses. The book is an excellent summary of the state of the art and contains much valuable information. The text is very comprehensive and thorough, and undoubtedly contains more information than any other current source.

Most of their work is broadly applicable to the large angle zone plate, with focal lengths comparable to the plate diameter, as well as the small-angle examples developed in the past for optical use. It is interesting to note that although the Fresnel zone plate concept has been used at optical wavelengths for over one hundred years, only the small-angle configuration (that is, with large focal length compared to diameter) has been employed, and the analytical expressions that have been derived previously are often not accurate for millimetre wavelengths.

The technology of millimetre-wave diffractive optics has seen great activity in recent years. The first millimetre-wave phase-correcting Fresnel zone plate lens antennas were developed in 1960, but there was little activity until about 1985. Since then nearly 100 publications about the subject have appeared in the literature. The field of millimetre-wave optics has been used in many system applications, including radar, radiometry, point-to-point telecommunications, missile terminal guidance seekers, and field instrumentation tests of atmospheric effects, generally where ease of manufacturing, low attenuation, low weight, low volume, and low cost are considerations.

The first chapter gives an introduction to the theory of diffraction, starting from the Huygens–Fresnel principle of radiation of electromagnetic waves, with an application to zone plates. The appendix includes some of the complicated equations that support chapter 1. Chapter 2 deals with the properties of high-aperture zone plate lenses, including numerous results measured at millimetre wavelengths. Although millimetre waves are emphasized in this volume, the methods and analysis and even some measurements are valid at normal microwave frequencies. Chapter 3 deals with the principles of construction of diffractive optics for either planar or curved or conical surfaces, and the advantages of each are discussed. As before, the material includes general analysis, detailed calculations, and supporting measured results. Chapter 4 includes alternative methods of synthesizing diffractive elements for specialized focusing applications, such as a line focus instead of a point focus. Chapter 5 extends the options to scanning antennas or multiple-beam antennas, and considers applications such as the use in automobile anti-collision radars or in missile radomes or omnidirectional telecommunication systems, including cases for microwave frequency ranges. Chapter 6 deals with unusual applications of diffractional elements and chapter 7 covers the use of synthetic dielectrics.

The book is unusually good and should see wide application.

James C Wiltse
Georgia Tech Research Institute
Georgia Institute of Technology
Atlanta, Georgia, USA

Acknowledgments

We are grateful to Dr J C Wiltse (Georgia Institute of Technology, Atlanta, USA), Dr M H A J Herben, J M Van Houten and L C J Baggen (Eindhoven University of Technology, The Netherlands), Professor S K Barton (University of Manchester, UK), Professor W X Zhang (Southeast University, China), Professor T Onodera and Dr Takafumi Hoashi (Kumamoto Institute of Technology, Japan), Dr D N Black (Electromagnetic Science, Inc, USA), Dr Hristo D Hristov (Technical University of Varna, Bulgaria), Dr Y Jay Guo (University of Bradford, UK), Dr Aldo Petosa (Advanced Antena Technology Lab, USA), Dr G Webb (Innova Lab, USA), Dr Junji Yamauchi (Hosei University, Japan), Dr Juan Vassallo (Instituto de Fusica Aplicada, Spain), Dr Mark Gouker (Analog Device Technology MIT Lincoln Lab, USA), Professor D Sazonov (Russia), Professor Soon Yim Tan and Dr Chun Fei Ye (Nanyang Technological University, Singapore) for kindly sending us copies of their publications on mm- and microwave diffractive elements based on zone plates.

We are greatly indebted to Eric Pepper, Director of Publications International Society for Optical Engineering (SPIE), Paulette Goldweber, IEEE Intellectual Property Rights, and Jack Browne, Publisher/Editor Microwaves & RF for sending us copies of their publications and permission to use figures.

We are infinitely grateful to Dr J C Wiltse of Georgia Institute of Technology, Atlanta, USA, for his willingness to write the Foreword to this book and also to Tom Spicer, Senior Commissiong Editor, Institute of Physics Publishing, UK, for constant attention and help in our work with the book.

Our many thanks go to the translator Vitaly Kisin (Bristol, UK) without whose help the book could not have reached the publication stage.

Introduction

Quasioptics is a field on the border of two branches of physics; its topics overlap geometrical optics and diffraction [1]. More specifically, quasioptics studies phenomena based on geometro-optical propagation, refraction and reflection of very short electromagnetic waves in situations in which the wave nature of radiation is of paramount importance.

Elements of diffractive optics in the millimetre wavelength band (MWDO) use diffraction on the periodical or quasiperiodical structure of these elements, with characteristic length that is on the order of the wavelength of the incident radiation; they are not based on reflection and refraction of classical optics.

MWDO elements can be used for focusing (including focusing into an arbitrary area), filtration, polarization, formation, splitting and mixing of radiation beams. They can function in the wavelength bands from x-rays to microwave and also in acoustics, for instance for focusing of shock waves [2–10].

It seems that the first diffractive focusing elements were the Fresnel zone plate, whose ability to create an image was recognized by Charles Soret in 1875 [11], and the phase zone plate with rectangular profile [12] prepared by Robert Wood in 1898 [13]. In 1957 G G Slyusarev suggested modified zone plates [14] in which the length of the optical path connecting an object to its image becomes constant within each zone and changes jumpwise by 2π on the boundary between two zones. Research in artificial computer-synthesized holograms led to the creation of the kinoform [15, 16]. Both in focusators [17] and in kinoform optical elements a continuous relief of classical optics is replaced with a discontinuous one: the phase function is taken modulo 2π. Diffraction is the basis for the function of all the elements mentioned above. The optical thickness of these elements is within one wavelength. In this sense zone plates, kinoforms, focusators and computer optics elements [7] can be regarded as diffractive elements, that is, as a class of quasioptical focusing systems: indeed, according to the definition of quasioptics given in [18], they are typically computed following the laws of geometrical optics while the principles on which they function are those of diffraction.

Three main types of MWDO elements are recognized (the criterion is the principle of arrangement with respect to the direction of propagation of the electromagnetic wave) [19]: transversal (fabricated mostly on a flat surface), mixed longitudinal–transversal (fabricated on an arbitrary curvilinear surface) and longitudinal (built as a system of screens arranged along the direction of propagation of the electromagnetic wave).

Focusing diffractive elements may reflect radiation or they transmit it.

One specific feature of diffractive elements is that in comparison with classical optical elements they execute an essentially broader class of geo-metrical and wave transformations of fields. This property greatly expands the fields of practical applications, covering wavelength bands from x-rays to centimetre waves and including nonlinear acoustics [20].

The advantages of microwave and millimetre diffractive lenses and antennas are: small thickness on the order of radiation wavelength, low weight, high diffraction efficiency, high spatial resolution dictated by the diffraction limit, simple fabrication technology, and the feasibility of creating them on arbitrary surfaces.

By synthesizing those elements of diffractive optics that focus radiation into a predefined three-dimensional configuration and possess specific frequency-selective characteristics, it is possible to create new elements for integrated optics and optical computers [21], for systems of optical processing of information and also for spectroscopy, new devices for nondestructive testing and for radiovision of essentially three-dimensional arrangements, for various antenna systems etc.

MWDO elements can be used in various systems providing road traffic security, for example, in traffic flow control, in automobile collision preven-tion systems, in automotive radars, satellite television receivers etc. In such systems MWDO elements can be incorporated into car components, for instance, as a part of the bonnet, and not conflict with the overall design of the automobile [22].

Furthermore, diffractive focusing elements can be useful as multiple-beam or scanning antennas in diverse systems for navigation, control and communications, for detecting weapons and plastic explosives hidden on the body under clothes, in broadband wireless access systems, in perimeter fence security devices, in nondestructive detection of imperfections etc.

This book is devoted to describing the principles of construction of MWDO elements fabricated on an arbitrary surface. We discuss the focus-ing, frequency and information processing characteristics of such elements. Special attention is paid to high-aperture MWDO elements designed for the millimetre and sub-millimetre wavelength bands.

The microwave band occupies an intermediate position between the radio frequency band $f < 10\,\text{GHz}$ and optics $f > 100\,000\,\text{GHz}$. One should also keep in view the decisive advantages of milimetre waveband in comparison with longer and shorter wavelengths. Being the shorter part of the radiofrequency band, microwaves provide high spatial resolution, extremely wide transmission bands, high-rate all-weather information transmission in free space, minimal overall dimensions of antenna systems, maximum directivity, noise immunity and ease of concealing the communications channel.

Diffractive focusing elements for use in the microwave band were first mentioned in 1936 [23]. Research into zone-plate-based diffractive focusing elements in the millimiter and sub-millimetre wavelength bands began in the 1960s [24]. The diffractive microwave focusing elements fabricated over spherical and parabolic surfaces were first proposed in the 1970s [25]. Intense research into the application of diffractive focusing elements as antennas for satellite TV reception began in the mid-1980s. Antennas both with axial and with off-axial position of the focal point were considered. A two-component millimetre band diffractive lens generating radio images of essentially three-dimensional objects was first suggested at the same time [26].

Further development of microwave and millimetre diffractive antennas and lenses followed in the following directions [10]:

- 'three-dimensional' elements—antennas and lenses on curvilinear surfaces;
- creation of 'flat' reflector zone plates with phase inversion zones fabricated of an artificial metal–dielectric or frequency selective surfaces;
- development of flat zone plates—lens-type antennas with phase inversion structure involving piecewise-constant distribution of refractive index of the material along the zone radius;
- development of optically controlled millimetre-band diffractive elements;
- creation of multicomponent millimetre-band diffractive objective lenses, including objectives on non-planar surfaces.

Special interest in the physics of microwave radiation arose recently in connection with the need to solve a number of new problems that promised future applications: creation of global satellite communications network and multichannel television systems; systems of all-weather high- and superhigh-resolution radiovision to be used in, among other fields, automotive radars for safe driving in low visibility conditions (fog, rain etc.); ecological monitoring of environment; energy beaming to flying apparatuses via a microwave channel; and development of a new generation of computers.

The authors believe that the book will also be of interest to designers of optical systems because, with scaling effects taken into account, the characteristics of diffractive quasioptical elements are valid for diffractive focusing elements of integrated optics. Furthermore, fabrication of diffractive optical elements on three-dimensional surfaces is still a problem awaiting a technological solution. In this sense the progress in elements of diffractive quasioptics is 'ahead' of their counterparts in the optical range and points to future perspectives in optics.

Bibliography

[1] Valitov V A (ed) 1969 *Sub-millimeter Waveband: Devices and Techniques* (Moscow: Soviet Radio)

[2] Koronkevich V P and Pal'chikova I G 1992 'Modern zone plates' *Avtometriya* 1 85–100

[3] Bobrov S T, Greisuh G I and Turkevich Yu G 1986 *Optics of Diffractive Elements and Systems* (Leningrad: Mashinostroenie) p 223 (in Russian)

[4] Wiltse J C 1998 'Recent developments in Fresnel zone plate antennas at microwave/ millimeter wave' *Proc. SPIE* **3464** 146–154

[5] McGee J F, Hesser D R and Milton J W 1969 *X-Ray Reflection Optics* (*Recent Developments*) (Springer, New York).

[6] Hristov H D 2000 *Fresnel Zones in Wireless Links, Zone Plate Lenses and Antennas* (Boston, London: Artech House)

[7] Goncharskii P V, Popov V V and Stepanov V V 1991 *Introduction to Computer Optics* (Moscow: MGU) p 312 (in Russian)

[8] Minin I V and Minin O V 1992 *Diffractive Quasioptics* (Moscow: Research and Production Association 'InformTEI') (in Russian)

[9] Minin I V and Minin O V 1999 *Diffractive Quasioptics and its Applications* (Novosibirsk: SibAGS) (in Russian)

[10] Minin I V and Minin O V 2001 'Diffractive quasioptics in millimeter waveband: chronology of progress' *Mezhvuzovskii Sbornik. Estestvoznanie. Ekonomika Upravlenie. Samara IPO SGUU* 2 16–23

[11] Soret J L 1875 'Concerning diffraction by circular gratings' *Ann. Phys. Chem.* **24** 429–451

[12] Lord Rayleigh 1888 'Wave Theory' in *Encyclopaedia Britannica* 9th edn vol 24 pp 429–451

[13] Wood R W 1898 ' Phase reversed zone plates and diffraction telescopes' *Phil. Mag.* Series 5 **45** 511–522

[14] Slyusarev G G 1957 'Optical systems with phase layers' *DAN SSSR* **113**(4) 780–783

[15] Jordan J A, Hirsch P M, Lesem L B and Rooy D L 1970 'Kinoform-lens' *Appl. Opt.* **9** 8

[16] Lesem L B, Hirsch P L and Jordan J A 1969 'The kinoform: a new form of reconstruction device' *IBM J. Res. Develop* **13**(3) 150

[17] Golub M A, Karpeev S V, Prohorov A M *et al* 1981 'Focusing of radiation to a predefined space using computer-synthesized holograms' *Pis'ma v ZhTF* **7**(10)

[18] Katsenelenbaum B Z 1969 'Quasioptical methods of formation and transfer of millimeter waves' *UFN* **13**(3)

[19] Minin I V and Minin O V 1989 'Diffractive optics: achievements and prospects' *Abstracts of Reports to the USSR Seminar on Methodology and Techniques of Processing of Two-Dimensional Signals* (Moscow: TsNIINTIKPK) Part 2 pp 25–26

[20] Minin I V and Minin O V 1994 'Advantages of diffraction quasi-optics elements and their applications' *Proc. SPIE* **2211** 445–452

[21] Minin I V and Minin O V 1999 'Diffraction elements of the mm-waves planar integral optics' USNC/URSI National Radio Science Meeting, Orlando, Florida, USA, 11–16 July

[22] Minin IV and Minin O V 1999 'Low cost multibeam antennas in the mm-wave regime for radar systems of transport means' USNC/URSI National Radio Science Meeting, Orlando, Florida, USA, 11–16 July, B1-1-002

[23] Clavien A G and Darbord R H 1936 'Directional radio transmission system' US Patent 2,043,347

[24] King M, Rodgers J, Sobel F, Wentworth F and Wiltse J C 1960 Quasi-optical components and surface waveguides for the 100 to 300 GHz frequency range' Electronic Communications Inc. Report No. 2 on contract AF 19(604)-5475, November. Reprint in: 1961 *IRE Trans. Microwave Theory and Techniques* **MTT-9** 512–518

[25] Khastgir P, Chakravorty I N and Dey K K 1973 'Microwave paraboloidal, spherical and plane zone plate antennas: comparative study' *Indian J. Radio Space Phys.* **2** 47–50

[26] Minin I V, Minin O V, Skarbo B A *et al* 1985 'Application of holographic radio objectives for flaw detection and plasma diagnostics in microwave band' in *Abstracts 5th USSR Conf. on Holography, Riga*, 12–14 November, pp 233–234

Chapter 1

Theory of diffraction: brief exposition

1.1 Diffraction

The plane wave is a wave whose wavefront is unlimited in space and whose source lies at infinite distance. If we create a wave process bounded to a certain space, e.g. by sending a plane wave through a stop (aperture) in a screen opaque to radiation, the result is a bounded wave bundle. Sometimes this wave bundle can be treated as a beam, or ray, whose behaviour is described by the laws of geometrical optics. However, actual wave bundles propagate differently from rays. For instance, imagine moving a large-size metal sheet at right angles to the straight line connecting the antennas of a transmitter and a receiver; it will be noticed that the intensity of oscillations at the receiver will vary, growing higher or lower even before the edge of the sheet crosses this line, and will be quite appreciable even after the edge moves fairly far across it [1, 2]. The cause of this behaviour is the phenomenon of diffraction which Sommerfeld defined as any deviation of light rays from the straight line that cannot be explained in terms of reflection or refraction.

The theory of diffraction deals with wave processes in the cases when wave propagation is disturbed by obstacles, for instance, screens or holes in opaque or semi-opaque screens, or by inhomogeneities of the medium, that is, of a medium whose properties vary from point to point. The task of diffraction theory is to determine the wave perturbation that propagates away from the obstacle.

Sections that follow deal with the fundamentals of the theory of diffraction required to model the focusing properties of diffractive optics elements. As known direct calculation of the diffraction integral involves considerable difficulties, this chapter briefly discusses both efficient algorithms for computing it and a number of practically important particular cases for transforming the diffraction integral.

1

1.2   The Huygens–Fresnel principle

The wave nature of electromagnetic oscillations produced by radio waves or light becomes more appreciable as the wavelength increases and the sizes of the bodies we consider (obstacles) become comparable with wavelength (size on the order of several tens of wavelengths).

The theory of diffraction is based on the principle formulated by Huygens in 1690. It states that each point of a wave surface can be treated as an independent centre emitting elementary spherical waves. By constructing an envelope of all these waves one can find the resulting shape of the wave surface under consideration. This principle was not yet sufficient in itself to make possible a numerical simulation of diffractional phenomena. The solution of diffractional problems of various types became possible only on the basis of the joint Huygens–Fresnel principle, combining the Huygens principle with Fresnel's idea of wave interference.

The Huygens–Fresnel principle is essentially an assumption that the amplitude of oscillations at any point can be determined by summing up the fields emitted by all points of the wavefront. It is also assumed that each point is an elementary emission source whose phase of oscillation differs from any other by a constant value (in other words, the sources are coherent).

The plane wave is defined as a wave with flat surface of equal phase and with amplitude independent of distance; hence, the wave propagates without damping [3]. Mathematically it is expressed by

$$U = U_0 \exp(t - R/c) = U_0 \exp(\mathrm{i}\omega t) \exp(-\mathrm{i}kR)$$

where U_0 is the initial amplitude, ω is the oscillation frequency and $k = \omega/c = 2\pi/\lambda$. Here R is the distance from the source to the point under consideration and c is the propagation velocity of the electromagnetic wave.

The field of a spherical wave can be written as

$$U = U_0 \frac{\exp[\mathrm{i}\omega(t - R/c)]}{R} = U_0 \exp(\mathrm{i}\omega t) \frac{\exp(-\mathrm{i}kR)}{R}.$$

A spherical wave's characteristic is that its amplitude decreases in an inverse proportion to distance, $1/R$, and the equiphase surface is spherical.

The mathematical expression for the Huygens–Fresnel principle is

$$\mathrm{d}U = U_0 K_n \frac{\exp(-\mathrm{i}kR)}{R} \, \mathrm{d}S$$

where K_n is a proportionality coefficient.

We see from this formula that the field emitted by an elementary segment of the surface is proportional to its area $\mathrm{d}S$ and the field amplitude U_0 on this segment. The total field created by all sources equals the integral of $\mathrm{d}U$ taken over the entire surface S of the given surface (figure 1.1).

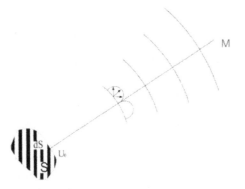

Figure 1.1. The diffractional field of a body in terms of the Huygens–Fresnel principle.

Let us consider how to explain the rectilinear propagation of a plane electromagnetic wave in terms of the Huygens–Fresnel principle, if each point of the wave surface emits spherical waves.

Let us divide the wave surface into smaller segments by applying the following general rule: all points of each segment emit the field of the same phase, while the phase of the field emitted by the neighbouring segment is the opposite of the former field. In order to satisfy these requirements, we choose a certain point P in which we determine the field. Let us draw concentric spheres around the selected point P. The radius of each sphere is greater by one half of the wavelength λ than that of the smaller sphere, that is

$$r_1 = r_0 + \lambda/2, \ldots, r_n = r_0 + n\lambda/2, \qquad n = 1, 2, 3, \ldots .$$

These spheres intersect the plane M along curves that form a set of circles with radii $a_0, a_1, \ldots, a_n$ (figure 1.2(a,b)).

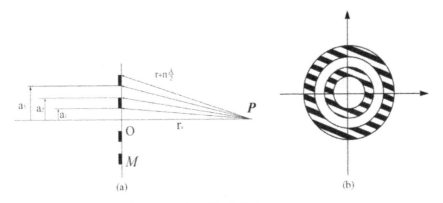

(a)   (b)

Figure 1.2. (a, b) Fresnel zones numbered by 1, 2, 3, . . .

The entire surface is thus divided into annular zones (the Fresnel zones). Each zone generates oscillations of the same phase because the distance to P is constant, while each subsequent zone generates oscillations at P with phase that is opposite to that of the preceding zone. This follows from the geometry of our construction: the distances from zone boundaries to the point P differ by $\lambda/2$ and hence, the phases differ by $180°$.

The radius of the nth zone is found from the triangle Pa_nO as $a_n = (nr_0\lambda + (n\lambda/2)^2)^{1/2}$ where $n = 1, 2, 3, \ldots$ enumerate the zones. Obviously, the excitation at P is due to the sum of actions of each zone.

The amplitude of oscillations at P is proportional to the size of the element dS and also depends on the angle φ between the normal n to the zone and its radius r_n. This factor is taken into account by the coefficient K_n.

Using here the Huygens–Fresnel principle, we obtain the following expression for the amplitude of oscillations at P:

$$dU_{np} = K_n \frac{\exp[i\omega(t - R/c)]}{R} dS_n.$$

The action of the entire zone is found as the integral over the area of this zone:

$$U_{np} = K_n \int_S \frac{\exp[i\omega(t - R/c)]}{R} dS_n.$$

As $r_n^2 = a_n^2 + r_0^2$, then an area element in cylindrical coordinates is $dS_n = a_n \, d\alpha \, da_n = r_n \, d\alpha \, dr_n$ and we finally obtain

$$U_{np} = K_n \exp(i\omega t)\lambda i(-1)^n \exp(-ikr_0), \qquad k = 2\pi/\lambda.$$

In order to take into account the effect of all zones at the point P, it is necessary to sum up their actions and obtain as a first approximation that the field amplitude at P equals half of the field amplitude at the central Fresnel zone [3].

As a result we have

$$U_p = K_1\lambda \exp(i\omega t) \exp(-ikr - i\pi/2).$$

This expression coincides with that for the plane wave if we assume that $K_1\lambda = U_0$. Therefore it proved possible, using the Huygens–Fresnel principle, to demonstrate that the plane wave maintains its plane phase wavefront as it propagates.

1.3   Methods of computation of the Fresnel–Kirchhoff diffraction integral

Diffraction of a scalar wave by a diffractional element can be described mathematically using the Fresnel–Kirchhoff integral that in the general case has the form [4]

$$U(P_0) = \frac{A}{i\lambda} \iint_S \frac{\exp[-ik(r_{21} + r_{01})]}{r_{21}r_{01}} \chi(r)\,dS \qquad (1.1)$$

where P_0 is the observation point and r_{01} and r_{21} are the distances from a point source to a certain point within the aperture and to the point source of emission, respectively; $\chi(r)$ is the inclination factor which is given by the Kirchhoff theory as

$$\chi(r) = \tfrac{1}{2}[\cos(\mathbf{n}, \mathbf{r}_{01}) - \cos(\mathbf{n}, \mathbf{r}_{21})]$$

where $\mathbf{n}$ is the normal to the screen surface. As the divergent spherical wave transforms itself to the plane wave, the factor $\chi(r)$ tends to K_n.

Although the mathematical expression (1.1) is fairly simple, practical computations using this formula encounter considerable difficulties because the integrand in the double integral contains a rapidly oscillating phase factor $\exp[-ik(r_{21} + r_{01})]$. As a result, direct computations of (1.1) using conventional techniques require unrealistically large computation times; these become completely unacceptable in the case of diffraction elements, the simplest of which is the zone plate. It proved necessary therefore to develop specialized high-efficiency algorithms for computing integral (1.1) both in its general form and in a number of specific cases.

Gravelsaeter and Stammes [5] suggested an algorithm for computing the Fresnel–Kirchhoff-type diffractional integral in the case of the diffraction of a scalar wave on a zone plate. As far as the efficiency of computing (1.1) is concerned, the advantage of the algorithm suggested in [5] in comparison with already familiar algorithms is that we do not need to compute the original double integral (1.1) but can replace it with computing two single integrals. But it has a shortcoming too: this single integral diverges as we approach the shadow boundary of the *i*th ring. Despite the presence of singular points, this method of calculating a double integral of a rapidly oscillating function of the type $\exp[if(\Phi)]$ makes it possible to reduce the overall computation time. This is only possible, however, in the near-axis case, that is, when the emission source and/or its image lie close to the optical axis: $\Delta x/D \ll 1$, where D is the zone plate diameter and Δx is the displacement of the source (image) at right angles to the optical axis. If the emission source is located essentially off-axis, so that the condition $\Delta x/D \ll 1$ does not hold, the size of the integration subregion in which the integrand varies slowly becomes comparable with $\lambda/2$. Therefore, in the case of the off-axis location of the emission source the number of divisions of a single *i*th Fresnel zone into subsegments grows substantially and we cannot expect any significant reduction of the computation time.

A more universal and efficient method of computing integral (1.1) was considered in [6, 7] in the case of problems of diffraction of scalar waves on diffractional elements. Let us consider, in correspondence with the

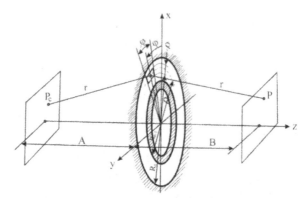

Figure 1.3. Diagram for computation for a diffraction element [6].

symmetry of the problem, a cylindrical coordinate system (ρ, φ, z); we place its origin over the centre of the diffraction segment while the axis Oz points along the normal to it (figure 1.3).

Assume now that l pointlike emitters are placed in the region of space $Z < 0$ at points with coordinates P_i^j. In the region $Z > 0$ the field $U(P)$ diffracted by the diffraction element is determined at the observation point $P(\rho, \varphi, z)$. To simplify the presentation, we limit the case to a single emission source $(l = 1)$ and a single Fresnel zone n between the radii R_{n-1} and R_n; we also use the rectangle method, which is acceptable from the point of view of computation accuracy of (1.1), and can now rewrite the initial integral (1.1) in the form

$$\sum_{m=1}^{M} \Delta S_m f_m \exp(\mathrm{i}2\pi\psi_m) \tag{1.2}$$

where f_m, ψ_m are functions of (ρ_m, φ_m), that is, functions of polar coordinates of the central point of the sub-area $\Delta S_m = (\rho\,\Delta\rho\,\Delta\varphi)_m$. This shows, among other things, that for the integration step $\delta \approx \lambda/4$ the number of terms in the sum (1.2) is $M \approx 2\pi(D/\lambda)^2$ and a direct computation of (1.2) requires a considerable amount of computer time. To reduce the computation time (the number M) we resort to an obvious identity

$$\exp(\mathrm{i}2\pi\psi) = [\mathrm{i}2\pi\,\mathrm{frac}(\psi)], \qquad \psi > 0$$

where $\mathrm{frac}(\psi)$ is the fractional part of the real function ψ. Hence, this function can always be reduced to the interval $[0, 1)$. On this interval, we can introduce a discrete grid whose nodes are enumerated by an integral index

$$t = \mathrm{entier}(\psi T + 0.5), \qquad 0 < \psi \le 1, \quad t = 1, 2, \dots, T$$

where $\mathrm{entier}(x)$ is the largest integer not exceeding x. Let ψ assume on this interval one of $t \in [1, T]$ possible discrete values of ψ_t that differ among themselves by $\Delta\psi = 1/T$. Then the phase $2\pi\psi$ and the exponential term in (1.2) assume not more than T different values. If we prescribe integration steps $\Delta\rho = \sigma$ and $\Delta\varphi = \sigma/\rho$ then the number of terms in the sum (1.2) at $T \approx D/\lambda$ is $M \approx 2\pi T^2$. Therefore, for each t we can form groups of terms of the type

$$\Delta S_{kt} f_{kt} \exp(\mathrm{i}2\pi\psi_t), \qquad k = 0, 1, \ldots, k_t$$

where k_t is found by counting the number of times that ψ falls within the interval t as (ρ_m, φ_m) runs through the entire integration range. Then we find

$$g(\psi_t) = \sum_{k=1}^{k_t} \Delta S_{kt} f_{kt}, \qquad t = 1, 2, \ldots, T \tag{1.3}$$

and finally the sum

$$\sum_{t=1}^{T} g(\psi_t) \exp(\mathrm{i}2\pi\psi_t^*). \tag{1.4}$$

The result of this calculation is the same as in (1.2) but the number of computations, $\exp(\mathrm{i}2\pi\psi_t)$, decreases in comparison with the direct computation (1.2) by a factor of $M/T = 2\pi T$.

The computation of the sum (1.2) is thus reduced to building a histogram $g(\psi_t)$ by (1.3) and then calculating the sum (1.4).

Further improvement of the efficiency of the algorithm is achieved using the fact that the terms $\exp(\mathrm{i}2\pi\psi_t)$ on a uniform grid ψ_t can be found by consecutive multiplication of complex numbers. By denoting $E_t = \exp(\mathrm{i}2\pi\psi_t)$ we obtain, instead of (1.4),

$$\sum_{t=1}^{T} g(\psi_t) E_t, \qquad E_t = \Delta E\, E_{t-1}, \qquad E_0 = 1 + \mathrm{i}0, \qquad \Delta E = \exp(\mathrm{i}2\pi/T). \tag{1.5}$$

Therefore, the determination of $U(p)$ is independent of the number l; also, it is sufficient to calculate ΔE only once.

A technique similar to (1.5) is used to determine the centre of ΔS_m. Let us have an m-paired index (k,j), $k = 1, 2, \ldots, k_n$, in the region of a ring defined by the radii from R_{n-1} to R_n in (1.2): $k_n = \max(1, (R_n - R_{n-1})/\sigma)$. Then the step along the radius is $\Delta\rho = (R_n - R_{n-1})/k_n$ and the current radius is $\rho_k = (k - 0.5)\Delta\rho$. The number I of divisions with respect to angle φ is $I = \max(4, 2\pi\rho_k/\sigma)$ If a_{nj} is a complex number whose components are the coordinates of the centre of ΔS_{kj}, then the sequence a_{kj} is found from the relation

$$a_{kj} = \Delta a_k a_{k,j-1}, \qquad j = 1, 2, \ldots, I_k$$

where

$$\Delta a_k = \exp(\mathrm{i}2\pi/I_k), \qquad a_{k,0} = \rho_n + \mathrm{i}0.$$

The efficiency of the algorithm was evaluated in the following manner. Program modules were developed to implement the computation of the Fresnel–Kirchhoff integral (1.1) by several methods: (a) using the algorithm suggested above, (b) with the rectangular grid and (c) using the Waddle method [8]. In a zone plate with an aperture ratio $D/F \approx \frac{1}{2}$ the focal field intensity was computed for a single point-like emission source and computation times for each of the three methods for one computation point were compared. It was obtained that for half-wavelength integration step of $\lambda/2$ and the zone plate aperture $D/\lambda = 250$ the suggested algorithm provided the gain of 2.45 in comparison with method (b) and 53.8 in comparison with (c), and for the one-wavelength integration step λ, the gain was 4.45 in comparison with method (b).

For the off-axis position of the emission source removed by $\Delta x \approx 50\lambda$ from the optical axis, and for the integration step equal to one wavelength (computation error $\sim2\%$), and the zone plate aperture $D/\lambda \approx 250$, the computation time for one computation point, using algorithm (a) was less by a factor of ~20–25 compared with the results for algorithm (b). This difference in computation time increases substantially if several pointlike emission sources are introduced and/or the zone plate aperture increases.

In the case of diffraction of plane electromagnetic wave on a zone plate, the initial integral (1.1) is transformed to

$$U(p') = \iint U_0(p) \, \frac{\exp[-ik\rho(p,p')]}{\rho(p,p')} \, \mathrm{d}S. \qquad (1.6)$$

We will now change to the coordinate system whose origin is placed at the observation point which lies at a distance H from the optical axis. Let us discuss the geometry of the problem as it appears from the observation point [7, 9]. For this, we draw circles around the selected point; the phase of the wave is identical on them because the distances r_1 and r_2 are identical (provided the illumination of the zone plate is uniform in amplitude and phase). We introduce now a one-dimensional function $\varphi(r)$ that describes the length of the intersection half-arc of two circles of radii R_i and $r' = (r^2 - F^2)^{1/2}$, whose centres are at a distance equal to the distance H from the optical axis to the observation point. Then (figure 1.4):

$$U(p) = \iint_S \frac{\exp(ikr)}{r} \cos(\gamma_1 + \cos\gamma_2) \, \mathrm{d}S = \iint_S \frac{\exp(ikr)}{r} \, 2\cos\gamma \, a \, \mathrm{d}a \, \mathrm{d}\varphi. \qquad (1.7)$$

In view of

$$a = (r^2 - F^2)^{1/2}, \qquad \cos\gamma = F/r, \qquad a \, \mathrm{d}a = r \, \mathrm{d}r$$

we have

$$U(p) = 2F \int_r \frac{\exp(ikr)}{r} \, \varphi(r) \, \mathrm{d}r \qquad (1.8)$$

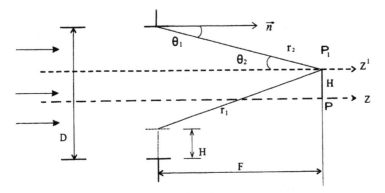

Figure 1.4. Computation of the diffraction integral for the planar incident wavefront [9].

where

$$\varphi(r) = \begin{cases} 0, & \text{if } |H - r| > a \\ \pi, & \text{if } r \le a - H \\ \tfrac{3}{2}\pi, & \text{if } x < 0, \ x = (r^2 + H^2 - a^2)/2rH \\ \tfrac{1}{2}\pi, & \text{if } x = 0 \\ \text{arctg}[(1 - x^2)^{1/2}/x], & \text{otherwise.} \end{cases}$$

Therefore, the initial integral (1.6) can be transformed to a single integral of the type of the Fourier transform of an analytical function $\varphi(r)$.

Note that if the incident wave front is planar, the Fresnel–Kirchhoff integral at the geometric focal point (at the point P) gives the exact analytical result [9]:

$$U(p) = \iint_S \frac{\exp(ikr)}{r} (\cos \gamma_1 + \cos \gamma_2)\, dS$$

$$= \iint_S \frac{\exp(ikF)}{F} (\cos \gamma_1 + \cos \gamma_2)\, dS.$$

Taking into account that $\cos \gamma_1 = 1$, $dS = 2\pi a\, da$, $a = F \sin \gamma_2$, we have

$$U(p) = 2\pi F \exp(ikr) \int_0^y (1 + \cos \gamma_2) \sin \gamma_2\, d\gamma_2$$

$$= 2\pi F \exp(ikr)\{1 + D^2/8F^2 - [1 - (D/2F)^2]^{1/2}\}$$

where $y = \arcsin(D/2F)$.

In the paraxial approximation, the field distribution across the optical axis is described within the region of the zone plate focal point (in the approximation of the Fresnel diffraction) by the Airy function $2J_1(x)/x$ for the ring formed between the Fresnel zone radii R_{n-1} and R_n. Here J_1 is the Bessel function of the first kind. At the same time, the zone plate is a set of concentric rings each having a different surface area; correspondingly, a more precise description of the field intensity distribution in the neighbour-hood of the focal point was suggested in [9], namely to use a 'modified' Airy function 'weighted' over the areas of the Fresnel zones:

$$U(p) = U_0 \left\{ \frac{\left(\sum_{n=0}^{N-1} (-1)^{n+1} R_{n-1} \frac{2J_1(x_{n-1})}{x_{n-1}} \right)^2}{\sum_{n=0}^{N-1} (-1)^{n+1} R_{n+1}^2} \right\}.$$

1.4 Diffraction of electromagnetic waves

Reliable determination of the diffraction field in an arbitrary area behind a stop can be achieved by applying the physical theory of diffraction (PTD) in which the results of the rigorous theory are used to prescribe the boundary conditions. It was shown in [10, 12] that the solution of the diffraction problem on the basis of Maxwell's equations with Kirchhoff–Kottler bound-ary conditions, using the Rubinowicz transformation, provides sufficient accuracy and reliability in the calculation of field components of the electro-magnetic wave for ideally conducting circular stops; the components have the polarization and direction of propagation of the initial wave. This solution makes it possible to evaluate the cross-polarization component for the absolute black aperture. The Rubinowicz transformation allows one to obtain an analytical expression for the field along the axis of the aperture. It also makes it possible to calculate the depolarization of the diffracted radiation, that is, the angular dependences for the electromagnetic field components, something we cannot achieve using traditional scalar methods.

Expressions for the components of electromagnetic filed are given in the Appendix at the end of the book.

In order to better understand the limits of applicability of the scalar and vector methods, figure 1.5 shows the results of simulation of the field intensity distribution in the focal region of a diffractionally constrained system [27]. To clarify the field structure in the focal region, it is possible in a number of cases to ignore the specific device which generated the converging wave front. The problem then reduces to investigating the convergent spherical wave in a given geometry. The data shown in these figures were obtained for the lens aperture $D/\lambda \approx 120$, with various values of aperture ratio D/F. The quantity ds on the plots is the relative displacement of the observation plane from the geometric focal point of the system. The vector-method computations were run using the results published in [25, 26].

The data show that the diffraction curves calculated using the scalar and vector techniques for the focal plane practically coincide for aperture ratios $1:4$ and $1:2$. Substantial differences are observed for larger values of the

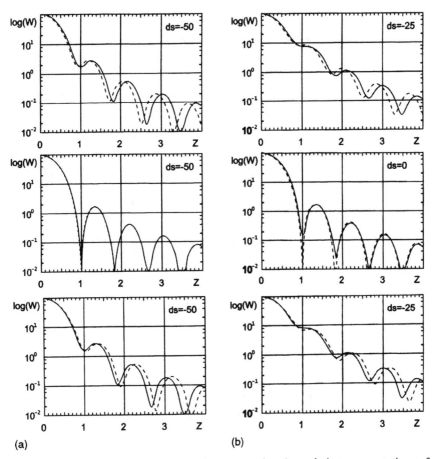

Figure 1.5. Comparison of the results of vector- and scalar-technique computations of field intensity distribution for different values of aperture ratio D/F: $A = 1/4$, $B = 1/2$, $C = 1/1$, $D = 1/0.66$. Solid curves: vector method; dashed curves: scalar method.

aperture ratio: the central maximum is somewhat widened and the side lobes are less pronounced in the distribution generated using the vector technique. The differences become more important in planes that are shifted off the focal plane. Therefore the scalar technique can provide sufficient accuracy for designing systems with low numerical apertures (1:2 and lower), provided no data are needed on wave polarization.

1.5 Asymptotic behaviour of the diffraction integral

Assume that an object at a distance $Z_0 = A$ from the plane of the zone plate is illuminated with a plane monochromatic wave. If the object is pointlike and

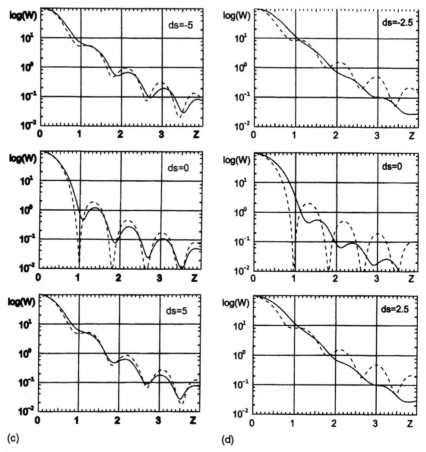

Figure 1.5. *(Continued)*

its coordinates are $(\varepsilon, \eta, z = A)$, then the wave incident on the zone plate is a divergent spherical wave which can be written in the paraxial approximation as [13]

$$E^{(-)} = \frac{1}{iz_0} \exp\left\{i\frac{\pi}{\lambda z_0}[(x-\varepsilon)^2 + (y-n)^2]\right\} \qquad (1.9)$$

where λ is the wavelength (we dropped the time factor).

We can write the transmittance function of the zone plates as a Fourier series [13]

$$T(x,y) = 2\sum_{n=-\infty}^{n=+\infty} \frac{\sin(n\pi/2)}{n\pi} \exp\left[i\frac{\pi n}{\lambda F}(x^2 + y^2)\right] \qquad (1.10)$$

where F is the principal focal distance of the zone plate. After the electromagnetic wave crosses the plane of the zone plate, the distribution changes to

$$E^{(t)}(x, y) = E^{(-)}(x, y) T(x, y). \tag{1.11}$$

In the Fresnel diffraction approximation, the momentum response of the zone plate in the focal plane $(u, v, z = B)$ is given by the integral [13]:

$$h(\varepsilon, \eta, u, v) = \frac{\exp(i2\pi z/\lambda)}{i\lambda z} \iint_S E^{(t)} \exp\{(i\pi/\lambda z)[(u - x)^2 + (v - y)^2]\}. \tag{1.12}$$

Substituting (1.18)–(1.20) into (1.21) and making use of the circular symmetry of the zone plate, we obtain

$$h(\varepsilon, \eta, u, v) = \frac{4kN^2}{(k + 1)^2 b^2} \sum_{n=-\infty}^{n=+\infty} \frac{\sin(n\pi/2)}{n}$$

$$\times \int_0^r rJ_0\left(\frac{2\pi Nwr}{(k + 1)b}\right) \exp\{iNr^2(n + F/z_0 + F/z)\}\,\mathrm{d}r \tag{1.13}$$

where $k = z/z_0$, $N = b^2/\lambda F$, $w = ((k\varepsilon + u)^2 + (k\eta + v)^2)^{1/2}$ and $r = [(x^2 + y^2)^{1/2}]/b$.

Owing to the rapid oscillation of the factor $\exp[i\pi Nr^2(n + F/z_0 + F/z)]$, all integrals in (1.22) are negligibly small if the number of Fresnel zones N that can be drawn within the zone plate of radius b is sufficiently large. Only if $n + F/z_0 + F/z = 0$ for $n = -1$ (the case when the image is considered within the principal focal plane), we have

$$h(\varepsilon, \eta, u, v) = \frac{2kN^2}{(k + 1)^2 b^2} \frac{2J_1(x)}{x} \tag{1.14}$$

where $x = (2\pi Nw)/(k + 1)b$.

Expression [1.14] shows that the momentum response of the zone plate in the Fresnel approximation and for $b \gg (\lambda F)^{1/2}$ (i.e. for the zone plate radius that is much greater than the diameter of the first Fresnel zone) coincides, to within a constant factor, with the momentum response of a convex aberration-free lens. A similar conclusion was obtained as far back as 1946 by Malyuzhints [14] who wrote integral (1.13) in the form

$$U(y, r) = \frac{kF}{\pi} \exp(ikF) \int_0^{\alpha_m} \tan\alpha \exp(iky\cos\alpha) I_0(kr\sin\alpha)\,\mathrm{d}\alpha \tag{1.15}$$

where α_m is the angle subtended by the radius r from the point $(F, 0)$. It was pointed out [14] that integral (1.15) resembles the Debye integral for diffraction near the focal point and differs from it in only two respects:

- the factor $1/\pi$ (which reflects specific properties of the zone plate),
- the integrand in (1.15) includes a factor $\tan\alpha$ instead of $\sin\alpha$ because in this particular case the mean amplitude (but not the mean phase) is constant not on a spherical front but on the plane $x = 0$ within the circle $r < R_0$.

Later Khastgir *et al* [18] analysed a long-focus zone plate of acoustic frequency range with triangular in-plane profile of Fresnel zones. Calculations for the diffraction of plane waves on such a zone plate showed that in this case the paraxial-approximation distribution of field intensity at the focal point in the plane perpendicular to the optical axis resembles the Debye integral with $\sin \alpha$ replaced with $\tan \alpha / \cos \alpha$ [18].

Therefore, the field intensity distribution in the focal region of the zone plate across the optical axis is described, in the Fresnel diffraction approximation, by the Airy function $2J_1(x)/x$ for the ring formed between the Fresnel zone radii R_{n-1} and R_n.

1.6 Computation of the diffraction integral of the Fraunhofer zone

The term 'Fresnel diffraction' is known to be used if a point source is close to the diffracting object or the observer is close to the diffracting object. The term 'Fraunhofer diffraction' is used if the diffracting object is illuminated with a plane wave and the distance between the diffracting object and the observer is large compared with the object size. Let us consider how to find the field at the observation point when a scalar wave is diffracted by a diffraction element on an arbitrary surface of rotation of second order [7, 16]. The Huygens–Kirchhoff integral for the diffraction wave is [17]

$$U_2(\mathbf{R}_2) = \frac{1}{4\pi} \int_\Sigma \left\{ U_1(\mathbf{R}_1) \frac{\partial \exp[-ik(|\mathbf{R}_2 - \mathbf{R}_1|)]}{\partial \mathbf{n}|\mathbf{R}_2 - \mathbf{R}_1|} \right.$$
$$\left. - \frac{\partial U_1(\mathbf{R}_1) \exp[-ik(|\mathbf{R}_2 - \mathbf{R}_1|)]}{\partial \mathbf{n}|\mathbf{R}_2 - \mathbf{R}_1|} \right\} \mathrm{d}\mathbf{S} \qquad (1.16)$$

where $U_1(\mathbf{R}_1)$ is the field created by the emitter on the surface Σ and $\partial/\partial n$ is the derivative along the normal to the surface Σ.

Let us write the emitter's field as that of a pointlike isotropic source, that is,

$$U_1(\mathbf{R}_1) = a_0 \frac{\exp[-ik(|\mathbf{R}_1 - \boldsymbol{\rho}|)]}{|\mathbf{R}_1 - \boldsymbol{\rho}|} \Phi(\theta_1).$$

Let us denote $\Delta R_1 = |R_1 - \rho|$, $\Delta R_2 = |R_2 - R_1|$. Then the original Huygens–Kirchhoff integral (1.16) changes the form to

$$U_2(\mathbf{R}_2) = \frac{a_0}{4\pi} \int_\Sigma \left[\frac{\exp(-ik\,\Delta R_2)}{\Delta R_2} \left(-\frac{ik}{\Delta R_1} - \frac{1}{\Delta R_1^2} \right) \exp(ik\,\Delta R_1)\mathbf{r}_{10} \right.$$
$$\left. + \frac{\exp(-ik\,\Delta R_1)}{\Delta R_2} \left(-\frac{ik}{\Delta R_2} - \frac{1}{\Delta R_2^2} \right) \exp(-ik\,\Delta R_2)\mathbf{r}_{20} \right]$$
$$\times \Phi(\theta_1)\,\mathrm{d}\mathbf{S} \qquad (1.17)$$

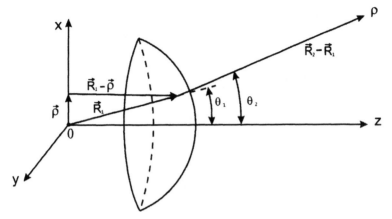

Figure 1.6. Algorithm of computation of the Fresnel–Kirchhoff integral in the Fraunhofer zone.

where $\mathbf{r}_{10}$ and $\mathbf{r}_{20}$ are unit vectors for the vectors $(\mathbf{R} - \boldsymbol{\rho})$ and $(\mathbf{R}_2 - \mathbf{R}_1)$, respectively (see figure 1.6).

Since $(k\,\Delta R_1) \gg 1$ and $(k\,\Delta R_2) \gg 1$, expression (1.17) can be simplified to

$$U_2(\mathbf{R}_2) = \int_{\Sigma} \left[\frac{\exp(-ik\,\Delta R_2)}{\Delta R_2} \left(ik + \frac{1}{\Delta R_1} \right) \frac{\exp(ik\,\Delta R_1)}{\Delta R_1} \mathbf{r}_{10} \right.$$
$$\left. + \frac{\exp(-ik\,\Delta R_2)}{\Delta R_2} \frac{\exp(-ik\,\Delta R_1)}{\Delta R_1} ik\mathbf{r}_{20} \right] \Phi(\theta_1)\,\mathrm{d}\mathbf{S}.$$

The argument of the Green's function is $\Delta R_2 = (R_2^2 + R_1^2 - 2R_1 R_2 \cos\gamma)^{1/2}$, where γ is the angle between the vectors $\mathbf{R}_1$ and $\mathbf{R}_2$, and $\cos\gamma = \cos\theta_1 \cos\theta_2 + \sin\theta_1 \sin\theta_2 \cos(\varphi_1 - \varphi_2)$, so we apply the Taylor series expansion in powers of R_1, that is, $\Delta R_2 = R_2 - R_1 \cos\gamma + R_1 R_1 / 2R_2 + \cdots$ where the third and subsequent terms are of the order of R_1/R_2, $(R_1/R_2)^2 \cdots$ and can thus be ignored in the Fraunhofer zone. Note also that for $R_1/R_2 \ll 1$ the difference between ΔR_2 and R_2 must be taken into account only in the exponent of the exponential expression.

As a result, in the Fraunhofer zone we have

$$U_2(\mathbf{R}_2) \approx \frac{ika_0}{4\pi} \frac{\exp(-ikR_2)}{R_2} \iint \frac{\exp(ik\,\Delta R_1)}{\Delta R_1} \exp(ikR_1) \cos\gamma$$
$$\times \left[\left(1 - \frac{1}{R\,\Delta R_1} \right) \mathbf{r}_{10} + \mathbf{r}_{20} \right] \Phi(\theta_1)\,\mathrm{d}\mathbf{S}. \tag{1.18}$$

We will write this integral in its explicit form.

We denote $A \equiv A(R_2) = (ika/4\pi R_2)\exp(-ikR_2)$, where $k = 2\pi/\lambda$ is the wave number, $\Delta R_1 = (R_1^2 + \rho^2 - 2R_1\rho \sin\theta_1 \cos\varphi_1)^{1/2}$ and

$$\beta = kR_1 \sin\theta_1 \sin\theta_2,$$

$$\frac{U_2(\mathbf{R}_2)}{A} = I(\theta_2, \varphi_2)$$

$$= \iint_\Sigma \frac{\exp(ik\,\Delta R_1)}{\Delta R_1} \exp(ikR_1 \cos\theta_1 \cos\theta_2)$$

$$\times \exp[i\beta\cos(\varphi_1 - \varphi_2)]\Phi(\theta_1)R_1^2 \sin\theta_1$$

$$\times \left\{ \left(\cos\theta_1 + \frac{dR_1}{d\theta_1}\frac{1}{R_1}\sin\theta_1 \right) \cos\theta_2 + \frac{R_1}{\Delta R_1}\left(1 - \frac{1}{k\,\Delta R_1}\right) \right.$$

$$+ \sin\theta_2 \left(\sin\theta_1 - \cos\theta_1 \frac{dR_1}{d\theta_1}\frac{1}{R_1} \right)\cos(\varphi_1 - \varphi_2)$$

$$\left. + \frac{\rho}{\Delta R_1}\left(1 - \frac{i}{k\,\Delta R_1}\right)\left(\frac{1}{R_1}\frac{dR_1}{d\theta_1}\cos\theta_1 - \sin\theta_1\right)\cos\varphi_1 \right\} d\theta_1\,d\varphi_1.$$

$$(1.19)$$

The integral in (1.19) can be rewritten in the form

$$I(\theta_2, \varphi_2) = \iint \frac{\exp[-ik(\Delta R_1 - F)]}{\Delta R_1/F} \exp[ik(R_1 \cos\theta_1 \cos\theta_2 - F)]$$

$$\times \exp[-i\beta\cos(\varphi_1 - \varphi_2)]$$

$$\times \left\{ f_1(\theta_1, \theta_2) + \frac{F}{\Delta R_1} f_2(\theta_1)\cos(\varphi_1) + f_3(\theta_1, \theta_2)\cos(\varphi_2 - \varphi_1) \right.$$

$$\left. + f_4(\theta_1)\frac{F}{\Delta R_1} \right\}$$

$$\times \frac{R_1^2}{F_1^2} \sin\theta_1 \Phi(\theta_1)\,d\varphi_1\,d\theta_1 \qquad\qquad (1.20)$$

$$f_1(\theta_1, \theta_2) = \cos\theta_2 \left(\cos\theta_1 + \frac{dR_1}{d\theta_1}\frac{1}{R_1}\sin\theta_1 \right)$$

$$f_2(\theta_1) = \frac{\rho}{F}\left(\frac{dR_1}{d\theta_1}\frac{1}{R_1}\cos\theta_1 - \sin\theta_1 \right)\left(1 - \frac{1}{k\,\Delta R_1}\right)$$

$$f_3(\theta_1, \theta_2) = \sin\theta_2 \left(\sin\theta_1 - \frac{dR_1}{d\theta_1}\frac{1}{R_1}\cos\theta_1 \right)$$

$$f_4(\theta_1) = \frac{R_1}{F}\left(1 - \frac{i}{k\,\Delta R_1}\right).$$

$$(1.21)$$

Here F is a parameter converting formulas to dimensionless form, such that $\Delta R_1/F \approx R_1/F \approx 1$, for example, we may have $F \approx (R_1^{\max} + R_2^{\min})/2$.

We denote

$$B = \exp[ik(R_1 \cos\theta_1 \cos\theta_2 - F)]\Phi(\theta_1)\frac{R_1^2}{F^2}\sin\theta_1.$$

Then the inner integral in (1.20) over φ_1 becomes

$$I(\varphi_1) = B\int_{-\pi}^{\pi}\frac{\exp[-ik(\Delta R_1 - F)]}{\Delta R_1/F}$$

$$\times \exp[i\beta\cos(\varphi_1 - \varphi_2)]\left(f_1 + f_2\frac{F}{\Delta R_1} + f_3 + f_4\frac{F}{\Delta R_1}\right)d\varphi_1 \quad (1.22)$$

and can be rewritten via the sum of the integrals $I(\varphi_1) = B(I_1 + I_2 + I_3 + I_4)$ where

$$I_1 = f_1\int_{-\pi}^{\pi}\frac{\exp[-ik(\Delta R_1 - F)]}{\Delta R_1/F}\exp[i\beta(\varphi_1 - \varphi_2)]\,d\varphi_1$$

$$I_2 = f_2\int_{-\pi}^{\pi}\frac{\exp[-ik(\Delta R_1 - F)]}{(\Delta R_1/F)^2}\cos\varphi_1\exp[i\beta\cos(\varphi_1 - \varphi_2)]\,d\varphi_1$$

$$(1.23)$$

$$I_3 = f_3\int_{-\pi}^{\pi}\frac{\exp[-ik(\Delta R_1 - F)]}{\Delta R_1/F}\exp[i\beta(\varphi_1 - \varphi_2)]\cos(\varphi_1 - \varphi_2)\,d\varphi_1$$

$$I_4 = f_4\int_{-\pi}^{\pi}\frac{\exp[-ik(\Delta R_1 - F)]}{(\Delta R_1/F)^2}\exp[i\beta\cos(\varphi_1 - \varphi_2)]\,d\varphi_1.$$

Therefore, the inner integral over φ_1 in (1.19) reduces to the sum of four circular convolutions of the following type:

$$I_1 = f_1\left(\frac{\exp[-ik(\Delta R_1 - F)]}{\Delta R_1/F}\exp[i\beta\cos(\varphi_1 - \varphi_2)]\right) \quad (1.24)$$

$$I_2 = f_2\left(\frac{\exp[-ik(\Delta R_1 - F)]}{(\Delta R_1/F)^2}\cos\varphi_1\exp[i\beta\cos(\varphi_1 - \varphi_2)]\right) \quad (1.25)$$

$$I_3 = f_3\left(\frac{\exp[-ik(\Delta R_1 - F)]}{\Delta R_1/F}\exp[i\beta\cos(\varphi_1 - \varphi_2)]\cos(\varphi_1 - \varphi_2)\right) \quad (1.26)$$

$$I_4 = f_4\left(\frac{\exp[-ik(\Delta R_1 - F)]}{(\Delta R_1/F)^2}\exp[i\beta\cos(\varphi_1 - \varphi_2)]\right). \quad (1.27)$$

The following property of Fourier transforms is used to calculate the integrals (1.24)–(1.27): if $F_1(U_1)$ and $F_2(U_2)$ are Fourier transforms of the functions U_1 and U_2 and also

$$U = U_1U_2 = \int_{-\pi}^{\pi}U_1(\varphi_1)U_2(\varphi - \varphi_2)\,d\varphi$$

then $F(U_1U_2) = F(U_1)F(U_2)$, and therefore $U_1U_2 = F^{-1}(F_1F_2)$.

At the first stage, Fourier spectra of the following three functions included in the integral (1.24)–(1.27) are calculated:

$$\exp[-ik(\Delta R_1(\varphi_1,\theta_1) - F)]F/\Delta R_1(\varphi_1,\theta_1)$$
$$\exp[-ik(\Delta R_1(\varphi_1,\theta_1) - F)][F/\Delta R_1(\varphi_1,\theta_1)]^2 \qquad (1.28)$$
$$\exp(-i\beta \cos\varphi_1).$$

At the second stage, the spectra of the remaining functions in the expressions (1.24)–(1.27) are calculated using the formulas of cyclical shift of the already known spectra of the functions (1.28). Indeed, if we know, for example, the expansion $U(\alpha) = \sum_0^N C_n \cos n\varphi$ on the interval $-\pi \le \varphi \le \pi$, then the series for the function $U(\varphi)\cos n\varphi$ can be written as

$$U(\varphi)\cos n\varphi = \sum_0^N \tfrac{1}{2}C_n[\cos(n+1)\varphi + \cos(n-1)\varphi] = \sum d_n \cos n\varphi$$

where $d_n = (C_{n-1} + C_{n+1})/2$.

At the third stage, the spectra of the obtained Fourier transforms of the corresponding functions multiplied by the coefficients $f_j, j = 1,\dots,4$, (1.21) are cyclically added up:

$$\tilde{F} = \sum_{j=1}^4 F_j f_j. \qquad (1.29)$$

Finally, at stage four, one inverse Fourier transform of the expression (1.38) is calculated.

This method of calculating the integrals (1.24)–(1.27) makes it possible to diminish the number of the Fourier transforms of the corresponding functions and to trim the process down to calculating three direct transforms and one inverse fast Fourier transform (FFT).

Note that since with this algorithm of calculating the Fresnel–Kirchhoff integral in the far zone one finds the values of the real and imaginary field components simultaneously for all discrete values of the angle φ_2, an appropriate sparsing of the spectrum was carried out before the inverse FFT was calculated.

The sample size N required for the direct FFT is found from the condition $I_{N/2}(\beta_{\max}) \ll I_0(\beta_{\max})$. Or,

$$I_{(N-2)/2}\left\{ \frac{k}{[(R_1 + \rho^2)/(1 + q^2)]^{1/2}}\bigg|_{\max} q_{\max}^{(N-2)/2} \right\}$$
$$\le J_0\left\{ \frac{k}{[(R_1 + \rho^2)/(1 + q^2)]^{1/2}}\bigg|_{\max} q_{\max}^{(N-2)/2} \right\}$$

where $q = [1 - (1 - x^2)^{1/2}]/x$ and $x = 2R_1\rho/(R_1^2 + \rho^2)\sin\theta_1$.

For example, for the characteristic parameters of the problem with $\rho \le R_1/2$, the sample size N was taken to be 64 for the real and imaginary

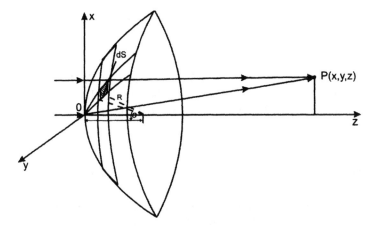

Figure 1.7. Computation of the Fresnel–Kirchhoff diffraction integral.

components of the functions. For the inverse FFT the sample size was increased to 1024.

Let us consider the calculation of the Fresnel–Kirchhoff diffraction integral in the three-dimensional case (figure 1.7) (here we follow [18]).

Let the generatrix of the surface of an MWDO element be described by an equation of the type $r = f(z)$. Then an element of the area of the surface can be written in the form

$$dS = r[1 + (dr/dz)^2]^{1/2} \, dz \, d\theta. \tag{1.30}$$

The diffraction integral for the observation point P with the coordinates (x_1, y_1, z_1) is written for a plane wave front that is incident on the diffraction element as

$$U_n = \int_{z_{n-1}}^{z_n} \int_0^{2\pi} \frac{\chi(Q)}{Q} \exp\left(-i\frac{2\pi Q}{\chi}\right) dS \tag{1.31}$$

where $\quad Q = [(\beta - z)^2 + (r\cos\theta)^2 + (r\sin\theta - x_n)^2]^{1/2} \quad$ and $\quad \chi(Q) = 1 + (\beta - z)/Q$.

We assume that the generatrix of the surface of the diffraction element is described by the square root of a quadratic trinomial,

$$r = [\alpha z^2 + \beta z + c]^{1/2}. \tag{1.32}$$

This presentation form is convenient because equation (1.32) describes:

an ellipse if	$c - \beta^2/4 < 0,$	$\alpha < 0$
a cone if	$\beta = c = 0,$	$\alpha > 0$
a hyperbola if	$\alpha c - \beta^2/4 < 0,$	$\alpha > 0$
a parabola if	$\alpha = c = 0,$	$\beta \neq 0.$

Taking into account (1.32), we calculate an element of the area surface using (1.30) and obtain

$$dS = 0.5[(2r)^2 + (2\alpha z + \beta)^2]\, d\theta\, dz.$$

Substituting this expression into (1.31), we arrive at the final form of the Fresnel–Kirchhoff integral, which describes the diffraction of a plane scalar wave on an MWDO element, selected on an arbitrary surface of rotation of second order.

1.7   Geometric modelling of diffraction elements

In a number of cases it is sufficient, when modelling the focusing properties of diffraction elements, to use the calculation of ray paths through this element. Thus for axisymmetric elements we use the array equation written in vector form [19]:

$$\mathbf{S'r} = \mathbf{Sr} + \frac{m\lambda}{d}\mathbf{g}$$

where $\mathbf{S'} = (L', M', N')$ is the vector of the diffraction ray, $\mathbf{S} = (L, M, N)$ is the vector of the incident ray, $\mathbf{r}$ is the vector normal to the diffraction element at the point of intersection with the ray, number m stands for the diffraction order, λ is the current wavelength, d is the array constant ($d = 2\Delta r_n$ is the unit vector parallel to the tangent), $\mathbf{q} = -\mathbf{pr}$ is the unit vector perpendicular to the tangent, $\mathbf{p} = (u, v, w)$ is a unit vector perpendicular to the tangent to the zones at the point of intersection with the ray in the plane of the element ($w = 0$) (figure 1.8(a)).

The general solution for the array is $\mathbf{S'} = \mathbf{S} - \Lambda\mathbf{p} + Q\mathbf{r}$, where $\Lambda = m\lambda/2\Delta r_n$ and Q is a numerical coefficient depending on the type of

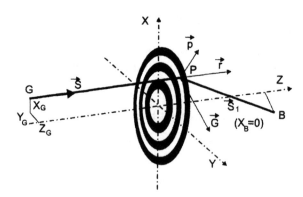

Figure 1.8. (a) Geometric modelling of a diffraction element.

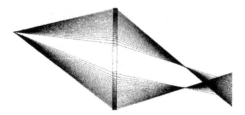

Figure 1.8. (b) Ray paths across an MWDO element that focuses the incident radiation of an off-axis radiation source into an annular area [20].

the diffraction element:

$$Q = (A_2^2 - B)^{1/2} - A_2, \qquad A_2 = N, \qquad B_2 = \Lambda^2 - 2\Lambda(Lu + M\nu).$$

The vector relations given above can be used for a rapid analysis of frequency and focusing properties of diffraction elements in various program complexes that are similar to those described in [7, 19–22] (first of all, to analyse aberration characteristics).

As an example, figure 1.8(b) shows the paths of rays through the diffraction element that focuses radiation from a point-like source into a ring, for a displaced position of this point-like source.

It is generally advisable to run a numerical experiment for any theoretical study of synthesis and for any analysis of the physical properties of MWDO elements.

A numerical experiment makes it possible to forecast the characteristics of the synthesized element, to carry out its optimization and for subtle analysis of the distribution intensity in the area under study until the very moment of physically manufacturing the diffraction element.

An interactive system of designing MWDO elements [22, 23] should be based on the modular principle of design and development of application programs, which operates by splitting a problem into individual subproblems that allow a relatively independent development and use of each module. This principle is a natural component of the general 'from top down design of software' principle of creating software complexes.

The following basic components are included in a numerical experiment. Stage one—the stage of input of the initial information—includes an analysis establishing the existence of a solution and checking the format of the initial data for correctness. At the second stage the boundaries of the computation region are requested, a discrete phase function of the element to be synthesized is created, its type is determined and the focusing and/or frequency properties are computed. At the last stage the results obtained are visualized as various relations and curves, or computer animations composed of display files are created.

If necessary, the information required for manufacturing a diffraction element is output by the computer.

1.8   On optimization of parameters of diffraction elements

Multiparametric optimization of parameters is the only way of reaching the limits in the characteristics of diffraction elements. Among numerous optimization algorithms, the so-called evolution simulation algorithms recently have become increasingly widespread. This methodology was first proposed by Holland in 1975 for artificial systems [28]. By now this is a well-known optimization methodology based on an analogy to processes of natural selection in biology. The biological basis for adaptation processes is the evolution from one generation to another; it operates by discarding 'weak' elements and retaining optimal or quasi-optimal elements. In [28] Holland gave a theoretical analysis of the class of adaptive systems in which structural modifications are represented by sequences (strings) of symbols selected from a certain (usually binary) alphabet. The search in the field of such representations is driven by the so-called genetic algorithms (GAs).

The main feature of GAs is that they analyse not just one solution but a certain subset of quasi-optimal solutions, known as 'chromosomes', or 'strings'; these solutions also have the ability of escaping from 'local' optimums. This subset is called the 'population'. For each chromosome, one has to calculate the criterion function $F(n)$ known as the evolution function, where n is the number of elements in a chromosome. Such functions calculate the relative weight of each chromosome. In each population, chromosomes are subject to various operators. The processes that take place are similar to those that occur in natural genetics. Among the main operators are: crossing-over (OK), inversion (OI), mutation (OM), translocation (OT), segregation (OS), and cross-mutation (OKM) [28–32].

The most optimal application of genetic algorithms is the optimization of multiparametric functions. Many realistic problems can be formulated as a search for optimal value, where the value is a complicated function of a number of input parameters. In some cases it is of more interest to find those values of parameters for which the best accurate value of the function is reached. In other cases the exact optimum is not required: any value which is better than a certain preset value is regarded as a solution. In this situation genetic algorithms are frequently the best among the acceptable methods of searching for 'good' values. The strength of a genetic algorithm lies in its ability to manipulate many parameters simultaneously.

A traditional view of a genetic algorithm is the code shown below.

```
BEGIN /* genetic algorithm */
    Create initial population
    Evaluate fitness of each individual
    stop := FALSE
    WHILE NOT stop EXECUTE
    BEGIN /* create population of new generation */
```

```
        REPEAT (size-population/2) TIMES
        BEGIN /* reproduction cycle */
             Select two highly fit individuals out of the
                  preceding generation for breeding
             Cross-breed the selected individuals and obtain
                  two offspring
             Evaluate fitness of offspring
             Place offspring into the new generation
        END
        IF population converges THEN stop := TRUE
    END
END
```

Estimates show that the application of genetic algorithms to diffraction antennas makes it possible to greatly reduce the size of the first side lobes of scattering (by 5–10 dB) while at the same time retaining amplification and the width of the direction diagram.

It is clear in this connection that, as the complexity of problems grows, straightforward increase in computer power will not always be adequate, even with the persistent trend of growing power of modern computers. Therefore, the development of efficient algorithms for computing diffraction integrals is an urgent and demanded task. These aspects become especially acute both since the aperture of diffractional elements D/λ increases, and since the width of the peripheral Fresnel zone for high-aperture elements becomes comparable with wavelength.

1.9 A complex of CAD programs for designing diffraction antennas

Modelling of the properties of diffraction lenses and antennas will be logically incomplete if we do not use special software complexes created for automatic manufacturing of designed and optimized elements.

Practically all developers of CAD systems claim to have used parametrization techniques. However, having been designed long before the concept of parametrization was born, these systems have to use parametrization to sustain their own internal data structures which were not originally meant for such parametrization. The resulting solutions are either inefficient or have limited scope. Among the CAD systems known to the authors, only the system of [24] (T-FLEX 3D CAD of the Russian company 'Top System') implements both the novel approach to the idea of parametrization and the requirement that a parametric model must form the basis of the drawings. The system uses the geometric kernel Parasolid v. 14 (supplied by EDS).

T-FLEX 3D CAD possesses unique capabilities of designing assembly parametric drawings; these can be complicated drawings whose parts may be interconnected. The interconnection can be formulated both as a geometric dependence and via the values of its parameters. If individual parts of a drawing overlap, the system removes invisible lines. The level of nesting of parts of the drawing is not restricted. Finalized drawings of a new designed part can be generated in seconds by varying the parameters of the assembly drawing. Together with a modified assembly drawing, a user also obtains drawings of its components as well as all the accompanying documents.

This system is sufficiently well suited to designing diffraction elements, antennas among them, because it allows maximum reduction in the time required to produce a complete set of drawings (provided the pilot drawing was correctly designed). In this case a *correctly designed* 'pilot drawing' is defined as a design in which the drawing reflects the true structure of antennas for all allowed values of variable parameters, and creates no distortions or unforeseen modifications of design. Using these capabilities of T-FLEX 3D CAD, it is possible to generate a drawing for the next modification of the antenna by simply resetting the antenna parameters to the new requirements; after this the new drawing is generated automatically.

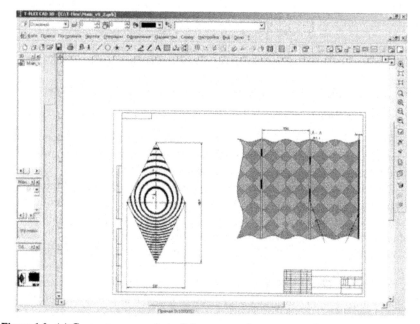

Figure 1.9. (a) Computer screenshot of the system of computer-assisted design of diffraction antennas. Diamond-shaped antenna.

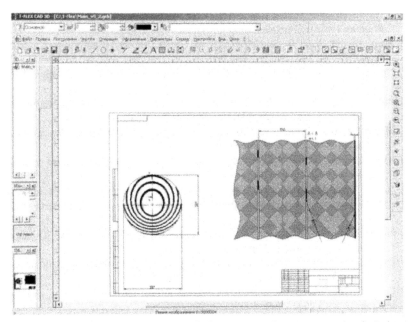

Figure 1.9. (b) Computer screenshot of the system of computer-assisted design of diffraction antennas. Circular antenna.

It is also worthy of note that our choice of T-FLEX 3D CAD [24] as the programming medium for antenna design was influenced by the fact that software products under this brand are offered by Top System as versions in either Russian or English, which allows the application of the resulting systems in English-speaking countries.

This system of automatic design of diffraction antennas operates in dialog mode. A user prescribes the antenna parameters (shape, number of levels of phase quantization, antenna type, type of dielectric etc.). The drawing of the antenna is generated automatically for this set of data. Correspondingly, the database of the system contains such parameters as, for instance, the properties of dielectric materials (type of dielectric, refraction and absorption indices in the millimetre wave range), skin layer thickness for several types of metal and so on.

Figures 1.9(a,b) display screenshots that demonstrate how the selected system works and some of its capabilities.

Bibliography

[1] Gorelik S G 1959 *Oscillations and Waves* (Moscow: Gos. Izd. FML)
[2] Vinogradova M B, Rudenko O V and Suchorukov A P 1979 *Theory of Waves* (Moscow: Nauka)

[3] Peresada V P 1961 *Visibility of Objects in the Sea for Radiolocation* (Leningrad: Ship-Building Industry Publishers)

[4] Born M and Wolf E 1999 *Principles of Optics* 7th edn (Cambridge University Press)

[5] Gravelsaeter T and Stammes J J 1982 'Diffraction by circular apertures' *Appl. Optics.* **21**(20) 3644–3651

[6] Baybulatov F Kh, Minin I V and Minin O V 1985 'Study of focusing properties of Fresnel zone plates' *Radiotekhnika i Elektronika* **30**(30) 1681–1688

[7] Minin I V and Minin O V Diffraction 1992 *Quasioptics* (Moscow: Research and Production Association 'InformTEI')

[8] Korn, G A and Korn T M 1968 *Mathematical Handbook for Scientists and Engineers* (New York: McGraw-Hill)

[9] Minin I V and Minin O V 1981 'Investigation of the field structure in the focal zone of the Fresnel zone plate' *19th Student Research Conference* (Novosibirsk: NGU Publ.) pp 54–59

[10] Pyatakhin M V and Suchkov A F 1988 'Large-angle diffraction in the Kirchhoff–Kottler approximation' Preprint No. 32 (Moscow: FIAN)

[11] Pyatakhin M V and Suchkov A F 1988 'Diffraction of plane electromagnetic wave by a circular aperture' Preprint No. 254 (Moscow: FIAN)

[12] Ganci S 1986 'Maggi–Rubinowicz transformation for phase apertures' *J. Opt. Soc. Amer. A* **3**(12) 2094–2100

[13] Shchukin I I 1973 'Formation of radio images by phase-inverting zone plates' in *Aspects of Scattering and Optimal Reception of Electromagnetic Waves* (Voronezh: VGU Publ.) pp 96–103

[14] Maluzhinetz G D 1946 'Diffraction near the axis of zone plates' *DAN USSR* **54**(5) 403–406 (in Russian)

[15] Vereszhagin V V and Lopatin A I 1981 'Theory of profiled zone plate' *Acoustic J.* **27**(6) 841–847

[16] Minin I V and Minin O V 1991 'Systems of millimeter wavelength antennas based on elements of diffraction quasioptics elements. Radio systems in millimeter and submillimeter wave ranges' in *Collection of Publications* (Kharkov, Institute of Radiophysics and Electronics of the Ukrainian AN) pp 120–127

[17] Kün R 1964 *Mikrowellenantennen* (Berlin: Veb Verlag Technik)

[18] Khastgir P and Bhowmick K N 1978 'Analysis of the off axis defocus of microwave zone plates' *Indian J. Pure Appl. Phys.* **16** 96–101

[19] Schmahl G and Rudolph D (eds) 1984 *X-Ray Microscopy* Series in Optical Sciences vol 43 (Berlin: Springer)

[20] Zherebin V V, Minin I V and Minin O V 1989 'A system of geometric modeling of diffraction optics' in *Abstracts of papers at 5th USSR Conf. on Computer Graphics' Computer Graphics-89* (Novosibirsk: Computer Centre of the Siberian Branch of AN SSSR) p 87

[21] Minin I V, Minin O V and Zherebin V V 1989 'Focusing of laser radiation onto an arbitrary 3D curve—a new principle of designing diffraction optics' *Abstract of report to 3rd USSR Conf. on Application of Lasers in National Economy* (Shatura, Research Centre of TL AN USSR) pp 213–214

[22] Minin I V and Minin O V 1987 'Interactive system for designing computer optics' *Reports to 4th USSR Conference on Computer Graphics Problems* (Serpukhov: High Energy Physics Institute of AN SSSR) p 160

[23] Minin I V and Minin O V 1990 'Interactive system for automating the design of elements of diffraction quasioptics DIK-M' *Computer Optics* 7 89–96

[24] Kuraskin S A *et al* 1999 'T-FLEX CAD: New capabilities—new horizons' *CAD Systems and Graphics* **9**

[25] Wolf E 1959 'Electromagnetic diffraction in optical systems. I. An integral representation of the image field' *Proc. Roy. Soc. A* **253** 349–357

[26] Richards B and Wolf E 1959 'Electromagnetic diffraction in optical systems. II. Structure of the field in an aplanatic system' *Proc. Roy. Soc. A* **253** 358–380

[27] Minin I V 1982 *Numerical Analysis of Diffraction of Electromagnetic Wave by Fresnel Zone Plates* (Novosibirsk: NGU)

[28] Holland J 1975 *Adaptation in Natural and Artificial Systems* (Ann Arbor, MI: University of Michigan Press)

[29] Goldberg D E 1989 *Genetic Algorithm in Search, Optimization and Machine Learning* (Addison-Wesley)

[30] Davis L (ed) 1991 *Handbook of Genetic Algorithms* (New York: Van Nostrand Reinhold)

[31] Michalewicz Z 1992 *Genetic Algorithms + Data Structures = Evolution Programs* (Berlin: Springer)

[32] Johnson J M and Rahmat-Samii Y 1997 'Genetic algorithms in engineering electromagnetics' *IEEE Antennas and Propagation Mag.* **39**(4) 7–25

Selected bibliography

Baggen L C J and Herben M H A J 1995 'Calculation the radiation pattern of a Fresnel-zone plate antenna: a comparison between UTD/GTD and PO' *Electromagnetic* **15** 321–345

Barakat R 1980 'The calculation of integrals encountered in optical diffraction theory' *Comput. Opt. Res. Math. Appl. Berlin A* 35–80

Cornbleet S 1976 *Microwave Optics. The Optics of Microwave Antenna Design* (London: Academic Press)

Fresnel O 1866 *Oeuvres Complètes* (Paris) vol 1, note 1, pp 365–372

Guo Y J and Barton S K 1995 'Analysis of one-dimensional zonal reflectors' *IEEE Trans. Antennas Propagation* **43**(4) 385–389

Hristov H D 2000 *Fresnel Zones in Wireless Links, Zone Plate Lenses and Antennas* (Boston: Artech House)

Hristov H D and Herben M H A J 1995 'Millimeter-wave Fresnel-zone lenses and antenna' *IEEE Trans. Microwave Theory Techniques* **43**(12) 2779–2785

Kirchhoff G 1882 *Berl. Ber.* 641; 1883 *Ann. der Physik* **18**(2) 663

Leyten L and Herben M H A J 1992 'Vectorial far-field analysis of the Fresnel-zone plate antenna: a comparison with the parabolic reflector antenna' *Microwave Optical Technology Lett.* **5**(2), 49–55

Lit J W Y and Tremblay R 1970 'Fresnel zone plate solved by the boundary-diffraction-wave theory' *Canadian J. Phys.* **48**(15) 1799–1805

Minin I V and Minin O V 1999 *Diffraction Quasioptics and Its Applications* (Novosibirsk: SibAGS) (revised edition in preparation)

Slujter J, Herben M H A J and Vullers O J G 1995 'Experimental validation of PO/UTD applied to Fresnel-zone plate antennas' *Microwave Optical Technol. Lett.* **9**(2) 111–113

Soret J L 1875 'Uber die durch Kreisgitter erzeugten Difractions phanomene' *Leipzig: Pogg. Annalen der Physik und Chemie* **156** 94–113

Van Houten J M and Herben M H A J 1994 'Analysis of a phase-correcting Fresnel-zone plate antenna with dielectric/transparent zones' *J. Electromagnetic Waves Applic.* **8**(7) 847–858

Wood R W 1898 'Phase-reversal zone plates and diffraction telescopes' *Phil. Mag.* Series 5 **45** 511–523

Chapter 2

Lenses based on high-aperture Fresnel zone plate: information properties

2.1 Focusing and frequency properties of the Fresnel and Soret zone plates

The quality of a lens is usually evaluated by quantizing the image of a point. The radiation energy distribution on this image is, according to the definition, the scattering function because the distribution function of complex amplitude provides the most complete data on a pointlike image. This distribution is obtained by computing the Fresnel–Kirchhoff integral using as a basis the wavefront generated by the quasioptic system at the exit pupil.

In principle the scattering function can be characterized by a single number, that is, the diameter of a circle within which certain energy is concentrated. However, this number does not tell us anything about the shape of the scattering function and is largely meaningless if it is asymmetrical and extends in one direction much farther than in the other. Furthermore, phase relationships in this distribution are important in a number of problems, such as direct quasioptic radiovision, tomography, holographic radiovision and some others, because interference phenomena will take place when adjacent pointlike images overlap.

It is common knowledge that in the absence of aberrations and axisymmetric aperture, the field intensity distribution in the focal region transversely to the optical axis is described by the so-called Airy function. The centre of this distribution lies at the point of the Gaussian image and the distance from the centre to the first minimum is known as the Rayleigh resolving power of the system:

$$r_a \approx 1.22\lambda\frac{F}{D}.$$

It can be mentioned that this equation assumes phase and amplitude to be uniform across the aperture. The value of r_a imposes limits on the capabilities

Figure 2.1. Soret zone plate.

of aberration-free (that is, diffractionally constrained) systems. We also need to take into account that the Fourier transform of the scattering function (the so-called optical frequency response function) describes optical systems as filters of spatial frequency.

In view of this, we will show in what follows the form of the diffractional spot produced as a result of diffraction of radiation, emitted by a pointlike source, on a diffractive element both longitudinally and transversally relative to the optical axis. Moreover, when analysing the formatting properties of an MWDO element based on a Fresnel zone plate (FZP) with an essentially off-axis position of the radiation source, the object of the study should be the shape of the diffraction solid of the image of this radiation source.

The Soret or Fresnel zone plate (figure 2.1) used as a focusing element began to attract the attention of designers in recent years in connection with problems of radiovision, satellite antennas, automobile locators etc.; indeed, the zone plate has certain advantages in the millimetre wave band of electromagnetic waves compared with dielectric radio lenses and mirror reflection antennas [1, 2]. The frequency properties of Fresnel zone plates have tremendous potential in applications, for example, for radiovision and tomography, since they make it possible to scan space depthwise by varying radiation wavelength (Fresnel zone plates can be used as radio lenses with variable focal length as it depends on the wavelength of the radiation employed). In addition, the Fresnel zone plate is the simplest diffractive element. Therefore, a study of its best achievable focusing and frequency properties is a way to a profound understanding of potential capabilities of more complex diffractive elements.

The idea of the zone plate came to Fresnel when he was calculating the intensity of radiation passing through a circular hole (stop or diaphragm) of diameter D placed at a distance A from the axis that goes through the centre of the hole at right angles to its plane. The scalar wave field at a point $(0, 0, F)$ can be calculated using the Kirchhoff formula

$$U(0, 0, F) = \frac{K}{2\pi i} A \iint \frac{\exp(ikr)}{r} (1 + \cos \varphi_0) \, dx \, dy,$$

$$\cos \varphi_0 = \frac{F}{(x^2 + y^2 + F^2)^{1/2}}, \qquad r = (x^2 + y^2 + F^2)^{1/2}.$$

Transforming this integral to the polar coordinate system, we arrive at the expression

$$U(0, 0, F) = \frac{KA}{i} \int_0^D \frac{\exp[ik(\rho^2 + F^2)]^{1/2}}{(\rho^2 + F^2)^{1/2}} \left(1 + \frac{F}{(\rho^2 + F^2)^{1/2}}\right) \rho \, d\rho.$$

Denoting $S = (\rho^2 + F^2)^{1/2}$ and correspondingly changing variables, we obtain

$$U(0, 0, F) = \frac{KA}{i} \int_F^{(F^2 + D^2)^{1/2}} (1 + F/S) \exp(ikS) \, dS.$$

Owing to the exponential containing imaginary exponent, the integrand above is rapidly oscillating so that the integral cannot reach large values. We split the integration range into rings, each consisting of two parts, and transform the expression for field to the form

$$U(0, 0, F) = \frac{KA}{i} \sum_{n=0}^N \left\{ \int_{F+2n\lambda/2}^{F+(2n+1)\lambda/2} (1 + F/S) \exp(ikS) \, dS \right.$$

$$\left. + \int_{F+(2n+1)\lambda/2}^{F+(2n+2)\lambda/2} (1 + F/S) \exp(ikS) \, dS \right\}.$$

The imaginary parts of the integrands in each of the terms have identical sign, and each term can reach large values. The division of the region used above corresponds to dividing the real circular aperture into annular zones (Fresnel zones) whose radii are given by the formula

$$r_n = (F\lambda n + n^2\lambda^2/4)^{1/2}.$$

Evidently the diffractive element described above focuses radiation like a lens.

In the general case, the expression for the radii of annular zones takes the form

$$r_n = \left(\frac{F\lambda n}{p} + \frac{n^2\lambda^2}{p^2}\right)^{1/2}$$

where p is the number of phase quantization levels in the Fresnel zone. This expression defines the so-called 'interference' zone plate which is free of spherical aberration [55]. In the optical wavelength band the quadratic term in the expression for calculating Fresnel zone radii is typically negligibly small. This, however, is not the case for diffractive elements with large relative diameter, as we find in the microwave frequency band.

2.2 Brief classification of zone plates

We will briefly describe the main types and characteristics of zone plates [3]. The progress in zone plates used as analogues of classical lenses and reflector antennas had a prevailing tendency to favour an increase in their diffractional efficiency. All diffractional focusing elements whose principle of construction was based on the Fresnel zone concept can be classified into two large classes:

- lens-type zone plates functioning in the transmission mode,
- mirror-type zone plates functioning in reflection mode.

Amplitude zone plates were first proposed by Fresnel [4]. The feasibility of creating them and the principle of construction followed from the Huygens principle formulated by Fresnel. An amplitude Fresnel zone plate consists of a number of alternating transparent and opaque concentric rings (known as the binary structure). The radii of the delimiting circles coincide with boundaries of Fresnel zones. For rays passing through the borders of two neighbouring zones, the optical path lengths from a point of an object to its image differ by one half of wavelength, $\lambda/2$.

A zone plate with opaque central zone is known as a Soret plate [5, 6]. It is in a way complementary to the Fresnel plate in which the first zone is chosen to be transparent to the incident radiation. As the radiation passes through a Fresnel or Soret plate, one half of the incident energy is lost while the remaining half is distributed over many orders of diffraction. The diffraction efficiency of Fresnel and Soret zone plates in the first diffraction order is 10.1%. The Fresnel and Soret zone plates can work both in transmission and reflection modes. The Fresnel and Soret zone plates function as transformers of a plane wavefront into a spherical one (they focus a plane wave into a 'pointlike' focal point) or as converters of a spherical divergent wave front to a spherical convergent one. Two Soret or Fresnel zone plates, placed parallel to each other and separated by a dielectric layer of thickness form a double Soret zone plate (figure 2.2), which possesses better focusing properties than a single zone plate [38].

A one-dimensional zone plate (with zone borders forming a set of parallel lines) converts a plane wave front into a cylindrical one and thus focuses a plane wave into a line transversal to the optical axis of the zone plate. This zone plate is known as a linear or cylindrical zone plate [39].

Figure 2.2. Double Soret zone plate.

Wood [7, 8], following Rayleigh's suggestion [9], eliminated the main drawback of Fresnel zone plates. Rayleigh suggested that the brightness of the image can be increased by increasing the diffractional efficiency of the main (focal) order of diffraction. To achieve this, one needs to stop blocking and removing the electromagnetic oscillations incident on opaque zones and reverse the phase in these zones instead. The diffractional efficiency of a zone plate is dictated by the shape and depth of the phase profile within a complete Fresnel zone (in a binary zone plate, the complete Fresnel zone comprises one transparent and one opaque zone). The phase zone plate on which the optical thickness of odd zones differs from the thickness of the even zones by $\lambda/[2(n-1)]$, where n is the index of refraction of the optical material, is known as the Rayleigh–Wood lens. The diffractional efficiency of the Rayleigh–Wood lens reaches 40.5%.

Figure 2.3 shows different modifications of the Rayleigh–Wood zone plate. In figure 2.3(a) the required change of phase on the zone plate is achieved by changing the thickness of the dielectric, in figure 2.3(b) it is achieved by using a dielectric with dielectric permittivity ε and air. In figure 2.3(c) the phase inversion structure of the zone plate is composed of two different solid dielectrics. This zone plate was suggested by Wiltse in 1976 [40].

Further increase in the efficiency of zone plates was achieved by increasing the number of quantization levels (discretization levels) of the phase profile: from half-wavelength zone plates to multilevel ones. The required phase increment can be implemented by varying dielectric thickness or by using dielectrics with different values of dielectric permittivity [41].

Still further increase in the number of quantization levels of phase-inverted zones gradually led to the so-called zoned Fresnel lenses (figure 2.4).

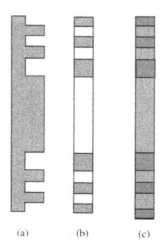

(a)   (b)   (c)

Figure 2.3. Wood zone plates: (a) with negative first phase zone, (b) with zones consisting of solid dielectric and air, (c) with zones consisting of two different solid dielectrics.

Figure 2.4. Fresnel lens.

Single zone plates can be combined to construct multicomponent millimetre wavelength-band lenses possessing better focusing characteristics [24]. In comparison with conventional lenses, zone plates for microwave and millimetre wavelength ranges have smaller thickness, smaller weight and greater efficiency (with radiation losses in the dielectric taken into account).

By design, reflection-type zone plates resemble the design of lens-type zone plates (figure 2.5). Soret or Fresnel zone plates can work both in transmission and reflection modes (figure 2.5(a)). A reflecting half-wavelength Wood zone plate is shown in figure 2.5(b); the design of a quarter-wavelength zone plate is very similar. An analogue of the Wood zone plate is obtained by

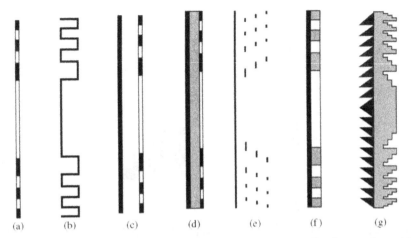

Figure 2.5. Reflection-type zone plates: (a) Soret zone plate, (b) Wood zone plate with profiled reflector, (c) Soret zone plate with flat reflector, (d) zone plate on a dielectric substrate with a reflector, (e) multilevel zone plate on 'freely suspended' zones, (f) half-wavelength Wood zone plate with dielectric reflector, (g) multilevel zone plate with dielectric reflector.

placing a planar metal screen at a distance of one quarter of the wavelength of the radiation used behind a classical Soret or Fresnel zone plate (figure 2.5(c)). A modification of a zone plate shown in figure 2.5(d), suggested by Huder and Menzel [41], consists in filling the air gap between the planar reflector and zone plate with a dielectric. A multilevel planar zone plate with freely suspended zones is shown in figure 2.5(e) [1]. Filling the air gap with a dielectric reduces the thickness of such zone plates [42] but their weight increases. Figure 2.5(f) shows a Wood zone plate with a flat reflector. The height of the phase inversion step is one-half of that for a similar zone plate which works in transmission mode. Figure 2.5(g) shows a completely dielectric reflecting zone plate. Various combinations of the described designs are possible in designing specific zone plates.

Note that a characteristic feature of all reflection-type zone plates discussed above is the presence of a planar or profiled metal reflector. A completely dielectric reflection-type zone plate was suggested in [43, 44], consisting of a multilevel dielectric zone plate and a dielectric diffraction grating (figure 2.6).

An amplitude-type zone plate obtained holographically by using interference of two spherical waves has zones with sine-wave transmissivity. This plate is known as the Gabor lens [10]. The boundaries of the Fresnel zones coincide with the extremums of the interference pattern. A phase plate with sine-wave profile is known as the Gabor phase lens [10]. The diffraction efficiency of the amplitude-type interference zone plate is 6.3% and that of the phase interference plate is 34%.

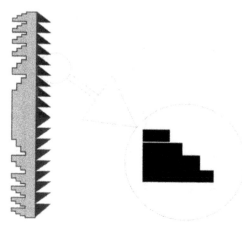

Figure 2.6. Non-metal or reflector zone plate with diffraction grating. © 2000 IEEE. Reprinted with permission from [43].

The shape of Fresnel zone boundaries may vary greatly; it is determined both by the shape of the surface on which they are created and by the required configuration of the focal area.

The kinoform lens differs from the Gabor phase lens in the shape of the zone profile. In a kinoform element the entire energy passing through the lens is concentrated into the main diffraction order. The maximum depth of the profiling is chosen so that the phase jump between two neighbouring zones at a given wavelength is 2π. The kinoform zone structure corresponds to full-period Fresnel zones. This means that the difference between the optical path lengths for rays passing through neighbouring zone boundaries equals an integral number of wavelengths, not of $\lambda/2$. The theoretical diffractional efficiency of a kinoform lens reaches 100%. Walsh was the first to suggest using such lenses as the echelette structure for working in reflected light [11], and Slusarev later proposed it for transmission light [12]. If a kinoform lens is prepared for the optical range using the photolithographic technology, the parabolic (within a zone) phase profile is replaced with a stepped one. The optical-thickness-dependent phase function of optical depth is divided into N levels within 0–2π. One of the pioneer publications on this was made by Goodman and Silvestri [13].

The efficiency of zone plates is further increased by replacing planar zone plates by diffractive elements on three-dimensional surfaces: spherical, parabolical, conical, cylindrical, ogival, pyramidal etc. [1, 39, 45–51]. The efficiency is increased because the number of Fresnel zones on a three-dimensional surface is greater than on the flat one (of identical diameter). Radiation can be incident both on the convex and concave side of the diffractive element if zone plates are created on curvilinear surfaces (figure 2.7). Note that the properties will be different in these two cases [20].

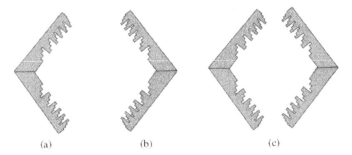

(a)   (b)   (c)

Figure 2.7. Diffractive elements on non-flat surfaces: (a) convex diffractive element, (b) concave diffractive element, (c) two-component diffractive element.

2.3 Fundamental properties of zone plates in the millimetre band

Zone plates on conical, ogival, parabolical and spherical surfaces were first investigated in detail at millimetre wavelengths by Minin and Minin in 1987–1992. Such diffractive elements were designed both for transforming a flat wavefront into a spherical one and for transforming a spherically divergent wavefront into a spherically convergent one.

One of the first papers [14] studied, experimentally and numerically, the focusing and frequency properties of the Fresnel zone plate in the microwave band for a plane incident wavefront at 3.2 cm wavelength. The scalar theory of diffraction was chosen as the apparatus with numerical modelling [15]. It was shown that a Fresnel zone plate possesses the same resolving power as a lens with identical aperture [14], that Fresnel zone plates retain this property as frequency changes slightly (~6%) and that the position of the focal point is approximately a linear function of frequency [14].[1]

What follows is a description of Fresnel zone plates with zone radii R_n calculated for the radiation source and the focal point position removed from the plate to finite distances

$$r_n = \{\tfrac{1}{4}[A + B + n\lambda_0/2 + (A^2 - B^2)/(A - B + \lambda_0 n/2)]^2 - A^2\}^{1/2} \quad (2.1)$$

where A and B are the front and rear segments or the calculated distances from the plate to the source and to the focal point, respectively, and λ_0 is the nominal wavelength for zone plate calculations. In the general case λ_0 may differ from the wavelength λ diffracting on the Fresnel zone plate. For a given diameter D of a Fresnel zone plate, the maximum number $n_{max} = N$ corresponds to the condition $rN \geq D/2$.

---

[1] It should be pointed out that [14] contains errors. The authors of [14] conducted measurements on a half-opaque and a phase-correcting zone plate, and claimed that the latter gave no improvement, which was a wrong conclusion. It was already explained in [61] that their phase-correcting zone plate was incorrectly designed, which makes the measurements and the comparison incorrect.

The following point must be emphasized. A familiar formula from optics is sometimes used to calculate the boundaries of Fresnel zones:

$$r_n = \sqrt{\frac{AB}{A+B}n\lambda_0}.$$

A comparison of this expression with the exact one, (2.1), shows the following.

The exact formula (2.1) for Fresnel zone radii given above does not become meaningless under passage to the limit, that is, for the values of A and B corresponding to diffraction angles of $\pi/2$ or tending to π. At the same time, the limit r_n found with the approximate formula $r_n = [n\lambda AB/(A+B)]^{1/2}$, tends to zero as the diffraction angle tends to $\pi/2$.

The boundaries of spatial zones determined from the approximate formula differ appreciably from ellipsoids. Note here that if $\lambda \neq 0$, the ranges of values of A and B under which both these formulas are physically meaningful, are different. For instance, it follows from the exact formula (2.1) that

$$\frac{-n\lambda}{4} \leq A \leq \infty \qquad \text{for } B \geq 0$$

and

$$\frac{-n\lambda}{4} \leq B \leq \infty \qquad \text{for } A \geq 0.$$

At the same time the approximate formula gives

$$0 < A \leq \infty \qquad \text{for } B > 0$$

and

$$0 < B \leq \infty \qquad \text{for } A > 0.$$

Furthermore, the values of A and B for which the zone radii have both minimal (zero) values and maximum values are different. Thus the exact formula gives the following expression for the maximum Fresnel zone radii:

$$r_{n,\max} = \sqrt{\left[\frac{n\lambda A}{2} + \frac{n^2\lambda^2}{16}\right]} \qquad \text{for } A = B > 0.$$

If the approximate formula is used, we have

$$r_{n,\max} = \sqrt{\frac{n\lambda A}{2}} \qquad \text{for } A = B > 0.$$

One of the conclusions that follows from the expressions above is that the familiar statement on the equality of Fresnel zone areas holds only approximately: the Fresnel zone areas only tend to equality as $A \to \infty$ and $B \to \infty$, with $A \approx B$ and n not very large. Therefore using the approximate formula for calculating the Fresnel zone boundaries is not correct in the millimetre and microwave wavelengths bands.

2.4   Computation of the phase-inversion profile

When calculating the phase profile of high-aperture phase-inversion diffraction, it is also necessary to take into account the height of the phase-inversion step (figure 2.8).

The optical length of a ray along the optical axis of the system is

$$HN + B$$

where H is the maximum height of the phase-inversion profile and N is the index of refraction of the material of the diffractive element.

The optical path of the ray passing through the nth zone is

$$hN + \{[B + (H - h)]^2 + r^2\}^{1/2} - n\lambda.$$

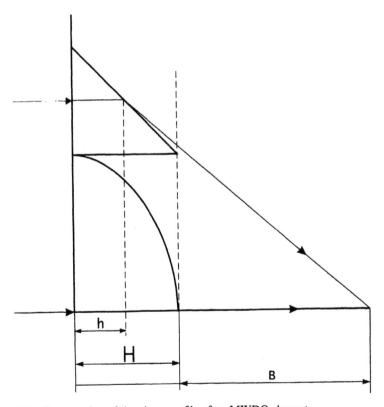

Figure 2.8. Computation of the phase profile of an MWDO element.

Then the condition of phase synchronization for radiation at point B can be written as

$$HN + B = hN + \{[B + (H - h)]^2 + r^2\}^{1/2} - n\lambda.$$

The relationship between the zone radius r and the profile height h is given by the expression

$$r = \{[B + (H - h)N + h\lambda]^2 - [B + (H - h)]^2\}^{1/2}.$$

If $h = 0$ we find the expression for the external radius of the nth zone:

$$r_n^2 = 2n\lambda B + n^2\lambda^2 + 2n\lambda^2/(N - 1), \qquad H = \lambda k/(N - 1), \qquad k = 1, 2, 3, \ldots .$$

(2.2)

In general, the choice of the maximum height of the phase profile H is fairly arbitrary. When choosing the integer k for the height of the phase-inversion profile, we control the number of Fresnel zones that fall within the aperture; the maximum height of profile can vary from zone to zone [16]. In fact in this case the diffractive element (or its part) is computed so as to make it work on a specific harmonic of the radiation.

Expression (2.2) differs from familiar expression for calculation of Fresnel zone radii but coincides with it if $N \to \infty$. Therefore, the expressions for calculation of the radii zone boundaries of a diffractive element are different for phase-inversion and amplitude types.

The internal radius of the nth zone is found for $h = H$:

$$r_n^2 = 2n\lambda B + n^2\lambda^2.$$

Therefore, for realistic values of the refraction index of the material, the internal radius of the $(n + 1)$th zone is smaller than the external radius of the nth zone, that is, zones may overlap. Consequently, 100% diffraction efficiency may not be achievable with high-aperture phase-inversion focusing diffractive elements.

We can evaluate the zone number from which zones start to overlap:

$$2(n + 1)\lambda B + (n + 1)^2\lambda^2 = 2n\lambda B + n^2\lambda^2 + 2n\lambda^2/(N - 1)$$

$$n = \frac{(2B + \lambda)(N - 1)}{(\lambda - N + 1)/2}.$$

The number of the zone boundary beginning with which we need to take zone overlapping into account is a function of the refraction index of the material and of the value of B/λ. The lower the quotient B/λ, the smaller the zone number where overlapping starts. Therefore, long-focal-length systems must be used to produce MWDO elements with minimum effect of overlapping.

A first approximation to the ideal structure of an MWDO element (figure 2.9(a)) is the binary phase structure with two quantization levels:

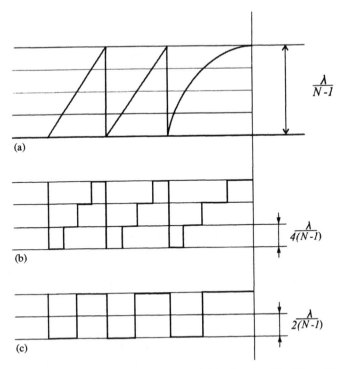

(a)

(b)

(c)

Figure 2.9. Phase profiles of MWDO elements: (a) kinoform profile, (b) profile with four phase quantization levels: 0, $\pi/4$, $2\pi/4$, $3\pi/4$; (c) with two levels of phase quantization: 0, π.

0 and π (figure 2.9(c)). The diffraction efficiency of the binary phase element is 40%. Four-level elements (figure 2.9(b)) are a good approximation to the theoretical profile; their calculated diffractional efficiency is 81%.

Note that diffractive optical elements used in the optical wave band have the number of zones at least $N \approx 200$ to 500 and aperture ratio $D/F \approx 0.002$ to 0.001 and are thus designed to work in a small-angle approximation. The angle subtended by a diffractive element from the focal point does not exceed several degrees of arc. The number of zones of MWDO elements for the millimetre and microwave wavelength ranges is of the order $N \approx 10$ to 50 and the aperture ratio is $D/F \approx 10$ to 0.2. The angle subtended by an MWDO element from the focal point now reaches tens of degrees of arc.

Let us consider how phase elements with the transfer function of the type $t(x, y) = \exp[ik\psi(x, y)]$ can be implemented [17]. Let the field $U_0(x, y, z)$ be the field created by the radiation source and $U_1(x, y, z)$ be the wave field creating a required image:

$$U_0(x, y, z) = A_0(x, y, z) \exp[ik\Phi_0(x, y, z)]$$

$$U_1(x, y, z) = A_1(x, y, z) \exp[ik\Phi_1(x, y, z)].$$

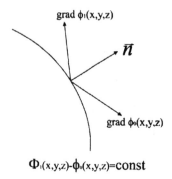

$$\Phi_1(x,y,z)-\Phi_0(x,y,z)=\text{const}$$

Figure 2.10.

The functions A_0 and A_1 describe the distributions of field amplitude and radiation intensity, respectively, and the functions Φ_0 and Φ_1 are the field phases that determine the direction of propagation of the radiation. We consider only phase elements that do not change the amplitude of the incident radiation and therefore can assume $A_0(x,y,z)=A_1(x,y,z)$ in the area of the transforming elements.

As an example, let us consider a reflecting optical element. We can assume that the familiar Snellius refraction law is valid; the incidence and refraction angles are measured off the normal **n** to the surface (figure 2.10).

The direction of propagation of the incident radiation is given by the vector grad $\Phi_0(x,y,z)$, and the direction of propagation of the reflected radiation, by the vector grad $\Phi_1(x,y,z)$. At each point of the reflecting surface the condition must be satisfied for the vector grad $\Psi_1(x,y,z)$ − grad $\Phi_0(x,y,z)$ to be orthogonal to the surface of the mirror element. The equations of the surfaces satisfying this condition can be found from the relation

$$\Phi_1(x,y,z) - \Phi_0(x,y,z) = \text{const} = C.$$

A specific feature of the problem of transformation of wave fronts, for example, by specular surfaces is that its solution is not unique. By choosing various values of the constant we can obtain various surfaces, each of which is a solution of the problem. Note that by changing the eikonal by a constant which is a multiple of $n\lambda$, we not only leave the field of rays unchanged but the values of the field also remain unchanged, that is, the value of the eikonal is in principle found only up to a constant which is a multiple of $n\lambda$.

Let the incident radiation be of the form $U_0(x,y,z) = A\exp(-ikz)$, that is, its eikonal is $\Phi_0(x,y,z) = -z$. We wish to focus this radiation on to a point with coordinates $(0,0,F)$. The eikonal of the reflected field is, to within a constant,

$$\Phi_1(x,y,z) = -(x^2 + y^2 + (z-F)^2)^{1/2}.$$

The equation of the equivalent reflecting focusing surfaces is

$$z - (x^2 + y^2 + (z - F)^2)^{1/2} = c + m\lambda, \qquad m = 0, \pm 1, \pm 2, \ldots .$$

By solving this equation for $z(x, y)$ we find

$$z = \frac{x^2 + y^2}{2(F - c - m\lambda)} + \frac{F + c + m\lambda}{2}.$$

All surfaces described by these expressions are paraboloids of revolution, with foci at the point $(0, 0, F)$. We are interested in the surfaces located close to the plane $z = 0$. Therefore, we set $c = -F$ and investigate the family of surfaces

$$z_m(x, y) = 0.5\left(\frac{x^2 + y^2}{2(F - m\lambda/2)} + m\lambda\right).$$

Approximate forms of these surfaces are shown in figure 2.11. Each of the surfaces shown in figure 2.11 generates an identical wave field that focuses radiation on to $(0, 0, F)$. The same wave field will also be created by a piece-wise-continuous surface shown in figure 2.11 by the solid curve (to within the accuracy of the shadow effects, provided the approximation of geometric optics is still valid, which allows us to neglect the scattering of radiation on the lines of discontinuity and at sharp angles of the surface). Note that ridge heights are not identical on composite surfaces.

From the point of view of technology it is essential that if an element is profiled as a piecewise-smooth surface, then the relative accuracy of manu-facturing this profile is not as good as in the case of a smooth surface.

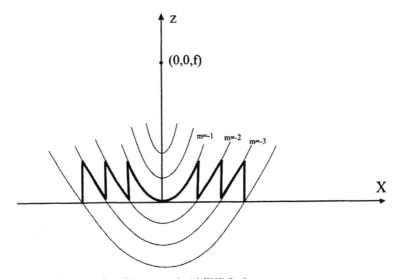

Figure 2.11. Phase profile of the 'specular' MWDO element.

2.5 Experimental setup

Experimental studies of focusing and frequency properties of MWDO elements based on Fresnel and Rayleigh–Wood zone plates (figure 2.12) were conducted on a setup whose principal diagram is shown in figure 2.13 [1].

A backward-wave tube (BWT) was used as a source of millimetre radiation. To set the working frequency of the BWT, radiation was fed to a calibrated cymometer through a coupler. The signal was received by a D-407 type diode and recorded on a selective nanovoltmeter. The output of the BWT was coupled via an attenuator to a pin-diode waveguide modulator. The modulator cut-off coefficient was about 10^5. The modulation frequency was imposed by the G6-15 generator and was chosen to be 1 kHz. Frequency control was carried out by monitoring the voltage on the retarding system of the BWT using a digital voltmeter.

A Fresnel zone plate in a metal frame was fixed on an optical bench at prescribed points; off-axes deviations did not exceed 6'. A pointlike source formed a spherical wave which illuminated the Fresnel zone plate. The illumination non-uniformity over the zone plate within its aperture did not exceed 8%, and non-planarity was not more than 1 mm. Computations showed that the effects of these factors on the structure and shape of the diffractional spot were negligibly small.

The receiver was fixed on an optic stage in the region of the Fresnel zone plate focal point and had micrometric screws for adjustment in various directions. Verniers were used to measure displacements of the receiver. The signal received by the receiver was fed to a second selective nanovoltmeter which worked in the $\pm7\%$ band at the 0.5 level of the modulation frequency. The shape of the signal was monitored on the oscilloscope.

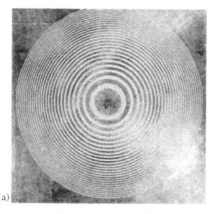
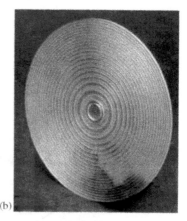

(a)   (b)

Figure 2.12. Zone-plate-based diffractive object lenses for the 140 GHz frequency band: (a) of amplitude type, (b) of phase-inversion type.

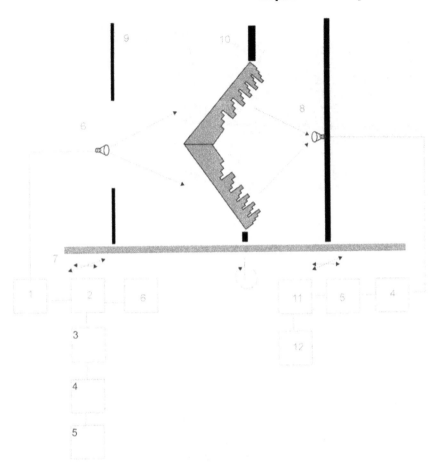

Figure 2.13. Principal diagram of the experimental setup: 1, attenuator; 2, directional coupler; 3, wavemeter; 4, microwave receiver; 5, selective nanovoltmeter; 6, HF signal generator; 7, optical bench; 8, open end of waveguide; 9, protection shields; 10, MWDO element; 11, digital recording oscilloscope; 12, continuous recording plotter.

To remove reflections from various metal objects and to protect the personnel from microwave radiation, metal screens coated with microwave absorbing material (5–10 mm thick Porolon saturated with Aquadag) were erected between the radiation source, the diffractive element and the receiver.

The radio image was visualized in two ways: by the method of movable probe and by thermal visualization of microwave radiation. In the first method, the diffraction field was mechanically scanned by an open end of a waveguide used as receiver connected to a detector head. The spatial distribution of the microwave field intensity was reconstructed from the obtained discrete data (figure 2.14(a)). In the second method, a thermal vision monitor recorded the thermal field using a screen absorbing the microwave radiation

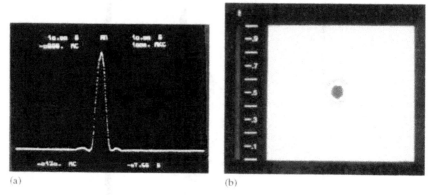

Figure 2.14. Radio image of a pointlike source obtained by (a) movable probe method and (b) thermal visualization of microwave field.

(figure 2.14(b)). The material absorbing the microwave energy radiated by the screen was a Lavsan film about $10\,\mu m$ thick, evaporation-coated with about $100\,\text{Å}$ thick aluminium layer.

The former technique has maximum sensitivity and is suitable for obtaining quantitative information. The latter technique has low sensitivity and considerable delay time and was used for qualitative rapid analysis of the shape of the focal region.

The calculated wavelength of the backward-wave tube was restored after each series of measurements. No drift of signal amplitude was observed at the focal point of the Fresnel zone plate. The stability of modulation by the pin-diode modulator was found from the stability of pulse generation by the G6-15 oscillator. Measurements revealed no drift in diode characteristics. Additional details about the experimental setup can be found in [18].

The factors contributing to the total experimental error were the measurements errors of independent quantities (receiver coordinates, signal amplitude, wavelength), deviations from the axial alignment of the system, non-planarity of the Fresnel zone plate, and non-pointlike dimensions of the receiver and source of microwave radiation.

On the whole, 600 measurements of field intensity were carried out in the experiments described below. For given values of arguments, the values were divided into groups of 15 to 20 measurements each. The average values $\langle W \rangle_j$ and $\langle W^2 \rangle_j$, were found within each group, j being the group number. The average values $W_j = \langle W \rangle_j$ are then presented as experimental values. The quantity $\varepsilon_e = \langle \langle W^2 \rangle_j - W_j^2 \rangle^{1/2}$, which is the mean square deviation from average values within groups over the entire set of data, was used as a measure of experimental error. The quantity $\varepsilon_t = \langle W_{tj}^2 - W_j^2 \rangle^{1/2}$, where W_{tj} were found by calculation of the particular values of arguments, were used as measure of deviation between the theory and the experimental data. The values ε_e, ε_t did not exceed 3 and 5%, respectively.

2.6   Field intensity distribution in the focal zone of the zone plate

The structure of the diffraction field in the focal region for the axial position of a pointlike radiation source was experimentally studied at wavelengths which are denoted below by λ_j, $j = 0, 1, 2, 3$. The experimental data are compared below with calculations based on the Fresnel–Kirchhoff scalar theory of diffraction.

The longitudinal distribution of field intensity for the actual position of the source reveals the maximum whose position on the OZ axis depends on λ. The dependence of the rear interval on wavelength can be found from expression (2.1):

$$B_\lambda(n) = \frac{A^2 + (A + n\lambda/2)^2 - 2(A^2 + r_n^2)^{1/2}(A + n\lambda/2)}{2(A^2 + r_n^2)^{1/2} - 2A - n\lambda}.$$

If B is averaged between its extreme values

$$B = 0.5(B_\lambda(1) + B_\lambda(N)) \tag{2.3}$$

then expression (2.3) coincides, within the accuracy of calculations and measurements, with calculations by the Kirchhoff scalar theory as $\Delta\lambda/\lambda_0$ varies within $\pm 20\%$ and with experimental values for all λ_j. These results are plotted in figure 2.15.

A similar dependence for a zone plate designed for focusing a wave with planar wavefront was obtained in [52]:

$$B_\lambda(p) = r^2 p/2N\lambda - N\lambda/2p.$$

Here p is the number of phase quantization levels and N is the total number of Fresnel zones.

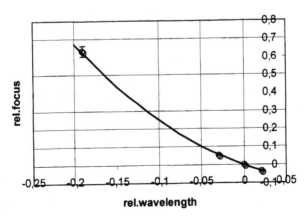

Figure 2.15. The position of the focusing area as a function of wavelength of the incident radiation [18].

As λ changes, the distance Δ_z from the maximum to the first minimum of field intensity distribution along the OZ axis changes too. The following two relations can be obtained from the in-phase and anti-phase conditions of waves at the geometric focal point and at the point corresponding to the first minimum of the distribution:

$$(A^2 + \rho^2)^{1/2} + (B_\lambda^2 + \rho^2)^{1/2} - (B_\lambda + A) = m\lambda$$

$$(A^2 + \rho^2)^{1/2} + (B_\lambda^2 + \rho^2)^{1/2} - (B_\lambda + A + \Delta_z) = m\lambda \pm \lambda/2$$

where ρ is a radius within the aperture of the Fresnel zone plate, and m is an integer. Solving these two equations simultaneously, we find

$$\Delta_z = \frac{\lambda^2/4 \pm \lambda(B_\lambda^2 + \rho^2)^{1/2}}{2(B_\lambda^2 + \rho^2)^{1/2} - B_\lambda \pm \lambda/2} \tag{2.4}$$

which gives results not very different from those given in [19] for large values of B:

$$\Delta_z = \pm 8\lambda(B/D)^2. \tag{2.5}$$

It must also be mentioned that expression (2.4) describes the field intensity distribution along the optical axes within the focal region of the zone plate better than (2.5) as far as the physics of the phenomenon is concerned because it shows the asymmetry of the given distribution relative to its maximum.

The best agreement of (2.4) with calculations based on the Fresnel–Kirchhoff integral and the experimental data is obtained if the mean radius of the middle point of all zones of the Fresnel zone plate is substituted into (2.4):

$$\rho = \tfrac{1}{2}N \sum_{n=1}^{N} (r_n + r_{n-1}).$$

The results of the calculations using the scalar diffraction theory and formula (2.4), and the results of measurements, are plotted in figure 2.16.

Figure 2.17 shows on the whole 370 values of field intensity W (measured and calculated via the Fresnel–Kirchhoff integral) in relative units as a function of dimensionless parameter $(z - B_\lambda)/\Delta_z$. We see from this plot that theoretical and experimental results practically coincide for $\lambda_0, \lambda_1, \lambda_2$, and only for λ_3 we do observe differences of up to 5–10% in the neighbourhood of the first minimum.

The following must be pointed out here. We know [12] that secondary foci are observed in the optical wave band for diffractional elements at distances $F/3, F/5, \ldots$. The field intensity at the point $F/3$ comes to about 10% of the intensity at the main focal point. Our studies showed that in the microwave band the field intensity in this region for zone plates with

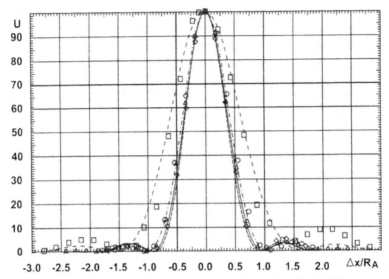

Figure 2.16. Profile of the diffraction spot for some values of $\Delta\lambda/\lambda_0$, % [18]:

	Theory	Experimental data
0.00	———	$\triangle$
1.90	— — —	$\diamond$
−4.64	—·—·—·—	$\bigcirc$
−19.8	— — —	$\square$

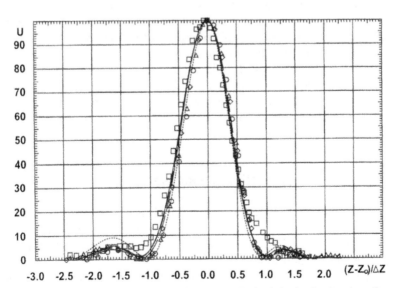

Figure 2.17. Field intensity distribution along the optical axis in the focal region of a zone plate (the notation is the same as in figure 2.16) [18].

$D/F \approx 0.5$ to 3 is of the same order as are the first side lobes of scattering. This is another principal difference between optical diffractional elements and those for the microwave and millimetre ranges.

2.7 Transverse field intensity distribution in the focal area of the zone plate

The spherical wave formed by an axially aligned radiation source was incident on the surface of the Fresnel zone plate and was then focused into a certain volume on the optical axis of the zone plate. A measurement probe scanned the plane located within the focal point area perpendicularly to the axis of the Fresnel zone plate. More than 200 measurements were carried out. The results were then compared with theoretical calculations.

The data of the experiments and normalized results of calculations are presented in figure 2.16. The ordinate axis is the intensity W in arbitrary units and the abscissa axis is the transverse coordinate x in units of the Airy radius $R_a = 1.21967\lambda B/D$. When a Fresnel zone plate is illuminated by radiation with wavelength differing from the calculated one by up to 20%, the experimental results coincide, within the accuracy of the experiment, with the calculated values obtained using the scalar Fresnel–Kirchhoff diffraction theory.

The experimentally obtained resolving power of the Fresnel zone plate for $\lambda \geq \lambda_0$ is somewhat better than the resolution found using the Rayleigh criteria. The resolving power deteriorates somewhat for $\lambda < \lambda_0$. The calculations and measurements for $\lambda = \lambda_3$ indicate broadening of the central maximum in the field intensity distribution in the focal region by a factor of 1.5 and to an increase in the intensity of side lobes.

We thus find from detailed studies of the focal properties of Fresnel zone plates that if frequency changes by up to 20% of the frequency for which a zone plate was designed, the zone plate maintains its focusing properties and thus provides resolving power not very different from the diffraction-restricted limit.

2.8 Information properties of zone plates

The geometry of the problem is clarified in figure 2.18. The centre of the Fresnel zone plate is juxtaposed with the origin of a rectangular coordinate system (x, y, z), with the OZ axis directed normally to the zone plate and coinciding with its optical axis.

The number L of coherent emitters are fixed in the plane $z = z_n$ along a straight line parallel to the axis OX, at points $P'_l(-h'_l, 0, z_n)$ where $l = 1, 2, \ldots, L$; h'_l is height measured in the negative direction of the OX

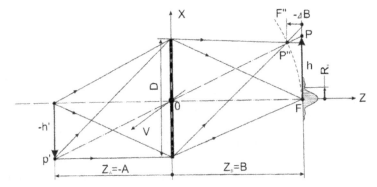

Figure 2.18. Geometry of the problem [20].

axis (only one source is shown in figure 2.17). The positions of images at the points P'_l are given in two ways: either at points $P_l(h_l, 0, z_B)$ conjugate to points P_l, where $h_l = B/Ah'_l$, or on an arc $F'F''$, which is the curve of best focusing. In the latter case a local rectangular coordinate system (x'', y'', z'') is chosen at the intersection point O of the arc $F'F''$ and the straight line $P'_l P_l$, so that the axis Oy'' points along the Oy axis, the axis Oz'' along $P'_l P_l$, and Ox'' along the tangent to the arc $F'F''$. Consequently, the image of the source P_l is the point $P''_l(x''_l, 0, z''_l)$.

2.9 Single pointlike source

A series of numerical and physical experiments was run to evaluate the field of view and the number of the elements in the frame for the objective lens based on the Fresnel zone plate. These experiments investigated the shape of the diffraction solid to evaluate the quality of the image of a pointlike source of radiation [20].

It is clear from the geometry of the problem that as the pointlike source of radiation moves away from the optical axis, its image moves not along a plane but along a certain surface $F'F''$; we will refer to it as the best focusing surface (BFS) (see figure 2.18).

In the first approximation, the profile of the generatrix of the BFS can be found from the condition of equality of the optical path of the rays passing from the radiation source to its image through the centre of the Fresnel zone plate and the boundary of the nth Fresnel zone. Taking also into account the geometry of the problem we can write:

$$[(x_1 - x_n)^2 + z_1^2]^{1/2} + [(x_2 - x_n)^2 + z_2^2]^{1/2}$$
$$= [(z_2 - z_1)^2 + (x_2 - x_1)^2]^{1/2} + 0.5n\lambda \qquad (2.6)$$
$$x_1 z_2 = x_2 z_1$$

where (x_1, z_1), (x_2, z_2) are the coordinates of the radiation source and its image, respectively, and x_n is the distance from the centre of the zone plate to a point at the boundary of the nth Fresnel zone. When writing a set of equations (2.6) we assumed that $y_1 = y_2 = 0$, which does not diminish the generality of the problem. The set (2.6) was solved numerically to achieve better agreement of computations and experimental data and the values for (x_2, z_2) were averaged for the extreme values of n ($n = 1$ and $n = N$).

For Fresnel zone plates with $D \geq 200\lambda$, $A = B = 2D$ experiments were run to find the cross sections of the diffraction solid formed by a single radiation source ($L = 1$) on the surface of best focusing for different ratios of $h/R_a = 0$, 16, 20, 24; hence, the diffraction solid was presented as a set of two-dimensional cross sections $W(x'')$ at $y = 0$ and $W(y'')$ at $x'' = x_0$.

The results of these experiments are shown in figures 2.19 and 2.20 where we also show the results of computations using the Fresnel–Kirchhoff scalar diffraction theory. The results are represented by a set of curves. The arguments of the functions $W(x'')$ or $W(y'')$ are measured off the position of the image centre $P_1'(x_0'', 0, z_0'')$, where x_0'' and z_0'' are the coordinates of the intersection of the line $P'P$ and the arc FF'' for a given $h' = B/Ah$.

It is clear from the results shown that the agreement between experimental data and computation results is quite satisfactory: the mean-square deviation of the entire series of measurements from the calculated curves is about 3%.

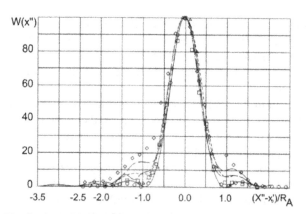

Figure 2.19. Distribution of field intensity source on the best-focusing surface for off-axis position of the radiation source [20].

x/R_a	Theory	Experiment
0	—·—·—·—	○
16	– – – –	□
20	—– —– —	◇
24	——————	△

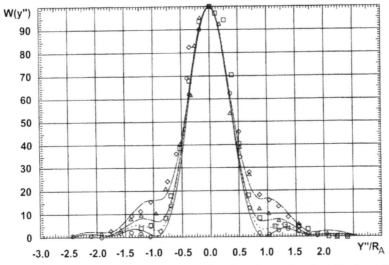

Figure 2.20. Field intensity distribution on the best-focusing surface for off-axis position of the radiation source [20] (notation identical to that of figure 2.19).

The field of view of the zone plate can be evaluated following [21]. The momentum response of the radio objective lens can be written in the form

$$h = \frac{\cos(\mathbf{n}, \mathbf{r})}{\lambda^2 rB} \exp(ikr) \exp[-ik(x^2 + y^2)/2F]$$

$$\times \exp\{ik[(x_1 - x)^2 + (y_1 - y)^2]/2B\} \, dx \, dy$$

$$P = \begin{cases} 1, & \text{if } |x|, |y| \le D \\ 0, & \text{if } |x|, |y| > D. \end{cases}$$

Taking into account that normally $A, B \gg x, y$ and expanding r into a series, we obtain

$$r = A\{1 + 0.5[(x_0 - x)/A]^2 + 0.5[(y_0 - y)/A]^2 - 0.125[(x_0 - x)/A]^4 - \cdots\}.$$

The quadratic term of the expansion gives the Fresnel approximation. The term proportional to x^4 takes into account third-order aberrations. Substituting r into expression for h, we can analyse the resulting cofactors. The term $\exp\{-ik[(x_0 - x)^4 + (y_0 - y)^4]/8B^3\}$ in the integrand dictates the contribution of aberration errors in the non-paraxial case. If $y_0 \approx y$ and $x_0 \approx x$, we have the paraxial approximation: the exponent in the exponential is much less than 1 and the momentum response of the system coincides with the response for the paraxial approximation h_0 in the case of off-axis radiation source $(x_0, y_0) > D/2$, so that

$$\exp\{-ik[(x_0 - x)^4 + (y_0 - y)^4]/8B^3\} = \exp[-ik(x_0^4 + y_0^4)/8B^3].$$

In this case the co-factor under discussion can be factored out of the integral. It then follows that

$$h = h_0 \cos^2 \varphi \exp[-ik(x_0^4 + y_0^4)/8B^3] = h_0 \cos^2 \varphi \exp[-i\pi B \tan^4 \varphi/4\lambda]$$

where $\tan \varphi = (x_0^2 + y_0^2)^{1/2}/B$ is the field of view of the image-forming system.

According to the Rayleigh wave criterion, the additional phase shift produced when the image is formed in a coherent system must not exceed $\pi/2$, that is, $\alpha = 2\pi(x_0^4 + y_0^4)/8\lambda B^3 \leq \pi/2$, whence

$$\tan \varphi \leq (2\lambda/B)^{1/4}.$$

This estimate of the field of view is valid in the Fresnel approximation provided $r < 2D^2/\lambda$, where D is the diameter of the Fresnel zone plate (FZP) objective lens. If the objective lens is located in the far-field zone, the field of view is determined by its directional diagram.

It is possible to use simple geometric-optics arguments to evaluate the effect of the transverse (relative to the optical axes) displacement of the radiation source on the quality of the image of a point in the assumption of infinitely thin diffractive objective lens [22]. The phase function of the lens can be written in the form

$$\varphi_0 = 2\pi/\lambda[z_1 + z_2 - (z_1^2 + \rho^2)^{1/2} - (z_2^2 + \rho^2)^{1/2}]$$

where z_1 and z_2 are the distances along the optical axis from the FZP to the radiation source and to its image, respectively, and ρ is the distance from the centre of the FZP to a point within its aperture. Now let the source be displaced by a distance Δx transversely to the optical axis Ox. The phase of the incident wave is then

$$\varphi_1 = 2\pi/\lambda\{[z_1^2 + (x + \Delta x)^2 + y^2]^{1/2} - z_1\}$$

and the phase of the wave emerging from the FZP is

$$\varphi_2 = \varphi_1 + \varphi_0.$$

Let us assume that the wave converges at a point $P(z_2, x_2, 0)$. If the aberrations introduced by the FZP are small, the phase of this wave is

$$\varphi_3 = 2\pi/\lambda\{-[z_2^2 + (x + x_2)^2 + y^2]^{1/2} + z_2\}.$$

The difference $\Delta\varphi = \varphi_2 - \varphi_3$ is the error of the wave with respect to the spherical reference wave whose centre lies at a point P. We assume that the coordinate of the reference wave is related to Δx by the formula $\Delta x_2 = -\Delta x z_2/z_1$. Taking now into account that the Strehl number (St) for small aberrations is

$$\text{St} \approx 1 - (\Delta\Phi)^2, \qquad (\Delta\Phi)^2 = (\Delta\varphi^2) - (\Delta\varphi)^2$$

expanding the expression for φ_2, φ_3 into a series, and averaging over $\Delta\varphi$, we obtain

$$\mathrm{St} \approx 1 - \left(\frac{2\pi}{\lambda}\right)^2 \left(\frac{1}{z_1^2} - \frac{1}{z_2^2}\right) \frac{\Delta x^2 D^2}{z_1^2 32}.$$

Figures 2.21 and 2.22 give the results that illustrate modelling of the image of a pointlike radiation source as it is moved transversely away from the optical

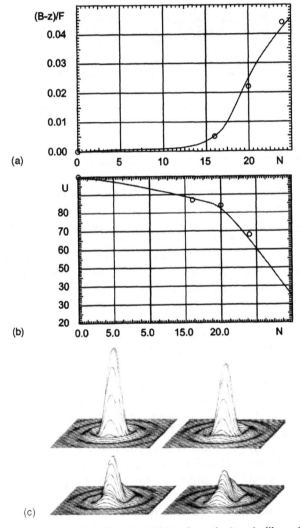

Figure 2.21. (a) Shape of the best focusing surface for a single pointlike radiation source: ———, modelling: O, experimental data. (b) Signal amplitude on the best focusing surface as a function of the positional number of the off-axis element: ———, modelling; O, experimental data. (c) Image of a pointlike off-axis radiation source [20].

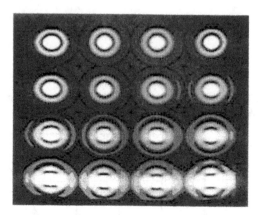

Figure 2.22. Nonlinear processing of the image of a pointlike off-axis radiation source: from left to right and from bottom upwards; the displacement increment between two successive images equals the Airy radius [20].

axes. Figure 2.21(a) shows the shape of the best focusing surface and figure 2.21(b) shows the signal amplitude on this surface as a function of the number indicating the position of the off-axis element. The distance from the emission source to the optical axis in figures 2.21(c) and 2.22 increases with increment $\Delta x = R_A$, from bottom upwards and from left to right on the figure. Figure 2.21(c) shows the image of a pointlike source in isometric presentation, and figure 2.22 shows images after high-frequency filtration for better perception of the structure of the side lobes on the image.

The results presented above show that on the surface of best focusing, an FZP with the aperture ratio $D/B \approx 1/2$ and lens aperture $D/\lambda \approx 200$ can resolve about 50×50 diffractive elements within the frame. If the image frame is flat, then the results demonstrate that the number of elements of the image falls off to $N^2 \approx 30 \times 30$ [18, 23].

Numerical and experimental studies were also conducted along similar lines but at wavelengths deviating from the calculated value λ_0 by $\pm 10\%$. It was shown that in accordance with the results of [18] the position of the focal plane (and of the best focusing surface) changed depending on the value of λ. All other properties of Fresnel zone plates were unchanged (if the data are properly presented in dimensionless format).

It was shown above that the best focusing surface of diffractive objective lenses is a sphere with the radius equal to the rear segment of the MWDO element at the chosen radiation wavelength. We can anticipate *a priori* that if, at a given wavelength λ, we have $\max(|\mathrm{d}s|) \ll \lambda$ where $\mathrm{d}s$ is the optical path length difference for the rays connecting the radiation source to the surface of the sphere cantered at a point on the source image and having the radius equal to the rare segment of the objective lens, then waves will be focused in the same way as the aberration-free wavefront is. In fact we will be able to show that for a quasi-ideal wavefront it is sufficient to meet a less rigid condition,

$$\max(|ds|) < \lambda/4.$$

To prove the condition shown above, we computed field amplitudes for high-aperture short-focus parabolic diffractive objective lens as functions of distance Δz from the geometric focal point along the optical axis and the transverse distance Δx for the cases of the ideal wavefront and a real wave-front. An analysis of these computations demonstrated that the maximum of field intensity on the optical axis lies not at the geometric focal point but nearer the objective lens: this shift is about 10 mm for the ideal wavefront with the lens D/F of about 2.5 and about 60 mm for the real wavefront and $D/\lambda = 150$. In the region of maximum field intensity for the ideal wave-front, the field intensity for a real objective lens reaches about 0.6 of the maximum value. As D/F diminishes, the displacement of the real focal point from the geometric focal point decreases too.

The field structure for the ideal and real objective lenses is shown in table 2.1 as a function of the displacement of the observation point transversely to the optical axis. The focal field was created by a pointlike source; calculation results for field intensity distribution cross section transversely to the optical axis are given for the geometric focal area of a test sphere ($\Delta_z = 0$), ideal wave-front ($\Delta_z = 10$, U_i) and real wavefront ($\Delta_z = 60$, U_a). For comparison we give the corresponding calculations using the Airy formula for distances $F + \Delta_z$. In all cases the intensities on the optical axis ($\Delta x = 0$) were normalized to 100.

The data shown in the table demonstrate that the real aberration wavefront differing from the spherical by not more than $\lambda/4$ generates the

Table 2.1.

$\Delta x/R_a$	$\Delta_z = 0$		$\Delta_z = -10$		$\Delta_z = -60$	
	U_i	U_a	U_i	U_a	U_i	U_a
0.00	100	100	100	100	100	100
0.25	79.32	79.22	79.00	78.89	76.20	76.20
0.50	37.15	36.93	36.51	36.24	30.81	30.72
0.75	7.44	7.28	7.11	6.84	3.97	3.83
1.00	0.003	0.00	0.133	0.002	0.53	0.30
1.25	1.46	1.52	1.61	1.57	2.03	1.75
1.50	1.23	1.24	1.19	1.17	0.76	0.58
1.75	0.09	0.09	0.07	0.06	0.05	0.03
2.00	0.19	0.20	0.23	0.23	0.37	0.41
2.25	0.39	0.40	0.39	0.39	0.18	0.19
2.50	0.10	0.10	0.08	0.07	0.01	0.01

diffractional image of a point which is practically indistinguishable from the ideal one.

Therefore, when the diffraction objective lens is optimized (especially a short-focus lens), it is advisable to use a criterion which minimizes the deviation of the wavefront behind the lens from the sphere whose radius and position of the centre are specially adjusted (this can be done by minimizing the mean-square deviations of the front from spherical shape).

However, the following aspects must also be taken into account. In the general case, the dependence of the quality of the generated image on geometric aberrations is dictated in the general case by the ratio of the radiation wavelength to the values of aberration. If the wavelength is small in comparison with aberrations, aberrations completely dominate the quality of images. This situation is typical in optics.

If, however, the wavelength is long, the behaviour is substantially different. In this case objective lenses can be successfully used which are completely unsuitable in the short-wavelength spectrum. Therefore, only limited minimization of geometric aberrations is advisable, say, in the millimetre wave band because better correction would anyway be masked by diffraction phenomena. In other words, if the objective lens already generates an image of almost the same quality as an ideal objective lens at the given wavelength, further reduction of geometric aberrations would serve no useful purpose.

2.10   A system of pointlike coherent sources of radiation

Numerical experiments were run to study the effect of interference between pointlike coherent emitters on the shape of the best focusing surface and on the number of resolved elements within the frame. Calculations using the Fresnel–Kirchhoff integral were run simultaneously for nine ($l = 9$) radiation emitters spaced by the same distance

$$h_l = B/A8R_a(l - 1), \qquad l = 1, \ldots, 9.$$

Computations of field intensity $W = |U|^2$ were carried out at different points on the XOY plane and were displayed on a raster greyscale screen. The result was a visualized set of converging waves shown in figure 2.23 (the bunch of rays from the axial source numbered by $l = 1$ is shown only in the half-space $x \geq 0$).

This study also confirmed the conjecture that that the field of view of FZP-based objective lenses can be extended by increasing D and at the same time maintaining constant aperture ratio, $D/B = 1/2$.

As follows from figure 2.23, if $D \approx 300\lambda$, the quality of the $l = 5$ image of the source is still satisfactory on a certain segment of the best focusing arc. Hence, the maximum number of image elements resolved on a curvilinear row (using the Rayleigh criterion) is $N \approx 64$.

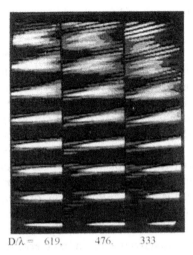

D/λ = 619,   476.   333

Figure 2.23. A visualized family of convergent beams [18] for lens apertures $D/\lambda = 619$, 476, 333.

As D increases by a factor of two, the number of elements in one row of the image increases to 80–90, that is, the diameter of the diffractive element increases faster than the number of elements resolved in a row. Computations carried out in the case of $D = 200\lambda$ provide about 50 image elements in a row.

As also follows from figure 2.23, the best focusing surface has substantial curvature and is concave towards the FZP. The profile of the FF'' arc (see figure 2.18) is approximately represented by the fractional-power curve:

$$\Delta B = Hh^{3/2} \tag{2.7}$$

where ΔB is the sag of the arc FF'' relative to the plane $z = z_B$ in millimetres and h is height in millimetres. Estimates of the constant H for $D/\lambda = 300$, 450 and 600, obtained from figure 2.23 using formula (2.7), give values close to 1.9×10^{-2}, 1.7×10^{-2} and $1.6 \times 10^{-2}\,\mathrm{mm}^{-1/2}$ for $l = 5$, 6 and 7, respectively.

2.11   The effective spectrum range of a zone plate

When a phase-inversion diffractive element is illuminated with radiation of wavelength that differs from the reference wavelength, then the shape of the generated wavefront differs from the calculated one. In the general case the errors stem from the following two factors:

• dispersion of the index of refraction changes optical path length differences through the diffractive element;

- the phase synchronization condition for the Fresnel zones of the diffractive element is disturbed in the calculated focal area.

Errors of the former type are similar to chromatic aberrations of ordinary transmissions lens elements. In the millimetre and sub-millimetre wavelength ranges, the dispersion of the refractive index of dielectrics is typically low.

The latter factor is specific precisely for diffractive elements in which the maximum optical pathlength difference must correspond to the phase $2\pi m$ for the selected wavelength ($m = 1, 2, \ldots$). As wavelength changes, even if the dispersion of the refractive index is negligible, the condition of phase synchronization is violated. Obviously, the phase synchronization of radiation from Fresnel zones will again set in if the maximum phase difference for the wavelength used is nearly $2\pi m$, or $4\pi m$, or $8\pi m$ etc.

It is not difficult to use expression (2.2) to evaluate the effective spectral range of the FZP. Let us consider an FZP designed for the transformation of a divergent spherical wavefront into a convergent spherical one. The minimum admissible wavelength λ_{min} is found from the condition $B \to \infty$; setting the numerator in (2.2) to zero, we find

$$\lambda_{min} = 2/n[(A^2 + r_n^2)^{1/2} - A]. \tag{2.8}$$

Likewise, $B \to 0$ at a maximum admissible wavelength λ_{max}, so that in view of (2.8), formula (2.2) gives

$$\lambda_{max} = \lambda_{min} + 4\{2A[A + (A^2 + r_n^2)]^{1/2}/n^2\}^{1/2}.$$

If we assume that $\lambda_{min} \ll \lambda_{max}$, then

$$\lambda_{max} \approx 2/n\{2A[A + (A^2 + r_n^2)]^{1/2}\}^{1/2}. \tag{2.9}$$

Formulas (2.8) and (2.9) yield the working spectral range of the Fresnel zone plate. If the FZP converts a plane incident wavefront into a convergent spherical front, we can easily obtain, in a similar way, that

$$\lambda_{min} = 0, \qquad \lambda_{max} = 2r_n/n.$$

This indicates that this type of FZP has very wide frequency band.

The following expression was derived in [53, 54] for the frequency band of the zone plate:

$$\Delta w = p/Nw.$$

Here Δw is the frequency band of the zone plate at 3 dB level, w is the calculated radiation frequency, p is the number of phase quantization levels and N is the total number of Fresnel zones. For instance, if we take a half-wave zone plate of 1.2 m in diameter, designed for the 12 GHz frequency, with the aperture ratio $D/F = 1.39$, then the total number of Fresnel zones is $N = 15$. The frequency band of such a zone plate is, in correspondence with the above expression, 1.6 GHz. An exact calculation using the Fresnel–Kirchhoff diffraction theory gives 1.5 GHz.

Therefore, the frequency band of the zone plate increases as the number of phase quantization levels increases, and decreases as the number of Fresnel zones on its surface increases.

We can evaluate the frequency band of a zone plate in paraxial approximation depending on its design parameters. To do this, we make use of the zone plate frequency properties. We find the focal length as a function of wavelength from the expression for the Fresnel zone radii. In the first approximation, the half-width of field intensity distribution along the optical axis in the focal area (defined in relation to its maximum) is

$\Delta z_+ \approx 8\lambda(F/D)^2$ for the positive direction (away from the zone plate)

$\Delta z_- \approx -8\lambda(F/D)^2$ for the negative direction (towards the zone plate).

The width of the field intensity distribution along the optical axis in the focal area can be found as

$$\Delta z \approx |\Delta z_+| + |\Delta z_-|.$$

Now we assume that as we go from one wavelength to another, the focal length changes by a value equal to the width of the field intensity distribution along the optical axis in the region of the focal point:

$$\Delta z \approx 16\lambda(F/D)^2.$$

By setting the focal lengths for two extreme wavelengths equal to the expression given above and doing some elementary transformations, we easily obtain the following estimate:

$$F(\lambda_1) - F(\lambda_2) = \frac{D^2}{4N}\left(\frac{\Delta\lambda}{\lambda^2}\right) = 16\lambda\left(\frac{F}{D}\right)^2$$

hence

$$\frac{2\Delta\lambda}{\lambda_0} \approx 64\left(\frac{\lambda}{D}\right)\left(\frac{F}{D}\right).$$

The estimate of the frequency bandwidth obtained above for the zone plate connects two of its most important parameters: the aperture ratio and the lens aperture.

Figure 2.24 shows how the frequency band of the zone plate depends on the ratio D/F for two values of aperture ratio. The main conclusion is that the zone plate frequency band decreases as the ratio D/F diminishes, and decreases as the aperture ratio increases.

Let us evaluate at what parameters of the zone plate its frequency properties become nonlinear. For simplification, we will consider the focusing of incident plane wavefront to a point. In this case the Fresnel zone radii are

$$r_n^2 = 2Fn\lambda_0/p + (n\lambda_0/p)^2$$

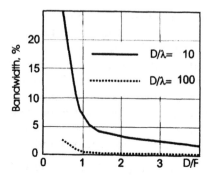

Figure 2.24. Zone plate frequency bandwidth.

where p is the number of phase quantization levels, $p = 2, 3, 4, \ldots, N$. Hence

$$F'(\lambda) = \frac{r_n^2 - (n\lambda/p)^2}{2n\lambda/p} = \frac{2Fn\lambda/p + (n\lambda_0/p)^2 - (n\lambda/p)^2}{2n\lambda/p}.$$

This equation can be rewritten in the form

$$F'(\lambda) = \frac{\lambda_0}{\lambda}\left(F + \frac{n\lambda}{2p} - \frac{n\lambda^2}{2p\lambda_0}\right).$$

This expression implies that

- $F'(\lambda)$ is linear in λ_0/λ;
- there is a constant term $n\lambda/2p$, that is, parallel translation of the linear dependence $F(\lambda)$;
- there is a nonlinear term $n\lambda^2/2p\lambda_0$.

Let us evaluate the number n of the zone at which the nonlinearity of the function $F'(\lambda)$ becomes significant.

Let the additional term to F be comparable with the focusing depth of the zone plate, that is, to the displacement of the focal point by ΔF in response to changes in λ.

The focal depth is $\Delta_z \approx 2\lambda(F/D)^2$. We denote $F/D = K$, then

$$4K \approx \frac{n\lambda_0}{2p} - \frac{n\lambda^2}{2p\lambda_0}$$

hence

$$n \approx 8K^2 p \frac{\lambda_0^2}{\lambda_0^2 - \lambda^2} = \frac{2p}{\lambda_0^2 - \lambda^2} F_0^2$$

where $F_0 = \lambda F/(D/2)^2$ is the Fresnel number.

In high-aperture optics, $F/D = 2, p = 2, \ldots, 4, n \approx (0.5\text{--}1.3) \times 10^2$, that is, the frequency properties are practically always nonlinear.

In long-focal-length systems $F/D \approx 10$, $n \approx (1.3\text{--}3.2) \times 10^3$, that is, approximately 20 times greater than in high-aperture systems.

And finally, we will find simple relations that connect the aperture ratio of the zone plate (D/F), its lens aperture (D/λ) and the number of Fresnel zones within its aperture. To do this, we assume that the diameter of the zone plate is delimited by a certain Fresnel zone with number N, that is, $D = 2r_N$. Then we substitute the expression for Fresnel zone radii, for instance for a plane incident wavefront, into $D = 2r_N$ and solve the quadratic equation; we arrive at the expression:

$$i = \text{entier}\left[\frac{1}{2}\left(\frac{D}{\lambda}\right) \left\{ \frac{1}{(F/D) + \sqrt{(F/D)^2 + \frac{1}{4}}} \right\} \right].$$

Here $\text{entier}(x)$ is the maximum integer of a real number. This simple formula shows the interrelation of the three fundamental parameters of the zone plate. Thus we can see that the number of Fresnel zones depends linearly on the relative diameter.

2.12 The Q factor of diffractive objective lenses

Let us consider ray paths from a pointlike radiation source on the optical axis to its image through the edge zone of the Fresnel zone plate.

The number of wavelengths on the path

$$m = m_1 + m_2 = (A^2 + D^2/4)^{1/2} + (B^2 + D^2/4)^{1/2}$$

is $n_0 = (m_1 + m_2)/\lambda_0$. If wavelength λ_0 is detuned by a small increment $\Delta\lambda$, the value of n_0 changes by Δn. Let $\Delta n = 1$, then

$$\lambda = \lambda_0 + \Delta\lambda$$

$$n = n_0 - \Delta n$$

$$n_0 - 1 = (m_1 + m_2)/(\lambda_0 + \Delta\lambda)$$

$$n_0 = (m_1 + m_2)/\lambda_0.$$

Hence, $\Delta\lambda/\lambda_0 = \lambda_0/(m_1 + m_2 - \lambda_0)$. Taking into account that $\lambda = c/f$ and $d\lambda = -c\,df/f^2$, we have

$$\Delta f/f = -c/[f_0(m_1 + m_2) - c].$$

Therefore, the frequency mismatch for the path difference of λ_0 is $\Delta f \approx c/(m_1 + m_2)$. Let $f_0 = 150\,\text{GHz}$, $m_1 = m_2 = 50\,\text{cm}$, then $\Delta f \approx 300\,\text{MHz}$. From this we have that the effective Q factor of the zone plate is

$$Q = f_0/\Delta f = f_0(m_1 + m_2)/c - 1 \approx (m_1 + m_2)/\lambda_0 = 500.$$

One consequence of the obtained expression is the obvious fact that the Q factor of the diffractive lens is proportional to its lens aperture $Q \approx (D/\lambda)$, because $m_{1,2} \approx D$. In addition, long-focal-length lens systems are preferable from this standpoint because $Q \approx A, B$.

Considering the case of the zone plate transforming a plane wavefront into a spherical one, we easily obtain after elementary transformations that

$$Q \approx \frac{D}{\lambda} \sqrt{\left(\frac{F}{D}\right)^2 + \frac{1}{4}}.$$

Therefore, the Q factor of a zone plate is proportional to the lens aperture and inversely proportional to the aperture ratio D/F.

Considering in the same way the Q factor for an off-axis position of the radiation source and of its image, we readily obtain, for example, that the ratio of the Q factors for the axial and off-axial positions of radiation sources is nearly unity for $A \approx B \approx 2D$ and for a displacement of a pointlike radiation source from the optical axis to the edge of the field of view ($\Delta x \approx D/4$). This leads to an important conclusion that the Q factor of a diffractive lens depends only slightly on the source position in the focal plane when the electromagnetic wave from a pointlike source is focused by the lens.

2.13   Two-component diffractive lens

A diffractive objective lens based on a zone plate has a small field of view and the number of resolved elements in the frame not more than 50×50 (with the parameters as indicated above). The information-carrying capability of this objective lens can be improved by, for instance, switching to multi-component systems. We will illustrate this statement by considering as an example a two-component objective lens suggested in [24] and consisting of two phase-inversion zone plates.

The components of the diffractive lens were calculated as follows. The front component (facing the object) is calculated for the position of the centre of the converging spherical wavefront coinciding with the centre of the scanned field (that is, with a field view centre). Another centre of the wave diffracted on the front component coincides with the apex of the cone whose generatrix touches on the edges of both components of the objective lens. The rear component of the diffractive lens is designed by analogy to the front one, that is, it is calculated for the position of the centre of the diverging spherical front that coincides with the centre of the device that visualizes the radiation. If the components of the diffractive objective lens have the same value of D/λ and the centre of the wave diffracted on the front component lies at infinity (that is, each of the lens components is

Figure 2.25. Two-component diffractive objective lens for the millimetre wave band and its refractive analogue [24].

designed for the transformation of a plane wavefront into a spherical one), then the Fresnel zone radii r_n are calculated using the familiar formula [19]

$$r_n = (n^2 \lambda_0^2/4 + n\lambda_0 B)^{1/2}$$

where n is the Fresnel zone number and λ_0 is the selected (reference) wavelength.

The general view of the diffractive radio objective and of its refraction analogue is shown in figure 2.25.

A pointlike source in the experiment on finding the shape of the focal surface was moved along a spherical surface with a radius equal to the focal length F ($B = F(\lambda)$, at $\lambda = \lambda_0$) and the centre coinciding with the centre of the front component of the diffractive objective lens. The shape of the image surface was found by identifying those regions, corresponding to the current position of the pointlike radiation source, where energy density reaches maximum. The results obtained show that the image surface found in this way is a sphere with an error not more that 2% and the centre lying at the midpoint of the rear component of the lens, with a radius equal to $B(\lambda)$. As wavelength deviates from the reference wavelength λ_0, the shape of the image surface remains spherical with a radius B, corresponding to wavelength λ.

Figure 2.26 gives the results of experiments on finding the frequency properties of two-component diffraction lenses. The abscissa axis is the relative change in the source wavelength $\Delta\lambda/\lambda_0$ and the ordinate axis is the relative displacement $\Delta B/F$ of the focal area. For comparison, this figure gives similar curves for single-component diffraction lenses of two types: one transforming a divergent spherical wavefront into a convergent one (type I), and one transforming a plane wavefront into a spherical one (type II). All three types of diffractive objective lenses had identical ratios D/B and D/λ. The results shown demonstrate that the dependence $B(\lambda)$ for a two-component objective lens practically coincides (to within the

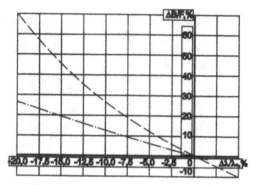

Figure 2.26. The position of focusing area $\Delta B/F$ as a function of relative detuning $\Delta\lambda/\lambda_0$ of the wavelength of millimetre-band radiation for three types of objective lens [24]: O, two-component; — —, single-component type I; – ·–, single-component type II.

accuracy of the experiment) with a similar dependence for the type-I objective lens.

Experimental data shown in figure 2.26 demonstrate that the sought dependence is nonlinear and that as the radiation wavelength changes by $\Delta\lambda/\lambda_0 = -17\%$, the focal area displaces by $\Delta B \approx 55\delta_{dp}$, where δ_{dp} is the resolution depth of a diffractive objective lens for the reference wavelength ($\lambda = \lambda_0$). Therefore, by controlling the frequency of the illuminating radiation, it is possible to scan space depthwise, which considerably increases the resolution depth of a diffractive objective lens.

As was shown above, the surface of the image of two-component diffractive objective lens is a sphere of radius $B(\lambda)$ formed as the radiation source displaces along the sphere. The position and shape of the diffraction spot from a pointlike radiation source were found on the thus established image surface for various values of distance from the radiation source to the optical axis. The resolution of the objective lens and thereby its field of view were found from the shape of the diffraction spot.

The results of experiments are shown in figure 2.27, where the abscissa axis is the relative distance in units of the Airy circle radius R_a measured off the generatrix of the image surface, and the ordinate axis is intensity U in arbitrary units normalized to 100.

As there are no universally excepted resolution criteria for off-axis radiation sources at the moment, table 2.2 gives the values of resolution using two of the most frequently utilized criteria for the axial radiation source (the detuning of the radiation wavelength was $\Delta\lambda/\lambda \approx 3.33\%$).

The experimental data show that a two-component objective lens possesses high resolving power which differs very little from the diffractive limit for a large field of view. We see from figure 2.27(a) that, using the Rayleigh criterion, the number of image elements resolved in one frame at

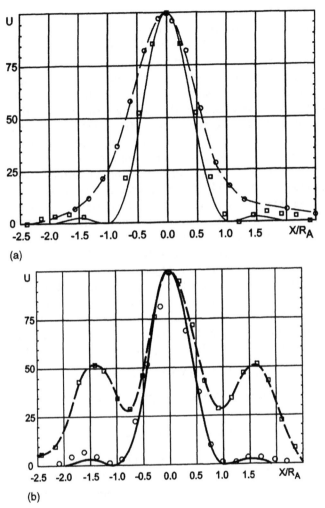

(a)

(b)

Figure 2.27. Diffractional image of a pointlike radiation source on the focal surface for: (a) $L = 100\lambda_0$, (b) $L = 47\lambda_0$ [24]. ———, at the centre of the field of view $N = 0$; O, off-axis source $N/2 = 10$; □, off-axis source $N/2 = 74$.

Table 2.2.

Field of view 2β (degrees)	Resolution, in units of R_a	
	At first minimum	At the 0.5 level
0	1.2	0.85
50	1.66	1.29
60	1.85	1.30

$L = 100\lambda_0$ and $2\beta = 60°$ is $N^2 \approx 160 \times 160$ (here L is the distance between the two components of the objective lens).

All the experimental results shown above characterize the focusing and frequency properties of a two-component diffractive objective lens and refer to the case when the distance L between the objective components was $L = 100\lambda_0$. The minimum distance between the objective lens components must be exactly as large because if the spacing is smaller then the resolution at the edges of the field of view deteriorates (the side lobes grow substantially bigger in the diffraction image of a pointlike radiation source). This conclusion is confirmed by the result given in figure 2.27(b).

The distance L between the components of the diffractive objective lens was in this case $L = 47\lambda_0$. The figure shows that the intensity of maxima in the diffractive image of a pointlike source substantially increases as the source moves away from the optical axis and reaches 50% of the intensity of the principal lobe at the distance $N/2 = 74$. This is explained by the fact that the wavefront is completely formed at a distance of at least $100\lambda_0$ from the surface of this component objective lens.

The upper bound L is imposed by vignetting, that is, the wavefront transmitted from one component to the other 'catches' the aperture of the second component only partially. If we assume that the signal amplitude diminishes by a factor of 2 at the edge of the field of view, the maximum distance between the objective lens components is found from the expression

$$L \approx D^2/1.22N0.5\lambda = D^2/0.61\lambda N.$$

Therefore the distance between the components of the objective lens may lie between the limits

$$100\lambda \leq L \leq D^2/0.61\lambda N. \qquad (2.10)$$

As for the resolution depth δ_{dp} of the objective lens described, and the longitudinal resolving power Δz, the experimental studies demonstrated that these values do not differ from the corresponding values for a single-component diffractive objective lens for which formula (2.4) works with sufficient accuracy.

2.14   Efficiency of diffractive objective lenses

A comparison of the efficiency for the Fresnel zone plate with that of lenses in the microwave band was carried out in [25]. Losses in a radio objective lens comprise diffractional losses and losses to reflection and absorption of radiation in the material of the lens. Diffractional efficiency η_{zp} (the ratio of power diffracted into the ith focal point to the entire power) for the Fresnel zone plate was determined in [25] on the assumption that one half of the incident power is transmitted without losses through open Fresnel zones while the second half of power is reflected (or absorbed) in opaque zones:

$$\eta_{zp} = 0.17[1 + (1 - R)(1 - \alpha_{ph})]$$

where $R = [(n-1)/(n+1)]^2$ and $\alpha_{ph} = \exp[-\pi n \tan \delta/(n-1)]$ are the coefficients of reflection and absorption of the incident power in the zone plate, respectively, and n and $\tan \delta$ are the refractive index and the tangent of loss angle of the material of the radio objective lens, respectively. Note that in the general case it is necessary to take into account the reflection of radiation by the two interfaces between materials of the diffractive element.

The materials employed in fabricating optical diffractive elements typically have sufficiently small loss angle tangent. However, this ceases to be true in the microwave and especially in the millimetre wavelength band.

The transmission coefficient $|T|^2$ and the additional phase shift in the transmitted wave ψ for a single-layer wall with losses as a function of an incident angle Θ can be expressed by the following relation [26]:

$$|T|^2 = \frac{A^2 + B^2}{1 - 2|r|^2 \exp(-2\beta) \cos(2\alpha + 2\psi_r) + |r|^4 \exp(-4\beta)}$$

where

$$A = [\cos \alpha - |r|^2 \exp(-2\beta) \cos(\alpha + 2\psi_r)$$
$$- |r|^2 \cos(\alpha + 2\psi_r) + |r|^4 \exp(-2\beta) \cos \alpha] \exp(-\beta)$$

$$B = [-\sin \alpha - |r|^2 \exp(-2\beta) \sin(\alpha + 2\psi_r)$$
$$+ |r|^2 \sin(\alpha + 2\psi_r) + |r|^4 \exp(-2\beta) \sin \alpha] \exp(-\beta)$$

$$\psi = \arctan(B/A) - (2\pi \cos \Theta)/\lambda$$

$$\alpha = (2\pi/\lambda)d[(\varepsilon - \sin^2 \Theta)/2]^{1/2}\{1 + [1 + \varepsilon^2 \eta/(\varepsilon - \sin^2 \Theta)^2]^{1/2}\}^{1/2}$$

$$\beta = (\pi d/\lambda)\varepsilon\eta/(\varepsilon - \sin^2 \Theta)^{1/2}$$

$$\psi_r = \arctan\frac{(\cos^2 \Theta(\varepsilon - \sin^2 \Theta))^{1/2} - \sin[0.5 \arctan(\varepsilon\eta/(\varepsilon - \sin^2 \Theta))]}{\cos^2 \Theta + (\cos^2 \Theta(\varepsilon - \sin^2 \Theta))^{1/2} \cos[0.5 \arctan(\varepsilon\eta/(\varepsilon - \sin^2 \Theta))]}$$

where $\eta = \tan \delta$, d is the wall thickness and $R_{1,2}$ are the Fresnel coefficients for the perpendicular and parallel-polarized waves relative to the wave incident plane:

$$R_1 = [\cos \Theta - (\varepsilon - \sin^2 \Theta)^{1/2}]/[\cos \Theta + (\varepsilon - \sin^2 \Theta)^{1/2}]$$

$$R_2 = [(\varepsilon - \sin^2 \Theta)^{1/2} - \varepsilon \cos \Theta]/[\varepsilon \cos \Theta + (\varepsilon - \sin^2 \Theta)^{1/2}].$$

The diffractive efficiency η_l of the lens, taking losses into account, was found from the expression

$$\eta_l = 0.84(1 - R_l)(1 - \alpha_l)$$

where

$$R_l \approx \left[\frac{(n-1)^2}{(n+1)^2} + \frac{(4n-n^2-1)}{16n^3(n^2-1)(kF)^2} \cdot \frac{1-(1+p^2+0.5p^4)\exp(-p^2)}{1-\exp(-p^2)} \right];$$

$\alpha_l \approx [n/(n-1)] \tan \delta (p^2 - 1)$ are the coefficients of reflection and absorption in the lens, respectively; $p = (\pi M)^2$; $M = b^2/\lambda F$; $k = 2\pi/\lambda$, and b and F are the radius and the focal length of the lens, respectively [25].

It is shown that at certain values of n, $\tan \delta$ and aperture D the application of the Fresnel zone plate in the millimetre band of the spectrum as a focusing device becomes expedient from the energy collection point of view [25].

A comparison of the focusing capabilities of three devices in the millimetre wave band—a lens, a phase-inversion and an amplitude zone plate—was carried out, for instance, in [27]. Experimental studies were conducted at the wavelength of $\lambda = 3$ mm using a zone plate with the aperture ratio $D/F = 1$. The lens and the zone plate were manufactured of Plexiglas ($n = 1.62$, $\tan \delta = 5.8 \times 10^{-3}$). It was shown that the maximum field intensity formed by the phase-inversion zone plate is greater by a factor 1.3 than that of the lens. This was caused by the smaller optical depth of the zone plate in comparison with the refractive lens. The width of the main peak of intensity distribution transversely to the optical axis in the focal zone is smaller for the zone plate than for the lens.

Estimates of the values of amplification coefficients of Fresnel and Soret zone plates were found in [28]. On the basis of the Huygens principle, the authors of this paper wrote the expression of the field intensity at the focal point of the zone plate in the form

$$I = ca_m(1 + \cos \Theta)/d_m$$

where c is the dimensional proportionality coefficient, a_m is the area of the mth zone and d_m is the mean distance from this zone to the focal point. In view of this the magnification coefficient of the zone plate with transparent central zone is

$$G_M = 20 \log \left\{ \frac{4F/\lambda + 1}{4F/\lambda + 0.5} \sum_{m=1}^{M} \frac{4F/\lambda + 2m - 3/2}{2F/\lambda + 2m - 3/2} \right\}$$

and for the zone plate with opaque central zone it is

$$G_M = 20 \log \left\{ \frac{4F/\lambda + 1}{4F/\lambda + 0.5} \sum_{m=1}^{M} \frac{4F/\lambda + 2m - 0.5}{2F/\lambda + 2m - 0.5} \right\}$$

where M is the number of Fresnel zones within the aperture of the zone plate. The authors found the aperture dependence of the magnification coefficients of the zone plate and of the parabolic reflection antenna. It was suggested that the zone plate may be a good choice as an antenna for a radio telescope in the centimetre wavelength band.

Another expression for the amplification coefficient of the zone plate with plane incident wavefront was given in [29]:

$$G = \frac{kF}{\pi} |\ln \cos \Theta_m|$$

where k is the wave number, F is the focal length and Θ_m is the angle under which the mth radius of the zone plate is subtended from the focal point.

If the area of FZP zones is constant (which is in fact only approximately correct) the expression for G can be rewritten in the form

$$2G/kD = \pi^{-1} |\ln(\cos \Theta_m) \cot \Theta_m|.$$

This expression implies that the function has a maximum at $\Theta_m \approx 63.5°$.

Losses due to scattering by interfaces on a profiled reflection-type zone plate that focuses radiation to a point were evaluated in [30, 31]. Thus the focusing coefficient χ due to losses at interfaces was found to be

$$\chi \approx 2/(2 + N\lambda/F)$$

and thus depend on the number N of zones and the focal distance that dictates the angle of reflection of rays from the surface of the element.

In reality, losses at zone boundaries are more significant but they are caused not by the 'shading' of parts of Fresnel zones but by the fact that owing to technological processes, the real zone profiles deviate from the calculated one.

As for the efficiency of flat elements transparent to radiation, we need to mention that significant losses occur due to reflection at their surfaces. To reduce these losses, it is necessary to 'bloom' the surfaces of diffractive elements.

However, in a number of cases it is possible to make blooming of flat elements unnecessary if the element is placed in a beam of polarized radiation. It is a known fact that at a certain angle between a flat dielectric and the incident beam (the Brewster angle) the polarized radiation passes through the plate completely, without being reflected at the surface. The phase structure of an element can be calculated in such a way that it will work at the incidence angle equal to the Brewster angle. In this case radiation is completely transmitted through the plate and there is no need for blooming. This effect cannot be achieved by traditional elements because their rotation produces aberrations that significantly distort the image.

To evaluate the efficiency of zone plates in an arbitrary diffraction order K when using a wavelength λ different from the reference wavelength λ_0, one can use the following expression [32]:

$$\eta \approx \left(\frac{\sin \pi(\lambda_0/\lambda - K)}{\pi(\lambda_0/\lambda - K)} \right)^2 100\%.$$

Amplitude-type Fresnel zone plates have a drawback: the fraction of energy gathered at the focal spot collects only about 10% of the energy incident on the plate. The reasons for this low efficiency are obvious: first, one half of the electromagnetic energy incident on the plate is absorbed by opaque zones, and second, the profile of the transparent zones does not provide for the maximum possible concentration of energy at the focal point. The first of these two reasons for low efficiency of the zone plate was identified by R Wood, who suggested replacing two opaque parts with zones inverting the phase. The second reason is counteracted by giving each zone a special profile calculated on the condition that the path length from the object to its image be constant over the entire zone [12].

The phase shifts required in the zones of a focusing element are created by using dielectric materials with variable optical thickness. Variable optical thickness is achieved by modifying the refractive index of the material according to a predetermined law or by creating a certain profile on the surface of the material transparent to radiation.

The diffractive efficiency η of the diffractive focusing element depending on the number of phase quantization levels is found from the expression [33]

$$\eta \approx \frac{\sin^2(\pi/p)}{(\pi/p)^2} \, 100\%.$$

In table 2.3 we give the values of the diffractional efficiency of zone plates depending on the number of phase quantization levels.

Three practically important conclusions follow from the data given above. Each diffractive element has an optimal number of phase quantization levels, so that the quality of a focusing element remains practically unchanged if this number is increased. For $p = 2$ we have the Rayleigh–Wood zone plate. In the space behind the diffractive element, the energy is mostly distributed between the +1 and −1 orders of diffraction, and in each of them the diffractional efficiency comes to about 40%. If $p = 4$, the diffractional efficiency of an element in the +1 order reaches 81% and then grows insignificantly as p increases. From the practical standpoint, this is the optimal version of a zone plate because further increase in diffractional efficiency does not pay off, owing to the complexity of its fabrication, especially in the short-wavelength part of the spectrum. As $p \rightarrow \infty$, the diffractional efficiency tends to 100% and the phase profile of the zone plate corresponds then to the profile with 'blazing'—a diffractive element of this type is usually known as *kinoform*.

Table 2.3.

p	2	3	4	5	6	7	8	9	10	20	∞
η (%)	40	68	81	87	91	93	95	96	97	99	100

Deviations from the ideal phase profile of a diffractive element reduce its diffractional efficiency and generate spurious diffractional orders (side lobes in the directional diagram). If a diffractive element is fabricated in the profiled-phase-inversion mode, the reduction of the diffractional efficiency due to technological errors of fabrication is described by the relation [34]

$$\Delta I = 1 - \exp[-(2\pi\sigma/\lambda)^2]$$

where σ is the mean-square deviation of the fabricated profile of the diffractive element from the prescribed one. It was shown [35] that the diffractional efficiency is much more sensitive to technological errors than the Strehl number.

The effect of fabrication errors in a kinoform lens on the quality of the image was evaluated in [36, 37]. The authors used the Maréchal criterion, which states that the system is well corrected if the normalized intensity in the diffractional focal maximum (the Strehl number) is greater than or equal to 0.8. It was shown that the admissible mean square random error in the position of the zones must not exceed the value $(\langle \Delta^2 \rangle)^{1/2} \leq 0.632(\lambda F)(\pi D)^{-1}$, and the ellipticity of the zones, $a < 1-1.1(4\lambda F)(\pi D)^{-1}$.

The effect of the depth of diffraction zones on efficiency is determined by the expression [56]

$$\eta = \text{sinc}^2 \left(\frac{h(n-1)}{\lambda} - p \right)$$

where p is the diffraction order. Obviously, when high-aperture MWDO are prepared, the demands to profile depth become more stringent.

When high-efficiency high-aperture MWDO elements are fabricated, sag-type errors in the shape of diffraction zones are frequently encountered. The effect of profile sag with phase depth of 2π on the diffractional efficiency is described by the expression

$$\eta = \text{sinc}^2(\Delta H/h)$$

where ΔH is the maximum sag.

For one-dimensional gratings with 'blazing', the maximum possible efficiency can be found from the formula [56]

$$\eta = 1 - \delta/T \qquad (2.11)$$

where δ is the linear dimension of the shadow in the tangential plane and T is the lattice period. The linear size of the shadow is $h\tan(\theta)$, where $\theta = \arcsin(\lambda/T)$, and h is the profile depth. Formula (2.11) can be written in terms of the inclination angle of the grating grooves, $\beta = \arctan(h/T)$, and the diffraction angle θ:

$$\eta = 1 - \tan(\beta)\tan(\theta).$$

However, when a real profile is considered, another and more important technological limitation arises, due to a finite value of the reverse slope

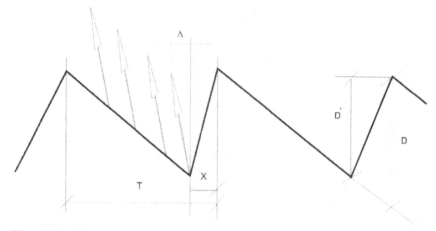

Figure 2.28. Joint action of inverse slope and 'shading' effect.

(figure 2.28). Reverse slopes may arise if MWDO elements are fabricated by punching or turning. The diffractional efficiency taking into account the reverse slope is found from the expression [56, 57]

$$\eta = (1 - x/T)^2$$

where x is the width of the reverse slope. The reverse slope effect mostly restricts the diffractional efficiency of elements with continuous profile. If the diffraction zones are narrow, both 'shading' and reverse slope effects play their role; in this case the resulting efficiency is found in the following way:

$$\eta = (1 - x/T)^2 \left\{ 1 - \frac{h}{T} \tan \left[\arcsin \left(\frac{\lambda}{T} \right) \right] \right\}$$

$$= (1 - x/T)^2 \left[1 - \frac{h}{T} \frac{\lambda}{\sqrt{T^2 - \lambda^2}} \right]. \qquad (2.12)$$

The formulas obtained for linear gratings can be used to evaluate the efficiency of more complex diffractive elements by summing up their contributions to efficiency over all zones. This approach was applied in [58] to calculate the efficiency of multilevel profile lenses. Let us consider a simplified approach to evaluation of diffractional efficiency of MWDO elements with continuous profile.[2] Its advantages lie in the fact that the method is independent of a specific transformation of the wavefront implemented by the diffractive element. This approach is expanded most simply to one-dimensional and

---

[2] This approach was developed by Dr V P Korolkov.

axisymmetric diffractive elements. The total diffractional efficiency is

$$\eta_{\text{tot}} = \sum_{i=1}^{N} \eta_i \frac{P_i}{P_{\text{tot}}} \qquad (2.12)$$

where i is the index that numbers the current diffraction zone, N is the total number of zones, P_i is the power of radiation incident on the ith zone, P_{tot} is the total power of radiation transmitted through the entire area of the element, and η_i is the diffractional efficiency of the ith zone calculated by the formula

$$\eta_i = \left(1 - \frac{x}{T}\right)^2 \left[1 - \frac{h}{T_i}\tan(\theta_i)\right].$$

The quantities P_i and P_{tot} are calculated by integrating the intensity distribution of the radiation incident on the element. The weight coefficient P_i/P_{tot} for uniformly distributed intensity equals the ratio of the area of the ith zone to the area of the entire element, S_i/S_{tot}. For the elements with linear zones (for instance, cylindrical lenses) and uniformly distributed intensity, the diffractional efficiency η_{zyi} can be evaluated using the formula

$$\eta_{zyi} = \sum_{i=1}^{N} \left[1 - \frac{x}{T_i}\right]^2 \left[1 - \frac{h}{T_i}\frac{\lambda}{\sqrt{T_i^2 - \lambda^2}}\right]\left[\frac{T_i}{Y_{\text{tot}}}\right]$$

where T_i is the width of the ith zone and Y_{tot} is the total width of all zones.

In the case of an MWDO element with circular zones and uniform illumination, expression (2.12) is transformed to the form

$$\eta_{\text{axial}} = \sum_{i=1}^{N} \left[1 - \frac{x}{T_i}\right]^2 \left[1 - \frac{h}{T_i}\frac{\lambda}{\sqrt{T_i^2 - \lambda^2}}\right]\left[\frac{r_i^2 - r_{i-1}^2}{r_N^2}\right]$$

where r_i and r_{i-1} are the outer and inner radii of the ith zone and $T_i = r_i - r_{i-1}$. This dependence shows that diffractional efficiency diminishes towards the periphery of the diffractive element.

2.15 Phase of the wave: behaviour in the focal region

In various problems of holography and tomography it is sometimes necessary to know the phase behaviour when the wave passes through the focusing region along the optical axis. In the geometric optics approximation, as the beam passes through the point of its tangency to the caustic at the focal point of the ideal aberration-free objective lens, the phase jumps abruptly from $+\pi/2$ to $-\pi/2$ [59].

We know, however, that the region of maximum density of energy has finite dimension and cannot be less than $\lambda/2$. Figure 2.29 shows the results

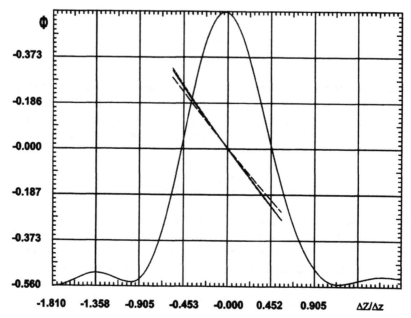

Figure 2.29. Variation of the phase of the wave within the focal area of the Fresnel zone plate [60]. ——, field intensity distribution; – – –, phase of the wave for $\Delta\lambda/\lambda_0 = 3.7\%$; — —, phase of the wave for $\Delta\lambda/\lambda_0 = 0\%$; ———, phase of the wave for $\Delta\lambda/\lambda_0 = -19.8\%$.

of computations of phase behaviour as the wave crosses the focal point; these results were obtained using rigorous use of the Fresnel–Kirchhoff diffractional integral.

The convergent spherical wave formed by the zone plate ($D/F = \frac{1}{2}$) was compared with the plane reference wave in phase with the spherical wave at the corresponding geometric focal point. The results given above show that the phase changes value from $+\pi/2$ to $-\pi/2$ as it crosses the focal point over a distance of about Δz [60]. As wavelength deviates from the reference value λ_0, the rate of change of the phase changes as well.

2.16 Off-axis zone plate

It is obvious from general physical arguments that if an arbitrary segment of surface is 'cut out' of a conventional zone plate, this element will also work as a focusing device. Figure 2.30 shows how to construct an off-axis zone plate of arbitrary shape.

The results of studying the properties of off-axis zone plates (circular- and diamond-shaped) with aperture ratio $D/F \approx 1$ in the millimetre

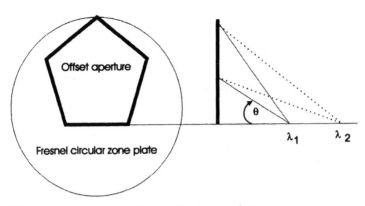

Figure 2.30. Principle of constructing an off-axis zone plate.

wavelength band show the following behaviour:

- Frequency properties of such zone plates are independent of their shape and coincide with the frequency properties of axial zone plates.
- The field intensity distribution in the focal zone along the optical axis is narrower than the similar distribution of an axial zone plate by a factor of $\cos(\theta)$, where the angle θ characterizes the degree of displacement of the centre of the off-axis zone plate from the centre of the axial one (see figure 2.30).
- The field intensity distribution transversally to the optical axis is dictated by the shape and size of the aperture of the off-axis zone plate.

The focusing properties of antennas based on off-axis zone plates will be discussed in chapter 6.

Bibliography

[1] Minin I V and Minin O V 1992 *Diffraction Quasioptics* (Moscow: Research and Production Association 'InformTEI')

[2] Minin I V and Minin O V 1994 'Elements of diffraction quasioptics, part 1. Fundamental properties' *Avtometriya* **3** 110–120

[3] Koronkevich V P and Palchikova I G 1992 'State-of-the-art zone plates' *Avtometriya* **1** 85–100

[4] Fresnel O 1955 *Selected Works in Optics* (Moscow: GITTL)

[5] Boivin A 1964 *Théorie et Calcul des Figures de Diffraction de Révolution* (Paris: Gauthier-Villars) p 511

[6] Soret J L 1875 *Archives Sci. Phys. Natur.* **28** 320

[7] Wood R W 1989 *Physical Optics* (Optical Society of America)

[8] Wood R W 1898 *Phil. Mag.* **45** 511

[9] Lord Rayleigh (Strut J W) 1902 *Scientific Papers* vol 111 p 87

[10] Gabor D 1949 *Proc. Roy. Soc. London A* **197** 454

[11] Walsh A 1952 'Echelette zone plates for use in for infrared spectroscopy' *J. Opt. Soc. Amer.* **42**(3) 213

[12] Slyusarev G G 1957 'Optical systems with phase' *Soviet Phys. Doklady* **113**(4) 780–783

[13] Goodman J W and Silvestri A M 1970 'Some effect of Fourier-domain phase quantization' *IBM J. Res. Dev.* **14**(5) 478

[14] Sanyal G S and Singh M 1968 'Fresnel zone plate antenna' *J. Inst. Telecomm.* **14** 265–281

[15] Kün R 1964 *Mikrowellenantennen* (Berlin: Veb Verlag Technik)

[16] Futhey J and Fleming M 1992 'Superzone diffractive lenses' in *Diffractive Optics: Design, Fabrication and Applications Technical Digest, 1992* vol 9 (Washington, DC: Optical Society of America) pp 4/MA2-1–6/MA2-3

[17] Goncharsky A V, Popov V V and Stepanov V V 1991 *Introduction to Computer Optics* (Moscow: Moscow State University)

[18] Baibulatov F Kh, Minin I V and Minin O V 1985 'Investigation of the focusing properties of Fresnel zonel plate' *J. Commun. Technol. Electron.* **30**(9) 1034–1039

[19] Born M and Wolf E 1999 *Principles of Optics* 7th edn (Cambridge University Press)

[20] Minin I V and Minin O V 1988 'Information properties of zone plates' *Kompyuternaya Optika* **3** 15–22 (in Russian)

[21] Andreev G A, Bazarsky O V, Kolesnikov A N and Khlyavich Ya L 1982 'Formatting properties of zoned Fresnel plates in millimeter wavelength band' *Radiotekhnika i Elektronika* **27**(10) 2027–2030 (in Russian)

[22] Dontzova V V and Lenkova G A 1990 'Effect of adjustment process on scattering function of the diffractional radio-lens' *Avtometriya* **6** 61–66 (in Russian)

[23] Minin I V and Minin O V 1987 'Diffractional radio-optics systems for the microwave band' *Abstracts of Talks at the 6th All-USSR Conf. on Techniques and Instruments for Measuring Electromagnetic Characteristics of Materials in RF and Microwave Ranges, Novosibirsk*, pp 169–170 (in Russian)

[24] Minin I V and Minin O V 1986 'A wide-angular multicomponent diffraction microwaves lens' *J. Commun. Technol. Electron.* **31**(4) 800–806

[25] Thchukin I I 1973 'On losses in lenses and zone plates. Solid-state radio-electronics' *Voronezh* 127–130 (in Russian)

[26] Kaplun V A 1974 *Radomes for Microwave Antennas* (Moscow: Soviet Radio) (in Russian)

[27] Shchukin I I 1976 'On possible applications of zone plates as antennas in microwave band' in *Selected Problems in Designing Systems for Optimal Data Processing* (Yaroslavl: Yaroslavl University Publ.) pp 111–116 (in Russian)

[28] Buskirk L F and Hendrix C E 1961 'The zone plate as a radiofrequency focusing element' *IRE Trans. Antennas Propagation* **AP-9**(3) 319–320

[29] Kanevsky I N 1977 *Focusing of Sound and Ultrasound Waves* (Moscow: Nauka) p 380

[30] Vereshchagin V V and Lopatin A I 1979 'Chromatic properties of profiles zone plates' *Optika i Spektroskopiya* **46**(5) 1002–1008 (in Russian)

[31] Vereshchagin V V and Lopatin A I 1980 'Chromatic properties of profiles zone plates' *Optika i Spektroskopiya* **47**(1) 159–165

[32] Burabli D A, Morris M and Robers I R 1989 'Optical performance of holographic kinoforms' *Appl. Opt.* **28**(5) 976–983

[33] Vereshchagin B B and Lopatin A I 1981 'Theory of profiled zone plate' *Akusticheskii Zh.* **27**(6) 841–847 (in Russian)

[34] Gan M A and Perveev A F 1988 'Kinoform optics—properties and problems of efficient application' *Soviet Phys. Izvestiya* **52**(2) 210–216 (in Russian)

[35] Palchikova I G 1991 'Pupil function, Strehl ratio and diffractional efficiency of phase zone plate' *Avtometriya* **5** 113–119 (in Russian)

[36] Palchikova I G 1985 'Effect of imperfections of kinoform lens on image quality' *Avtometriya* **6** 38–42 (in Russian)

[37] Palchikova I G and Ryabchun A G 1985 'On effects of imperfections of kinoform lenses on pupil function' *Avtometriya* **6** 38 (in Russian)

[38] Jiang G Z and W X Zhang 1997 'The effect of layer spacing on the properties of double-layer Fresnel zone plate lens' in *IEEE Int. Symp. Antennas and Propagation, Montreal, Canada*, vol 1, *Symp. Digest*, 13–18 July, pp 472–475

[39] Hristov H D 2000 *Fresnel Zones in Wireless Links, Zone Plate Lenses and Antennas* (Boston: Artech House)

[40] Wiltse J C 1985 'The phase-correcting zone plate' in *Proc. Int. Conf. Infrared Millimeter Waves, Lake Buene Vista, FL*, December, pp 345–347

[40] Hristov H D and Herben M H A J 1995 'Millimeter-wave Fresnel zone plate lens and antenna' *IEEE Trans. Microwave Theory Techn.* **43** 2770–2785

[41] Huder B and W Menzel 1988 'Flat printed reflector antenna for millimeter-wave applications' *Electronic Lett.* **24** 318

[42] Guo Y J and S K Barton 1992 'A high-efficiency quarter-wave zone plate reflector' *IEEE Microwave Guided Wave Lett.* **2**(12) 470–471

[43] Minin I V and Minin O V 2000 'The dielectric non-metallic reflecting FZP antennas' in *Proc. 25th Int. Conf. on Infrared and Millimeter Waves, Beijing, China*, 12–15 September, pp 461–462

[44] Minin I V and Minin O V 2000 'The dielectric non-metallic reflecting FZP antennas' in *Proc. 5th Int. Symp. on Antennas, Propagation, and EM Theory, Beijing, China*, 15–18 August, pp 164–166

[45] Bruce E 1939 'Directive radio system' US Patent 2,169,553

[47] Kasyanov D A 2001 'Investigation of focusing of cylindrical zones lens' XI Session of the Russian Acoustical Society, Moscow, 19–23 November, pp 365–368

[47] Ji Y and M Fujita A 1996 'Cylindrical fresnel zone antenna' *IEEE Trans. Antennas and Propagation* **44** 1301–1303

[48] Raisky S M 1952 'Zone plate' *UFN (Soviet Physics Uspekhi)* **47**(4) 516–536

[49] Khastanir P, Chakravorty I N and Dey K K 1973 'Microwave paraboloidal, spherical and plane zone plate antennas: comparative study' *Indian J. Radio Space Phys.* **2** 47–50

[50] Delmas J-J, Toutain S, Landrac G and Cousin P 1993 'TDF antenna for multisatellite reception using 3D Fresnel principle and multiplayer structure' in *IEEE Int. Antennas and Propagation Symp., Ann Arbor, MI, Symp. Digest*, 28 June–2 July, pp 1647–1650

[51] Cousin P, Landrac G, Toutain S and Delmas J J 1994 'Calculation of the focal field distribution and radiation pattern of a parabolic antenna with Fresnel zones' in *Proc. Int. Symp. Antennas, JINA 94, Nice, France*, 8–10 November, pp 489–492

[52] Malliot H A 1994 'Zone plate reflector antenna for applications in space' in *Proc. IEEE Aerospace Applications Conference, Vail Co., USA*, pp 295–311

[53] Garrett J E and J C Wiltse 1991 'Fresnel zone plate antennas at millimeter wavelengths' *Int. J. Infrared and Millimeter Waves* **12**(3) 195–220

[54] Wiltse J C 2002 'Bandwidth characteristics for the stepped conical zoned antenna' *Proc. SPIE* **4732**; Photonic and Quantum Techn. for Aerospace Appl. IV, 1–5 April, Orlando, FL

[55] Young M 1972 'Zone plates and their aberrations' *J. Opt. Soc. Amer.* **62**(8) 972–976

[56] Hessler T, Rossi M, Kunz R E and Gale M T 1998 'Analysis and optimization of fabrication of continuous-relief diffractive optical elements' *Appl. Opt.* 3 4069–4079

[57] Suleski T J and O'Shera D C 1995 'Gray-scale masks for diffractive optics fabrication: I. Commercial slide imagers' *Appl. Opt.* **34**(32) 7507–7517

[58] Levy V, Mendlovic D and Marom E 2000 'Efficiency analysis of resolution-limited DOEs' in *Diffractive Optics and Micro-Optics, OSA Technical Digest* (Washington, DC: Optical Society of America) pp 150–152

[59] Landau L D and Lifshits E M 1979 *Field Theory* (Oxford: Pergamon Press)

[56] Minin I V and Minin O V 1981 'Investigation of field structure in the focal area of the Fresnel zone plate' in *Conf. 'Physics', Novosibirsk, Novosibirsk University*

[61] J C Wiltse 1994 'Millimeter-wave Fresnel zone plate antennas' in *Millimeter and Microwave Engineering for Communications and Radar, Critical Reviews of Optical Science and Technology* (ed J C Wiltse), SPIE vol CR54, Bellingham, WA, ch. 11, pp 272–293

Selected Bibliography

Black D N and J C Wiltse 1987 'Millimeter-wave characteristics of phase-correcting Fresnel zone plates' *IEEE Trans. Microwave Theory Techn.* **35**(12) 1122–1128

Chen C, Shi S and Prather D W 2004 'Electromagnetic design of an all-diffractive millimetre wave imaging system' *Appl. Opt.* **43**(12) 2431–2438

Garrett J E and Wiltse J C 1991 'Fresnel zone plate antennas at millimeter wavelengths' *Int. J. Infrared and Millimeter Waves* **13**(3)

Garrett J E and Wiltse J C 1990 'Performance characteristics of phase-correcting Fresnel zone plates' *IEEE Int. Microwave Symp.* **11** 797–799

Hristov H D and Herben M H A J 1995 'Quarter-wave Fresnel zone planar lens and antenna' *IEEE Microwave and Guided Waves Lett.* **5** 249–251

Ji Y and Fujita M 1994 'Design and analysis of a folded Fresnel zone plate antenna' *Int. J. Infrared and Millimeter Waves* **15** 1385–1405

Jiang, G Z, Zhang W X and Kang K 1998 'The focusing fields of Fresnel zone plate lens' in *IEEE Int. Symp. Antennas and Propagation, Atlanta, GA, Symp. Digest*, vol 1, June, pp 166–169

Kearey P D and Klein A G 1989 'Resolving power of zone plates' *J. Modern Phys.* **36** 361–367

Kock W E, Rosen L and Rendeiro J 1966 'Holograms and zone plates' *Proc. IEEE* November 1599–1601

Minin I V and Minin O V 2000 'Fresnel zone plate lens and antennas for millimeter waves: history and evolutions of developments and applications' in *Proc. 25th Int. Conf. Infrared and Millimeter Waves, Beijing, China*, 12–15 September, pp 409–410

Sheikh U H 1974 'Focussing properties of zone plate lens' *Bull. Engineering, Univ. Tripoli* **2**(2) 191–196

Soifer V A (ed) 2000 *Methods of Computer Optics* (Moscow: Fizmatlit) (in Russian)

Stigliani D Y, Mittra R and Semonin R G 1967 'Resolving power of a zone plate' *J. Opt. Soc. Amer.* **57**(5) 610–613

Wiltse J C 1998 'Recent developments in Fresnel zone plate antennas at microwave/millimeter wave' in *Proc. SPIE Int. Symp., San Diego, CA*, 23 July, vol 3464, pp 146–154

Wiltse J C 1999 'History and evolution of Fresnel zone plate antennas for microwave and millimeter waves' *IEEE, APURSI*, pp 722–725

Wiltse J C 1999 'Second-generation zone plate antenna design' in *Proc. SPIE, Denver*, 19 July, vol 33795, pp 287–294

Wiltse J C 2000 'Advanced zoned plate antenna design' in *Proc. SPIE, San Diego*, August, vol 4111, pp 201–209

Wiltse J C 2001 'The stepped conical zone plate antenna' *Proc. SPIE, Orlando*, 17 April, vol 4386, pp 85–92

Young M 1972 'Zone plates and their aberrations' *J. Opt. Soc. Am.* **62**(8) 972–976

Zavyalov V V and Voronin V I 1976 'Scanner for visualization of transverse field distribution of sub-millimeter radiation' *Pribory i Tekhnika Eksperimenta* **6** 102–104 (in Russian)

Chapter 3

Principles of construction of elements of diffractive optics

3.1 Methods of synthesis of diffractive elements

The problem of synthesis of the amplitude-phase-type profile does not have a unique solution even for the simplest element—the Fresnel zone plate. As an example, we consider a diffractive element that allows focusing of radiation on to an annular region.

It appears that a diffractive element focusing laser radiation on to a ring was first described in [1]. It was meant to produce annular imprints on metal surfaces. The field structure in the focal region of this element of diffractive optics was later treated in the Fresnel approximation in [2, 3]. Specifically, it was shown [2] that there are two solutions to the problem of focusing on to a ring: one when rays intersect the optical axis and another when they do not intersect it. A similar analysis of elements with angular momentum response was given in [4] in relation to systems of geometric transformation of images. In addition, it was shown in this paper that field intensity has a maximum in the focal plane on the optical axis of the element of diffractive optics, produced by weak, spurious illumination along its optical axis [4]. By increasing the number of phase quantization levels P of a phase element of diffractive optics approximately to 16, it is possible to greatly reduce the field intensity at the optical axis, as shown in [5, 6] in the paraxial approximation. However, in the optical range this increases the technical difficulties in manufacturing the diffractive element and multiplies the number of masks required to generate the desired phase profile [5, 6]. Also, in a number of cases there are reasons that make this approach (calling for increased number of phase quantization levels) unsuitable. The highest quality phase profile is produced by the photolithographic etching technique [7]. Also, this method leads to high demands on the accuracy of juxtaposition of binary photographic templates whose number is usually not higher than 4 to 8.

In addition to increasing the number of phase quantization levels, there is another free parameter of diffractive optical element whose selection can regulate the field intensity distribution function in a prescribed region of space [8, 9, 66].

To simplify the explanation of the method, we consider here how monochromatic radiation from a point-like source is focused to a point using a flat diffractive element. The structure of zones on its surface can be easily obtained by considering equiphase contours (figure 3.1(a,b)); these lines are elliptical.

Obviously, to focus radiation to a point F it is necessary (for a binary element) that the difference between the eikonals of the incident and diffracted rays $|IOF|$ and $|IM + FM|$ be a multiple of $\lambda/2$ (see figure 3.1(a)). Since the phase shifts on the boundaries between the neighbouring zones have opposite

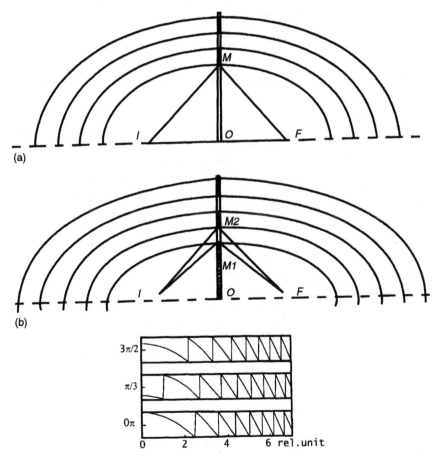

Figure 3.1. (a, b) Principle of construction of equiphase curves [9]. (c) Phase profile for various values of phase shift: 0π, $\pi/3$, $2\pi/3$.

signs, transparent and opaque zones alternate. Practically all known binary diffractive focusing elements built until recently followed this principle. On the other hand, the central zone of a diffractive element can be either transparent or opaque for the incident radiation. In the latter case the in-phase condition for the radiation at the point A is written differently: the difference between eikonals of the rays $|IM_2 + M_2F|$ and $|IM_1 + M_1F|$ must be a multiple of $\lambda/2$ (see figure 3.1(b)). We see from figure 3.1 that in this case the size R_0 of the central opaque zone $|OM|$ can be arbitrary. It is easily seen that it is expedient to choose $R_0 \leq |OM|$ since otherwise such effects as apodization will be observed. It is not difficult to apply similar reasoning to other types of diffractive focusing elements, such as focusing on to a ring.

In other words, another parameter that can be varied in calculating the modulation function of a diffractive element is a permanent phase shift. The diffraction zones can be displaced by adding an initial phase shift φ_0 to the phase function of the element:

$$\bar{\varphi}(r) = \beta \, \text{mod}[\varphi(x, y) + \varphi_0, 2\pi].$$

This means in fact that the zone is divided not by starting at the extremum of the transmission phase function but at an arbitrary value of transmission (figure 3.1(c)).

For a binary element that focuses radiation emitted by a pointlike radiation source to a ring of radius R_{ring}, the effect of R_0 on the quality of field intensity distribution function in the focal plane was evaluated by running a numerical experiment in scalar wave approximation.

We chose the following parameters for the focusing element: distance A from the point source of radiation to the diffractive element plane and distance B from this element to the focal plane were 7.5 cm, the diameter of the diffractive optic element $D = 2$ cm, the radius of the ring of focusing $R_{\text{ring}} = 0.1D$, and the wavelength $\lambda = 10.6$ μm. The results of the numerical experiment are plotted in figure 3.2 (the ordinate is the field intensity normalized to the largest maximum of the three distributions under study, and the abscissa axis is the coordinate along the ring radius in the focal plane). Figure 3.3 shows the corresponding shapes of focusing zones in isometric projection.

Curve 1 in figure 3.2 shows the distribution of field intensity in the focal plane along the ring radius of the known solution ($R_0 = 0$) while curve 2 gives the result of optimization of the binary element with the central maximum suppressed. The optimization consisted in minimizing the field intensity on the optical axis of the diffractive element R_0:

$$R_0: \quad I \approx \min_{R_0} |U(x = 0, y = 0, B) U^*(x = 0, y = 0, B)|.$$

In all cases shown in figure 3.2, the central zone of the diffractive optical element was opaque for radiation and the total number of zones on the surface of the focusing element remained unchanged.

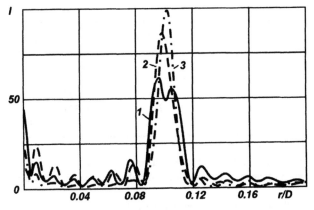

Figure 3.2. Field intensity distribution along the radius of the focusing ring.

As follows from figure 3.2, the stray central maximum of field intensity can be essentially 'suppressed' by choosing a suitable value of R_0. The gist of the effect observed is as follows. The boundary of each zone on the surface of a diffractive element is a source of toroidal convergent boundary waves, so that in-phase constructive interference of these waves can be observed not

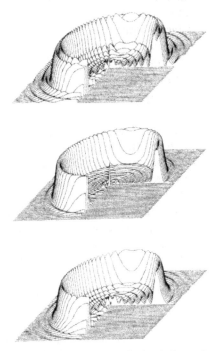

Figure 3.3. Field intensity distribution when radiation is focused on to a ring.

Table 3.1.

Curve in figure 3.2	R_0 [1]	I/I_{max} [2]	I_0/I [3] (%)	$\Delta r/R_{ring}$ [4]		
				0.1	0.5	0.7
1	0.00	0.69	65	0.302	0.238	0.198
2	0.47	0.81	21	0.270	0.143	0.103
3	0.64	1.00	25.5	0.248	0.132	0.094

[1] The value of R_0 is given in units of $(\lambda B)^{1/2}$.
[2] I/I_{max} is the normalized field intensity at the distribution peak.
[3] I_0/I is the intensity of the central (axial) maximum relative to the maximum intensity within the ring.
[4] Δr is the width of the ring at the indicated intensity levels.

only in a prescribed space (at a point R_k) but also in some different locality, for instance on the optical axis where the effect reaches its maximum. By redistributing the structure of zones on the surface of the diffractive element, the synchronization condition for boundary waves in a certain prescribed region can be changed by choosing the value of R_0. This makes it possible to considerably suppress the spurious central maximum even for a binary diffractive element and to increase field intensity in the ring itself through spatial redistribution of energy.

Similarly, it is possible to correct the shape of the focusing region. For instance, the problem posed by focusing radiation on to a ring with maximum intensity at a point R_k was solved in this manner, that is, through optimization of the type

$$I \approx \min_{R_0} |U(R_k, 0, B)U^*(R_k, 0, B)|.$$

The results of optimizing the focusing properties of diffractive optical elements are shown in figure 3.2 (curve 3).

Some comparative characteristics of the discussed variants of diffractive optical elements are shown in table 3.1.

The results shown in figure 3.2 and table 3.1 imply that, even for binary diffractive elements, the field intensity distribution function can be corrected in the focal area by choosing R_0. Similar effects are achieved also for elements that focus radiation on to an area of arbitrary shape, for instance, on to longitudinal or transversal segments, an ellipse, a transversal (relative to the optical axis) smooth curve etc. Furthermore, it is not difficult to show that the statements made above are also valid for phase-type diffractive elements.

The problem of synthesizing the structure of the focusing diffractive element thus does not have a unique solution and the field intensity distribution function can be corrected in the focal plane by choosing the free

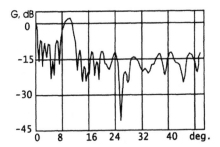

Figure 3.4. Directivity pattern of a binary diffractive antenna. One division of the angle scale is 2°. The maximum in directivity corresponds to the orientation of the principal lobe of scattering at 10°.

parameter R_0. Other, more complicated, techniques of constructing elements of diffractive optics are discussed in [10].

All this can be applied to elements of antennas not only to optimize their parameters but also to create directivity patterns of unconventional form. Figure 3.4 gives the directivity pattern for such an antenna for which gain reaches maximum within an angle of 10 degrees of arc. Taking into account that this directivity pattern corresponds to a binary amplitude-type diffractive element, the quality of focusing must be regarded as satisfactory. The efficiency of such an antenna can naturally be improved by going to a multilevel system.

The size of the first Fresnel zone can also be chosen by another, more general, method [8]. For example, the first Fresnel zone may not necessarily be axisymmetric, that is $R_0 = R_0(\varphi)$. Let us consider the design of a zone plate in this case. The radii of Fresnel zones are given by the expression

$$R_n(\varphi) = \{R_0^2 + [2n\lambda(R_0^2 + F^2)^{1/2}]/m + (n\lambda/m)^2\}^{1/2}.$$

Here n is the number of the zone, M is the number of phase quantization levels and F is the focal distance.

Let $F \gg R_0$. We expand the second term to a series and retain only the first order of expansion:

$$R_n(\varphi) \approx \{R_0^2 + 2nF\lambda/m + (n\lambda/m)^2\}^{1/2}.$$

It is usually advisable to choose R_0 to be not greater than the size of the first $r_{n=1}$ zone for the case of $R_0 = 0$ (otherwise apodization effects become significant). The size of the first zone is $R_1 \approx (\lambda F)^{1/2} \approx R_0$. Then if $n \gg 1$, $R_0^2 \ll 2\lambda Fn/m$ and therefore

$$R_n(\varphi) \to \{2n\lambda F/m + (n\lambda/m)^2\}^{1/2}.$$

The asymptotic expression for Fresnel zones obtained above is independent of φ and coincides with the Fresnel zone radii of a conventional zone plate. Therefore, the structure of zones on the surface of a diffractive element tends

Table 3.2.

D/λ	I_{ring}/I_c	$\Delta R/R_{ring}$			$\Delta R_c/R_{ring}$		
		0.7	0.5	0.1	0.7	0.5	0.1
100	0.37	0.38	0.50	0.82	0.57	0.91	1.43
200	1.67	0.18	0.25	0.43	0.33	0.43	0.67
400	2.70	0.10	0.15	0.23	0.09	0.12	0.29
800	3.13	0.05	0.06	0.10	0.05	0.09	0.19

asymptotically to the zone structure of ordinary elements (manufactured by familiar methods) and for $n \gg 1$ is independent of the form of $R_0(\varphi)$.

We have shown above that by choosing the value of R_0 we can control the field intensity distribution function in the area of focusing. The form of this function also depends, for obvious reasons, on the geometric size of the diffractive element. Let us consider as an example the problem of focusing radiation on to a ring. We know [1] that in this case a central spurious intensity maximum is formed on the optical axis of the element, as a result of interference of toroidal waves diffracted at the edges of Fresnel zones [9, 11, 66]. In general, the reduction of intensity of zero order beyond the main maximum (in this case, the maximum of annular shape) can be achieved in several ways. The simplest one is to reduce the size of diffraction pattern and increase the diameter of the element, that is, to increase the ratio D/λ. We shall consider now the effect of D/λ on the quality of focusing using as an example an element with $D/B = 0.5 = $ const and $R_{ring}/D = 0.1 = $ const, where B is the focal distance ($F = B$ for $\lambda = \lambda_0$) and R_{ring} is the radius of the focusing ring. The comparison will be carried out for two types of elements: type E1 where rays drawn from the boundaries of zones to the point R_{ring} do not intersect the optical axis; type E2 where the rays intersect the axis. The results of a numerical experiment are shown in tables 3.2 and 3.3, respectively (here I_{ring}/I_c is the ratio of the intensities within the ring (at the point R_{ring}) to the intensity on the optical axis; ΔR is the width of

Table 3.3.

D/λ	I_{ring}/I_c	$\Delta R/R_{ring}$			$\Delta R_c/R_{ring}$		
		0.7	0.5	0.1	0.7	0.5	0.1
100	1.00	0.35	0.51	0.80	0.57	0.76	1.23
200	2.78	0.18	0.17	0.40	0.29	0.33	0.57
400	3.45	0.10	0.10	0.21	0.16	0.19	0.33
800	8.33	0.05	0.06	0.10	0.09	0.11	0.17

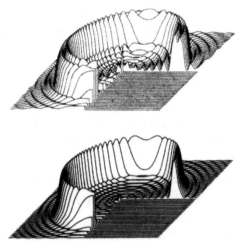

Figure 3.5. Field intensity distribution for radiation focused on to a ring, (a) for type-E1 and (b) type-E2 diffractive elements.

the ring at the corresponding intensity level and ΔR_c is the width of the axial intensity maximum). Figures 3.5(a,b) show the corresponding shapes of focusing regions in the axonometric projection.

Analysing the data given above we come to the following conclusions: since the problem of synthesizing the phase function of a diffractive optics element does not have a unique solution, we can control the distribution of field intensity in the region of focusing; with the aperture ratio D/λ exceeding 200 and $D/B \approx 0.5$ the geometric parameters (ΔR_c, ΔR) are in fact independent of the type of element (E1 or E2); an element of type E2 makes it possible to essentially reduce the central (spurious) maximum of intensity. For type-E2 elements the ratio I_{ring}/I_c increases faster with increasing D/λ than for type-E1 elements.

Another remark is in order. We already mentioned that two types of elements exist for focusing radiation on to a ring; we denoted them as E1 and E2. A natural question arises: is it possible to provide a focusing of radiation, for instance, on to a specific area for an arbitrary shift of angles $\Delta\varphi$ between a point at the boundary of the zone of the diffractive optical element and a point on the focusing ring? The answer is no because for large values of the parameter $\Delta R_{\text{ring}}/\lambda_B \gg 1$, the diffraction-defined width of the ring Δ [12] is given by the expression

$$\Delta \approx \lambda B/(0.5D\cos\Delta\varphi).$$

This expression implies that in order to provide focusing of radiation on to a prescribed region of space, we need to have $\Delta\varphi = 0; \pi$. Furthermore, if, for example, $\Delta\varphi = \pi/2$, this element will be an analogue of a zone plate and will focus radiation to a point. It is not difficult to obtain the corresponding restriction on $\Delta\varphi$ if radiation is focused on to an arbitrary curve. To achieve this, we need to replace the denominator in the expression for Δ by the size of the corresponding part of the diffractive element in the direction orthogonal to the tangent to the curve.

Another approach to correcting the field intensity distribution function in the region of focusing is to choose an optimal pupil function which ensures suppression of the required order of diffraction. For instance, it is expedient to use in the case we consider a phase step with modulation depth equal to π. The intensity distribution in the diffraction pattern of this function has the form [13]

$$I(u, v) \approx \text{sinc}^2(0.5UL)\,\text{sinc}^2(wV)\sin^2(0.5\pi UL)$$

where $U = \varepsilon/(\lambda F)$, $V = \eta/(\lambda F)$ and $\varepsilon = \pi UD/2$.

As a result of superposition of this phase function and the phase function of the diffractive element, the central intensity maximum is suppressed, as we easily see.

An analysis of the problem of reconstruction of a two-dimensional field $F(x, y) = (I(x, y))^{1/2} \exp[i\varphi(x, y)]$ from its intensity $I(x, y)$ in the Fresnel zone revealed the important role played by the vortex properties of the vector flux of electromagnetic energy [14]

$$j = \left(I\frac{\partial\varphi}{\partial x}, \ I\frac{\partial\varphi}{\partial y}, \ kI\right).$$

The following condition must hold for all electromagnetic fields:

$$\int_{R^2}(\text{curl }j)_1\,\mathrm{d}S = 0$$

where $(\text{curl }j)_1 = k^{-1}[\nabla I, \nabla\varphi]_1$. Note that the following expression holds at the points of isolated zeros of intensity:

$$|(\text{curl }j)|I = k^{-1}\left[\left(\frac{\partial^2 I}{\partial x^2}\right)\left(\frac{\partial^2 I}{\partial y^2}\right) - \frac{\partial^2 I^2}{\partial x\,\partial y}\right]^{1/2}.$$

When the field of this Fresnel zone is considered, the isolated zeros are formed in pairs, with opposite signs of $(\text{curl }j)_1$ in them. To illustrate how to take into account the vortex component when synthesizing electromagnetic fields, let us consider a problem of focusing laser radiation on to a ring of radius R_{ring} at a distance f [14, 15]. In the paraxial approximation, diffractive element of the type 'spherical lens + axicon' is a known solution. Its phase function is

$$\varphi_0(r) = -\frac{kr}{2f_0} + kr\frac{R_{\text{ring}}}{f_0}. \tag{3.1}$$

This solution is realized in the geometric optics approximation by the mapping $(\varepsilon, \eta) \rightarrow (x, y)$ (object electromagnetic field in the Fresnel zone) with zero Jacobean

$$x = \varepsilon + f_0 \otimes -\frac{k\varepsilon/f_0 + kR_{\text{ring}}\varepsilon/[f_0(\varepsilon^2 + \eta^2)^{1/2}]}{k} = R_{\text{ring}}\cos\alpha$$

$$y = \eta + f_0 \otimes -\frac{k\eta/f_0 + kR_{\text{ring}}\eta/[f_0(\varepsilon^2 + \eta^2)^{1/2}]}{k} = R_{\text{ring}}\sin\alpha.$$

The corresponding diffractive field has zero curl j and the following form in the focal plane:

$$F_0(\rho) = \frac{kE_0}{if_0}\exp\left[ik\left(f_0 + \frac{\rho^2 - R_{\text{ring}}^2}{2f_0}\right)\right]\int_0^a \exp\left(\frac{ikrR_{\text{ring}}}{2f_0}\right)J_0\left(\frac{k\rho r}{f_0}\right)r\,dr$$

where E_0 is the amplitude of the electromagnetic field in the (ε, η) plane, ρ is the polar radius and a is the radius of the diffractive element.

The appearance of a local intensity maximum at the centre is caused by the difference between the geometric-optics approximation and its implementation by wave functions. It is then clear that a stable zero intensity at the centre can exist only with non-zero curl.

The simplest solutions for focusing on to a ring with non-zero curl are the solutions of the type

$$\varphi_n(r, \alpha) = -\frac{kr^2}{2f_0} + kr\frac{R_{\text{ring}}}{f_0} + na \qquad (3.2)$$

where n is a non-zero integer and $a = \arg(\varepsilon + i\eta)$. The diffraction field is then

$$F_0(\rho, Q) = \frac{kE_0(-i)^n}{if_0}\exp\left[ik\left(f_0 + \frac{\rho^2 - R_{\text{ring}}^2}{2f_0}\right) + inQ\right]$$

$$\times \int_0^a \exp\left(\frac{ikrR_{\text{ring}}}{2f_0}\right)J_n\left(\frac{k\rho r}{f_0}\right)r\,dr.$$

If we now turn to a similarity between the behaviour of a liquid and the propagation of energy in electromagnetic field, the solutions (3.1) and (3.2) correlate exactly as two methods of creating a vortex on the surface of a liquid: by rotating it or by the centrally symmetric pressure on the liquid.

The element (3.2) implements a single-valued mapping (for $r \neq 0$) with non-zero Jacobian:

$$x_0 = r\cos\alpha - (nf_0\sin\alpha)/kr, \qquad y_0 = r\sin\alpha - (nf_0\cos\alpha)/kr. \qquad (3.3)$$

The single-valuedness (unambiguity) of mapping (3.3) signifies that the element (3.2) does not generate focusing in the sense of the stationary phase method. Moreover, the position of the intensity maximum r_{max} falls within the region of geometric shadow:

$$r_{\text{max}} < [R_{\text{ring}}^2 + (nf_0/ka)^2]^{1/2}.$$

The reason for this is that solutions (3.2) are wave solutions and thus cannot be obtained in terms of geometric optics.

3.2 Selection of harmonics of coherent radiation

The quantization of the phase function of a diffractive element results in new properties of the element. One of these manifestations of discreteness of the phase function of the elements that we discuss is the possibility of selecting harmonics of coherent radiation. Analytical expressions for a four-level diffractive element were obtained in [16–18] for the gain of the axisymmetric element G at the main focal point corresponding to the reference wavelength [8, 19, 20]:

$$G(N) = \left[\sum_{n=1}^{N} A(n) \left\{ \cos \left(2\pi M \, \frac{\lambda_0}{\lambda} [M \, \mathrm{Giv}(n) + 1] \right) \right. \right.$$

$$\left. \left. - \cos \left(2\pi M \, \frac{\lambda_0}{\lambda} [\mathrm{Giv}(n)] \right) \right\} \right]^2$$

$$+ \left[\sum_{n=1}^{N} A(n) \left\{ \sin \left(2\pi M \, \frac{\lambda_0}{\lambda} [M \, \mathrm{Giv}(n) + 1] \right) \right. \right.$$

$$\left. \left. - \sin \left(2\pi M \, \frac{\lambda_0}{\lambda} [\mathrm{Giv}(n)] \right) \right\} \right]^2$$

$$A(n) = \left[\frac{4MF}{\lambda_0} + 2n - 1 \right] \Big/ \left[\frac{4MF}{\lambda_0} + M(2n - 1) \right]$$

where N is the number of Fresnel zones within the aperture of the element, λ_0 and λ are the reference and the actual radiation wavelength, F is the focal length and $\mathrm{Giv}(n)$ is the function equal to the maximum integral value of its argument.

An analysis of this relation shows that the maximum gain (for constant signal-to-noise ratio) is achieved for the harmonic with the number $\lambda_0/\lambda = M/2$, while there is no focusing of radiation at the wavelength satisfying $\lambda_0/\lambda = M$, that is, this radiation is selected (figure 3.6(a)).

Spectrally selective properties of diffractive elements with a discrete phase function make it possible to use them for mixing a discrete set of wavelengths into the same focusing zone [8]. For instance, the following *theorem* is valid:

For radiation with harmonics $\Lambda_i = \lambda_0/i$, $i = 1, \ldots, N$ in the interval $Q = [\lambda_1 \ldots \lambda_2]$ to fall within this interval, that is, $\Lambda_i \in Q$, it is necessary and sufficient for the maximum number of the harmonic to be $N = i_{\max} = \mathrm{entier}(\lambda_2/\lambda_1)$, where $\mathrm{entier}(x)$ is the integral part of the real x.

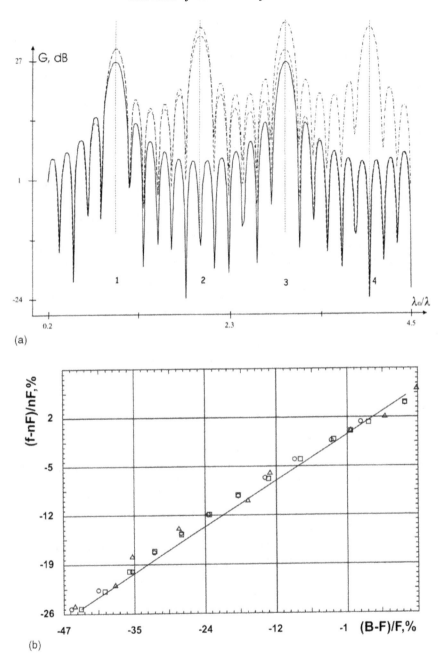

Figure 3.6. (a) Gain of a diffractive optical element for three numbers of phase quantization levels [20]: ——, two levels; – – –, four levels; –·–·–, six levels. (b) Frequency properties of a zone plate for the first three harmonics [20]. dn $F = (f - nf_0)/nf_0$. □, $n = 1$; △, $n = 2$; ○, $n = 3$.

The *proof* of this theorem is self-evident. The property of diffractive elements formulated above is also important for practical applications [20]. Thus it may be possible to use this principle and design elements for the x-ray range of wavelengths, to design novel optical elements for optical polychromatic computers etc.

As for the frequency and focusing properties of diffractive elements using radiation harmonics, their numeric and experimental analysis [8] revealed the following behaviour. When an element operates on a harmonic, the frequency properties expressed in arbitrary units are the same as the frequency properties of the diffractive element in the range of the main wavelength (reference wavelength)—see figure 3.6(b) (in this figure n is the number of the frequency harmonic). This is also true for the transversal and longitudinal resolving powers of a diffractive element if we appropriately replace the current (working) wavelength in the expression.

Another way of selecting harmonics of coherent radiation using Fresnel zone plates is available. It is based on an optimal selection of the number of Fresnel zones within the aperture. Such a space filter was proposed and investigated in [22, 23]. We will now give a brief explanation of the method.

Let two monochromatic waves λ and λ' be incident on a zone plate and the wave λ is to be selected. The wave amplitude close to the axis of the Fresnel zone plate illuminated with a plane wavefront and wavelength λ is [24]

$$I(z,r) \approx \sum_{m=0}^{N} I_0\left(\frac{2\pi}{\lambda} r \sin\gamma\right) \exp\left[i\left(\frac{2\pi}{\lambda} R_m + m\pi\right)\right]$$

where r is the distance from the axis of the Fresnel zone plate, I_0 is the Bessel function, γ is the angle subtended by the radius of the mth zone from the selected point on the zone plate axis, $R_m = (z^2 + \rho_m^2)^{1/2}$, and $\rho_m = (m\lambda F)^{1/2}$ is the radius of the mth zone of the Fresnel zone plate. The formula above is valid for small γ (the paraxial approximation) for the Fresnel zone plate with transparent first zone and odd m.

The wave amplitude distribution at a point $z = F$ for the wavefront with wavelength λ' will obviously be

$$I'(F,r) = \exp\left(i\frac{2\pi}{\lambda'} F\right) \sum_{m=0}^{N} I_0\left(2\pi r \frac{\lambda}{\lambda'}(m/\lambda F)\right)^{1/2} \exp\left[im\pi\left(1 + \frac{\lambda}{\lambda'}\right)\right]$$

(3.4)

or, in dimensionless units,

$$I'(F,r) = \exp\left(i\frac{2\pi}{\lambda'} F\right) \sum_{m=0}^{N} I_0\left(\pi r_A \frac{\lambda}{\lambda'}(m/N)\right)^{1/2} \exp\left[im\pi\left(1 + \frac{\lambda}{\lambda'}\right)\right]$$

so that at the main focal point F of the wave λ we have

$$I'(F,0) = \exp\left(i\frac{2\pi}{\lambda'}F\right)\frac{\exp[i(N+1)(1+\lambda/\lambda')\pi] - 1}{\exp[i(1+\lambda/\lambda')\pi] - 1}. \qquad (3.5)$$

As follows from expression (3.4) for $I'(F,r)$, the wave amplitude distribution for λ and λ' in the plane of the main focal maximum of the λ wave is determined by the quantities (λ/λ'), N and the diameter of the filtering aperture. The corresponding rule for choosing the number of zones N follows from expression (3.5) for $I'(F,0)$. Obviously, filtration will have maximum efficiency if the number N of zones of the Fresnel zone plate is such that $I'(F,0)$ has maximum value. If $\lambda/\lambda' = p/n$, where p, n are integers, then $I'(F,0)$ vanishes at $N = 2gn - 1$ ($g = 1, 2, \ldots$) for every $\lambda < \lambda'$ and all $\lambda > \lambda'$, except $\lambda/\lambda' \neq 2\varepsilon - 1$, $\varepsilon = 1, 2, \ldots$. If the condition $\lambda/\lambda' = 2\varepsilon - 1$ is met, the main focal point of the wave λ coincides with the additional focal point of order $(\varepsilon + 1)$ of the wave λ', and the amplitudes of both waves at this point equal $(N + 1)$. It is also obvious that the transversal dimensions of the focal areas are different for the wavelengths λ and λ' and are dictated by the function $I'(F,r)$.

If the original radiation is composed of discrete components related by a quotient of integers, then the amplitudes of all λ', except the minimum λ_{min} that needs to be singled out, vanish at the main focal point of the wave λ_{min} at $N = 2gS - 1$, where S is the lowest common denominator of the fractions (λ_{min}/λ').

Maluzhinets [24] proposed to fabricate a diffractive element with the maximum height of phase-inversion profile equal to an integral multiple of $h = \lambda/(n-1)$ or with the phase shift as an integral multiple of 2π. This signifies in fact that the diffractive element, or a part of it, is designed to work on one of the harmonics of the incident radiation. The advantage of such an element is a considerably lower number of zones within the aperture, which is important in the short-wavelength range. Table 3.4 lists the values of the Strehl number and the number of zones for various diffractive elements, depending on the phase shift.

Table 3.4.

Element	Strehl number	Number of zones
Kinoform	1.000	1557
Kinoform with triangular profile	0.999	1557
Diffraction lens $h = 10h_0$	0.958	156
Superdiffraction lens $h = 10h_0$	0.999	267
Diffraction lens $h = 25h_0$	0.790	63
Superdiffraction lens $h = 25h_0$	0.997	117

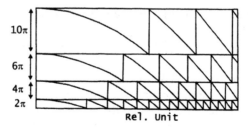

Figure 3.7. Example of coding the phase function of a diffractive element modulo 2π, 4π, 6π and 10π.

A diffractive element with a phase step equal to $h = Nh_0$ is calculated in such a way that the first several zones are phase-shifted by 2π, the subsequent several zones by 4π, and so forth up to the phase shift of $2\pi N$. The number of divisions for each value of the phase shift was equal for each group, even though this is not mandatory. Therefore, another possible free parameter (a degree of freedom) is available for optimizing the properties of diffractive elements: the height of the phase-inversion step. Furthermore, the frequency band with such a diffractive element also increases. This conclusion follows from the argument that the phase profile height for several wavelengths in the spectrum will be a multiple of 2π. Zone radii of the diffractive element are found from the expression (figure 3.7)

$$R_i = \sqrt{2i\lambda_M MF + i^2\lambda_M^2 M^2}$$

where M is the multiplicity coefficient for the wavelength λ_M. It is clear from this formula that the diffractive element computed for certain λ_M and M will

Figure 3.8. Diffractive element designed to work in the 75 GHz range with 6π-high phase profile.

have the same focal length for other wavelengths for which the product $\lambda_M M$ remains constant.

Figure 3.8 shows the diffractive element designed to work in the 75 GHz range and having the phase profile height equal to 6π.

3.3 Methods of controlling frequency properties of flat diffractive elements

We already mentioned that diffractive optical elements are qualitatively different from other types of focusing elements in their frequency properties. Golub *et al* [7] analysed the frequency properties of optical diffractive elements and concluded that owing to a higher chromatism of diffractive elements and to the impossibility of using here the standard techniques offered by optics for compensating chromatism (selection of materials of various dispersion, introduction of chromatic surfaces etc.), it must be considered inexpedient to design non-monochromatic objective lenses based on elements of diffractive optics. Furthermore, Slyusarev indicated [26] that the frequency properties of diffractive elements cannot be removed in principle, owing to the diffractive nature of these elements.

At the same time, a zone plate may be modernized in such a way as to compensate for its frequency properties. It appears that Raiskii [27] was the first to point to this possibility. The gist of the idea that he proposed was as follows. Let us consider two amplitude-type zone plates with identical focal distances, one of which was designed for the wavelength λ_1 and the second one for λ_2. Let us juxtapose them by making their optical axes coincide, and determine the positions of transparent zones common for the two zone plates of the stack. The zone plate produced in this way will be achromatic for the wavelengths λ_1 and λ_2. Such a diffractive element was indeed designed and its properties were investigated [28].

An approach similar to that described above can also be used to create flat diffractive optics with prescribed focusing properties. For instance, we can synthesize a zone plate that would focus radiation to several discrete focal points separated by certain distances along the optical axis. Thus if the neighbouring focal points are spaced by a distance not exceeding $\Delta_z/2$, this zone plate will have a greater depth of focus. If they are spaced by a distance greater than Δ_z, a multifocal zone plate is synthesized.

As an example, figure 3.9 shows the plot of field intensity distribution along the optical axis of a bifocal zone plate; its photograph is also shown. The two focal points were separated in space by a distance $B_1 - B_2 = D/4$. The studies conducted in 1983 in the 4 mm range demonstrated that in each of its foci, the bifocal zone plate provided resolving power very near the possible limiting value.

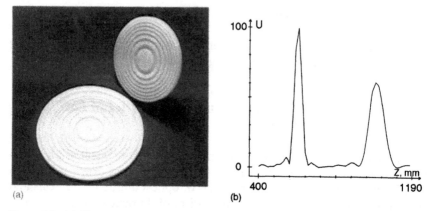

(a)   (b)

Figure 3.9. (a) The appearance of a bifocal phase-inversion zone plate [28] for the 4 mm wavelength range. (b) Field intensity distribution along the optical axis of a bifocal zone plate.

However, the algorithm above, as the one in the previous case, reduces the power efficiency of the zone plate even though it allows it to combine the properties of two or more elements in one diffractive optical element.

Multifocal elements are important for use, for example, in reading heads of laser optical disks in order to generate three-dimensional images. Such elements can be designed using, for instance, a diffractive element consisting of different segments, each segment having its own focal length. However, the difficulty lies in the control of energy distribution between different foci.

Golub *et al* [29] analysed a method of synthesizing highly efficient multi-focal optical elements with a prescribed number of foci and a prescribed intensity distribution between them (figure 3.10).

Let us find the phase function φ describing a multifocal diffractive element. This element must realize focusing of radiation on to N points located on the optical axis of the system, with specific ratios of intensities

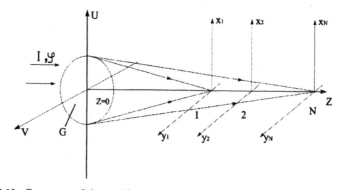

Figure 3.10. Geometry of the problem.

at these foci:

$$I_1, \ldots, I_N \left(\sum_{i=1}^{N} I_i = 1 \right).$$

Let us describe a multifocal diffractive element as a zone plate with the following phase function:

$$\varphi(\mathbf{u}) = \Phi[\varphi_1(\mathbf{u})] + \varphi_2(\mathbf{u}) - \varphi_0(\mathbf{u}), \qquad \mathbf{u} \in G. \tag{3.6}$$

Here $\varphi_0(\mathbf{u})$ is the phase function of the illuminating beam with aperture G, $\mathbf{u} = (u, v)$ are the Cartesian coordinates in the plane of the diffractive element, and

$$\varphi_1(\mathbf{u}) = \mathrm{mod}_{2\pi}(-k\mathbf{u}^2/2F_1), \qquad \varphi_2(\mathbf{u}) = -k\mathbf{u}^2/2F_2, \qquad k = 2\pi/\lambda \tag{3.7}$$

are the paraxial phases of lenses with certain focal distances F_1 and F_2. The term $\varphi_0(\mathbf{u})$ compensates for the phase of the illuminating beam. The introduction of the phase function of the lens $\varphi_2(\mathbf{u})$ in equation (3.6) allows us to select the coordinates of foci along the optical axis. The function $\mathrm{mod}_{2\pi}(x)$ maps the phase to the interval $[0, 2\pi)$. The function $\Phi(\varphi_1)$ lies in the interval $[0, 2\pi)$ and describes the nonlinear pre-distortion of the phase $\varphi_1(\mathbf{u})$. Therefore, a multifocal element is mathematically a superposition of a lens $\varphi_2(\mathbf{u})$ and a complex zone plate $\Phi[\varphi_1(\mathbf{u})]$.

Let the illuminating beam $[I_0(\mathbf{u})]^{1/2} \exp(i\varphi_0(\mathbf{u}))$ with intensity $I_0(\mathbf{u})$ be incident on a diffractive element that does not modify the intensity of the incident wave. The field behind this diffractive element can be presented in the form

$$w(\mathbf{u}) = [I_0(\mathbf{u})]^{1/2} \exp\{i\Phi[\varphi_1(\mathbf{u})] + i\varphi_2(\mathbf{u})\}. \tag{3.8}$$

After the function $\exp[i\Phi(\varepsilon)]$ is expanded to a Fourier series, equation (3.8) is transformed to the form [29]

$$w(\mathbf{u}) = [I_0(\mathbf{u})]^{1/2} \exp\left(-\frac{ik\mathbf{u}^2}{2F_2}\right) \sum_{l=-\infty}^{+\infty} a_l \exp\left[il\left(-\frac{k\mathbf{u}^2}{2F_1}\right)\right]. \tag{3.9}$$

Expression (3.9) can be interpreted as a superposition of a set of paraxial spherical beams with different focal distances. Here $l_1, \ldots, l_N$ are the corresponding diffraction orders.

To synthesize a multi-focal element, it becomes especially important to find the function $\exp[i\Phi(\varepsilon)]$, $\varepsilon \in [0, 2\pi]$. The phase function $\Phi(\varepsilon)$ can be treated as a phase function of an order-N diffraction grating with period 2π, producing intensities $I_1, \ldots, I_N$ in the orders $l_1, \ldots, l_N$.

As an example, let us consider a bifocal element with focal distances F_a, F_b and with intensity equally distributed between the foci. The aperture G is circular, of radius R. In this case we set $F_1 = F_a$, $F_{-1} = F_b$. Then we easily obtain [29]

$$F_1 = 2F_aF_b/(F_b - F_a), \qquad F_2 = 2F_aF_b/(F_a + F_b).$$

Now we define the function $\Phi(\varepsilon)$ as a phase function of a order-two diffraction grating:

$$\Phi(\varepsilon) = \begin{cases} 0, & \varepsilon \in [0, \pi) \\ \pi, & \varepsilon \in [\pi, 2\pi). \end{cases}$$

The Fourier coefficients of the function $\exp[i\Phi(\varepsilon)]$ are

$$a_1 = \begin{cases} 0, & l = 0 \\ (1 - (-1)^l/\pi il, & l \neq 0 \end{cases}$$

and $|a_1|^2 = |a_{-1}|^2 = 0.4053$. In this case more than 81% of the radiation incident on the diffractive element is focused into orders with numbers $l = 1$ and $l = -1$.

For the scalar field, the following expression can be obtained for the output of the diffractive element:

$$w(\mathbf{u}) = a_1 \exp(-ik\mathbf{u}^2/2F_a) + a_{-1} \exp(-ik\mathbf{u}^2/2F_b). \qquad (3.10)$$

Expression (3.10) describes the required process of focusing to two points in the paraxial approximation.

3.4   Elements of diffractive optics fabricated on surfaces of revolution of second order

Up to this point we were discussing the properties of diffractive optical elements fabricated on a flat surface. However, the introduction of a non-flat surface makes it possible to considerably broaden the range of problems they solve, to improve a number of properties of flat elements etc. Therefore, this parameter can be treated as an additional degree of freedom. As an example, we will consider diffractive elements fabricated on a parabolic surface (figure 3.11).

Presumably, Raisky [27] at the beginning of the 1950s and Gabor in mid-1960s [30] were the first to propose fabricating diffractive objective lenses on surfaces of revolution. Raisky pointed out that since the curvature of wavefronts has different signs for the source at a distance A from a zone plate and for the observer at a point B, the lengthening of the optical path by $\lambda/2$ can be achieved by a very small increment of the angle α (the angle between the optical axis and the direction to the nth Fresnel zone). As a result, the optical path difference expressed in the number of half-wavelengths $\lambda/2$ (i.e. the number of Fresnel zones) assumes very high values for the zone plate even at relatively small angles α [27]. To create high-speed diffractive optics with wide Fresnel zones, large aperture and

Figure 3.11. Phase-inversion zone plates for the 4 mm wavelength range on parabolic surface: half-wave $(D/\lambda = 44)$ and quarter-wave $(D/\lambda = 110)$ [28]. Material: polyethylene.

short focal distance, it was suggested in [27] to use a spherical surface which makes it possible to produce wavefront curvature of the same sign for the values of both A and B.

A theoretical study of relatively elementary focusing properties of radio objective lenses of the microwave range, carried out on flat and spherical surfaces, was reported by Dey and Khastgir [31]. They obtained a transcendental equation for finding Fresnel zone radii on a spherical surface. As an example, figure 3.12 gives the appearance of a phase inversion diffractive element with triangular zone profile [28], fabricated on a spherical surface and prepared for working in the 35 GHz range. In the cases where wavelength λ is much smaller than the width of the outermost Fresnel zone

(a)   (b)

Figure 3.12. (a) Element of mm-range diffractive optics with triangular zone profile fabricated on a spherical surface, and (b) a fragment of a zone [28].

(this is possible at microwave frequencies) Fresnel zone radii can be found from an approximate equation [31]

$$\frac{(A+B)^2}{c^2 r^2} R_n^4 + 2\left\{ \frac{A+B}{r} + \frac{(A-B)(A+B)^2}{rc^2} - 2\left(\frac{A}{r}+1\right) \right\} R_n^2$$

$$+ C^2 - 2(A^2 + B^2) + \frac{(A^2 - B^2)^2}{C^2} = 0$$

where r is the spherical radius of the diffractive lens and $C = A + B + n\lambda/2$. Using the geometrical optics approximation, these authors conducted an analysis of the effect of accuracy of placing the radiation source at the focal point on the amount of displacement of the focusing range along the optical axis. For instance, if a is a new distance from the source to the diffractive element, then the new focal length b is found from the equation

$$[\{a + [r - (r^2 - R_1^2)^{1/2}]\} + R_1^2]^{1/2} + [\{b - [r - (r^2 - R_1^2)^{1/2}]\} + R_1^2]^{1/2}$$

$$- (a + b + \lambda/2) = 0.$$

An analysis of this expression showed that if $a \gg D$, changes in this distance leave focal length unaffected.

Numerical computations carried out for $a_0 = 60$ cm, $b_0 = 10$ cm, $D = 80$ cm, $\lambda = 3.2$ cm showed that the half-width of the field intensity distribution along the optical axis for an element of diffractive optics on a spherical surface is somewhat narrower than for a zone plate. At the same time, the level of side lobes in the image of a point is lower than for a similar zone plate; the field intensity at a focal region of a 'spherical' diffractive element was found from the expressions of [31]:

$$A = I_0 \left[\sum_{p=0}^{N} \exp[-i(\alpha_{2p+1} - \alpha_1)] \right]$$

$$\alpha_n = 2\pi/\lambda \{ (R_n^2 + (a_0 + x_n)^2)^{1/2} + \cdots + (R_n^2 - (b - x_n)^2)^{1/2} \}$$

$$x_n = r - (r^2 - R_n^2)^{1/2}.$$

In later publications Dey and Khastgir [32] studied elements of diffractive optics on parabolic surfaces and Dey *et al* [33] compared focusing properties of 'parabolic', 'spherical' and 'flat' diffractive elements in the centimetre wave range.

Expressions for Fresnel zone radii for the corresponding types of diffractive optical element were given in [32]; it was shown that the narrowest half-width of field intensity distribution in the focal region along the optical axis is provided by the 'parabolic' radio lens. The authors explained these properties by indicating that the number of Fresnel zones on a 'parabolic' element within its aperture is greater than the corresponding number for 'spherical' and 'flat' elements.

Dey and Khastgir [32] also found an equation for calculating Fresnel zone radii on a 'parabolic' diffractive element when a spherical wavefront is incident on a convex side of the radio lens. It was shown that the focal length in the frequency range from 8075 to 10 875 MHz varied from 16 to 34 cm for a radio lens with $\lambda_0 = 3.2$ cm, $a_0 = 50$ cm, $b_0 = 24$ cm and $N = 12$.

Minin and Minin [34] published the results of experimental studies of focusing and frequency properties of 'parabolic' (one- and two-component) radio lenses fabricated for the millimetre wave range; the work was done between 1982 and 1987.

The computation of the discrete phase function of the objective lens (of its Fresnel zones) that converts a divergent spherical wavefront into a convergent spherical one was conducted using the expressions [32]

$$[R_n^2 + (a + b_n)^2]^{1/2} + [(b_0 - b_n)^2 + R_n^2]^{1/2} = a + b_0 + n\lambda/2, \qquad b_n = R_n^2/4F.$$

Let us consider the resolution of a parabolic objective lens. The shape of the diffractive spot formed by such a lens is shown in figure 3.13 for various frequencies of radiation. The solid curve traces the shape of the diffractive spot for a zone plate [35].

An analysis for the experimental data plotted in figure 3.13 shows the following features:

- parabolic radio lenses, just as the flat ones [35], show resolution close to the diffractional limit in a wide spectral range (+20.65% to −16.3%);
- for parabolic radio lenses, the level of side lobes in the field intensity distribution in a focal area is smaller, on average by 20–40%, than a similar value for flat radio lenses;
- the directivity of a zone plate on a parabolic surface is higher than that of its flat analogue.

The shape of the best focusing surface (figure 3.14(a)) is shown in figure 3.14(b) for two different radiation frequencies as a function of the angle of incidence of a plane wave on to the diffractive element. The data shown indicate that the generatrix of the surface can be described with high accuracy as a circle with a radius equal to the current value of the rear segment length. The following condition holds:

$$B_1(\lambda_1)/B_2(\lambda_2) = R_1/R_2.$$

Here B_i and R_i are the rear segment and the radius of the approximating circle for the corresponding wavelength, respectively.

The field intensity distribution along the optical axis was measured in [34] for various wavelengths of source radiation differing from the basic wavelength λ_0. The experimental results demonstrate that the field intensity distribution width along the optical axis, as determined from first zeros of intensity, is less by 15–20% than a similar characteristic of a 'flat' radio lens.

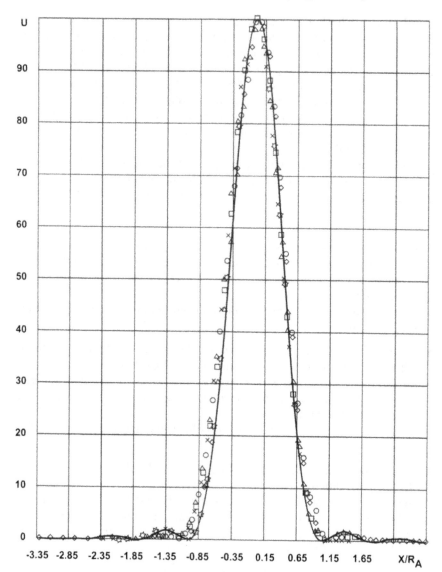

Figure 3.13. Diffraction spot for a diffractive optical element on a parabolic surface [34] for the following values of $\Delta\lambda/\lambda_0$: ×, 20.65%; □, 14.35%; ○, 8.69%; △, 5.41%; ◇, 13.01%; *, 16.30%.

Note another important property of diffractive optical elements fabricated on second-order surfaces of revolution: their non-reciprocity. The thing is, with a fixed position of a flat lens (zone plate), the positions of the front and rear segments can be swapped. For instance, if the zone plate is calculated for the front segment *A* and a rear segment *B*, then if these are

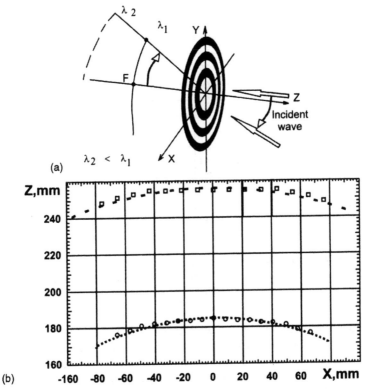

(a)

(b)

Figure 3.14. (a) Schematic shape of the best focusing surface. (b) Shape of the best focusing surface for a diffractive element on a parabolic surface as a function of radiation incidence angle [34] for two difference frequencies.

swapped (that is, if the radiation source is placed at distance B from the zone plate and the radiation receiver is placed at distance A), the radio lens will continue to function properly. This occurs because the values of zone radii for a Fresnel zone plate are independent of the swap $A \leftrightarrow B$. This property does not hold for diffractive optical elements on non-flat surfaces, so that the orientation of the apex of the diffractive element in space relative to the calculated values of A and B must be taken into account when calculating the parameters of such lenses.

3.5 Single-component 'parabolic' radio lens

A single-component parabolic radio lens (figure 3.11) was machined with a digitally controlled lathe. The material used was polyethylene, the radio lens aperture ratio was $D/\lambda_0 \approx 44$, the front segment $A/\lambda_0 \approx 87$, the rear segment $B/\lambda_0 \approx 43$, and the focal distance of the paraboloid $F/\lambda_0 \approx 14$.

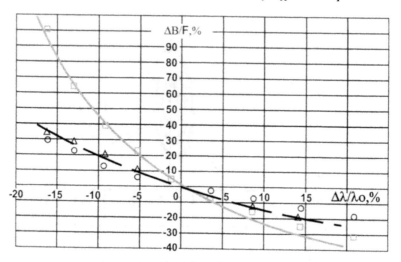

Figure 3.15. Frequency properties of diffractive objective lens [34]. – – –, zone plate; △, parabolic single-component lens; ○, parabolic two-component lens. Single component lens, with $A \Leftrightarrow B$ replacement: ——, theory; □, experiment.

The frequency properties of parabolic radio lenses are described by curves shown in figure 3.15. Such relations coincide to high accuracy in the range of negative detuning of wavelength from the calculated value λ_0 and differ by only 3–4% in the range of positive detuning. It appears that the identical behaviour of functions $B(\lambda)$ for the zone plate and diffractive 'parabolic' lenses is caused by the fact that almost the same number of Fresnel zones falls within the parabolic surface for the given parameters of the elements as the number of zones within the zone plate.

As for the resolving power of a parabolic objective lens for an off-axial position of the radiation source, the experimental study demonstrated that broadening of the main diffraction spot obtained from the first zeros at the edge of the field of view for a single-component radio lens was about 4% and that the level of side lobes (their intensity) was within 5–7% of the intensity of the main scattering lobe. The field of view of the single-component 'parabolic' radio lens with $D/\lambda \approx 44$ is not less than 30°.

In order to prove non-reciprocity of 'convex' diffractive elements, a number of experiments were run. For instance, the segment swap $A \Leftrightarrow B$ was carried out for a single-component 'parabolic' diffractive optic element. Then the field intensity distribution along the optical axis of the element was investigated. The study showed that in this case several secondary maxima of field intensity are formed along the optical axis in addition to the main maximum, and they are comparable with the amplitude of the main scattering lobe. Therefore, the efficiency of such diffractive elements is reduced because the radiation incident on the diffractive optical element is redistributed into several foci. In addition, when the radiation source is displaced off the optical axis by a distance $\Delta x \approx (2-3)R_A$, its image 'breaks into pieces'—meaning the element stops functioning, which is explained by the disparity of Fresnel zone radii and the corresponding radio lens segments (violation of the tautochronism).

3.6   Two-component 'parabolic' diffractive objective lens

A multi-component objective lens proposed in [36] is based on Rayleigh–Wood zone plates; its experimental studies (discussed in chapter 2) demonstrated that such systems make it possible to considerably enhance the informative properties of single-component focusing devices. Further increase of the information potential of diffractive objective lenses is possible through increasing, for instance, the number of lens components. However, this approach is not always acceptable, for the following reasons. First, the complexity of such devices gets too high: the overall longitudinal dimensions grow considerably, the relative adjustment of components becomes difficult, and so forth. Secondly, losses of incident radiation power increase, both owing to re-reflection of radiation among components and to the absorption of radiation in the lens material (e.g. within the microwave absorption band). We are of the opinion, therefore, that a more realistic approach is the design of two-component diffractive lenses on non-flat surfaces.

A two-component 'parabolic' diffractive lens (figure 3.16) had the same aperture as the single-component one; the length of the front segment of each component of the objective lens was $A/\lambda_0 = 47$. The distance L between the components of the lens was chosen in correspondence with the conditions for

Figure 3.16. Two-component parabolic objective lens for 4 mm wavelength range [34].

a flat two-component 'planar' diffractive objective lens. The Fresnel zone radii for each of the components of a 'parabolic' objective lens were found for the incident plane wavefront.

When a plane wave is incident on a convex side of a 'parabolic' diffractive lens, Fresnel zone radii are calculated using the expression [32]

$$R_n = \left(\frac{4b_0 Fn\lambda + Fn^2\lambda^2}{4F + n\lambda} \right)^{1/2}, \qquad R_n^2 = 4Fb_n$$

where b_0 is the distance from the apex of a paraboloid to the focal point along the optical axis, F is the focal distance of the paraboloid, and b_n is the projection of the nth Fresnel zone on to the optical axis.

The study of the properties of such an objective lens have shown the following. If a pointlike radiation source sits on the axis, the same resolving power is found (longitudinally and transversely to the optical axis) as for a single-component version. At the same time, the field of view of the two-component optical system is greater than that of a single-component one, coming to at least 40°. Figure 3.15 shows that the frequency properties of this diffractive objective lens coincide with the similar function for a single-component version.

The following conclusions can thus be drawn from the study of diffractive lenses fabricated on surfaces of revolution with a parabolic generatrix:

- non-flat surfaces of diffractive optical elements make it possible to expand the field of view of the lens and the number of image elements in it;
- two-component diffractive lenses possess a wider field of view than their single-component analogues;
- diffractive lenses can be fabricated on surfaces whose generatrix is described by a second-order curve of revolution; the focusing properties of such lenses are retained in a broad spectral range;
- zone plates on non-flat surfaces have higher gain than equivalent zone plates on a flat surface.

3.7   Invariant properties of diffractive optical elements

Let us find out what new possibilities are offered by diffractive optical elements created on non-flat surfaces [37]; we will also clarify the relation of their characteristics to the corresponding parameters of flat elements.

Without reducing the generality of the problem, we will consider, for visual clarity, an element of diffractive optics that converts a flat wavefront into a convergent spherical one. By writing expressions for the eikonals of the diffracted and reference waves, it is easy to show that for arbitrary phase shape the discrete phase transmittance function of the diffractive objective lens (its Fresnel zone radii) is found from the condition

$$[(B - x_n)^2 + y_n^2]^{1/2} = B - x_n + n\lambda_0/2 \qquad (3.12)$$

where B is the distance from the top point of the objective lens to the focusing area, (x_n, y_n) are the Cartesian coordinates of the boundary of the nth Fresnel zone, and λ_0 is the nominal radiation wavelength. If the electromagnetic wave is incident on a concave surface of the diffractive element, the sign in front of x_n must be reversed.

In order to find the frequency properties of lenses on non-flat surfaces, the quantity B must be found from expression (3.12):

$$B(\lambda) = \frac{y_n^2 - (n\lambda/2)^2}{n\lambda} + x_n. \qquad (3.13)$$

By substituting the values of Fresnel zone coordinates of the objective lens under consideration into (3.13), we find the relationship sought. The expression obtained, (3.13), leads to an important conclusion on the frequency response of diffractive elements fabricated on arbitrary surfaces of revolution. Indeed, we can find asymptotes for the bounds on the working spectral range for the objective lens on an arbitrary surface. The minimum wavelength is found from (3.13) by setting the denominator to zero:

$$\lambda_{min} \to 0. \qquad (3.14)$$

The maximum possible wavelength is found from the condition setting the numerator of (3.13) to zero:

$$\lambda_{max} \to 2(x_n + R_n')/n \qquad (3.15)$$

where $R_n' = (x_n^2 + y_n^2)^{1/2}$, which gives for $y_n \gg x_n$:

$$\lambda_{max} \to 2(x_n + y_n)/n.$$

The following conclusions can be made from comparing the estimates (3.14) and (3.15) obtained for the working spectral range of a diffractive optical element on an arbitrary surface of revolution:

- the maximum possible wavelength of an objective lens is independent of the direction of incidence relative to the apex of the objective lens surface and is determined by the flexure (degree of concavity or convexity) of the given surface;
- when electromagnetic radiation is incident on the side of the apex of the objective lens, its working spectral range is wider than in the opposite case;
- when radiation is incident on the side of the apex of the surface of the element, the focal length is less dependent on the relative detuning of the wavelength than in the opposite case.

The analytical dependences shown above clearly demonstrate that the frequency properties of diffractive optics discussed above differ substantially from similar properties of the zone plate (from flat optics behaviour) whose surface flexure is zero.

These specific frequency properties of diffractive optics on non-flat surfaces are an indication that these properties can be controlled—amplified or dampened—by choosing the shape of the lens surface and its orientation in space.

Let us consider now the longitudinal resolving power (depth of definition) of 'non-flat' diffractive lenses. It is not difficult, following [35], to obtain the following expression for the longitudinal resolving power, using the in-phase condition for the radiation at a point B and the anti-phase condition at a certain point $B + \Delta_z$:

$$\Delta_z^{(\pm)} = -\frac{(\lambda/2)^2 \pm \lambda[(B \mp x_n)^2 + y_n^2]^{1/2}}{2[((B \mp x_n)^2 + y_n^2)^{1/2} - (B \mp x_n) \pm \Delta/2]} \qquad (3.16)$$

where the sign $(\pm)$ at x_n refers to a convex $(+)$ and concave $(-)$ lenses, respectively. The sign $(\pm)$ for λ refers to the position of the first minimum of the field intensity distribution along the optical axis beyond the point $B(+)$ but closer than $B(-)$.

It also follows from the (3.16) that the longitudinal resolving power and depth definition can be controlled by choosing the flexure of the objective lens surface. Thus we obtain for a concave objective lens from (3.16) that for $y_n \gg (B - x_n)$ and $y_n \gg \lambda$

$$\Delta_z^{(\pm)} \to \lambda/2.$$

We correspondingly have for the convex objective lens for $y_n \ll (B + x_n)$ and $(B + x_n) \gg \lambda$ that

$$\Delta_z^{(\pm)} \to \lambda[(B + x_n)/y_n]^{1/2}.$$

This specific behaviour of definition depth of diffractive optical elements on a non-flat surface makes it possible to design systems that possess much higher gain than other known objective lenses.

3.8 Conical diffractive element

As an example, we consider a diffractive optical element fabricated on a conical surface (figure 3.17); it was studied by the authors in 1983. It was manufactured with a numerically controlled lathe, using optical-grade polystyrene with the following optical constants: diffractive index $n = 1.59$ and absorption coefficient $k \approx 10^{-3}$. The nominal radiation wavelength was $\lambda_0 = 4.6\,\text{mm}$, lens aperture $D/\lambda_0 = 44$ and the rear segment $B = D$. The maximum flexure of the diffractive lens surface was $\langle x \rangle / \lambda = 32$. The phase profile was machined on the inner surface of a shallow cone with

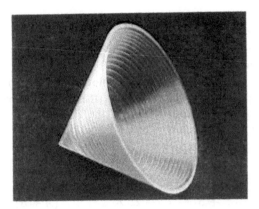

Figure 3.17. Diffractive optical element on a conical surface [37].

opening angle $\alpha = 70°$, and the phase step was calculated using the expression

$$h = [0.5\lambda/(n-1)](1 - \cos^2 \alpha/n^2)^{1/2}.$$

Fresnel zone radii were found from the formula

$$R = -\frac{K\lambda_0}{2\tan\alpha} + \left[0.25K\lambda_0\left(\frac{K\lambda_0}{\tan^2\alpha} + 4B + K\lambda_0\right)\right]^{1/2}, \qquad x_k = \frac{R_k}{\tan\alpha}.$$

$$(3.17)$$

In the general case, if a diffractive optical element is fabricated on a surface of revolution, one always has to deal with a technological substrate on which the necessary phase profile of the element is machined. The optical thickness of this substrate may be comparable with the height of phase profile, and the Fresnel zone radii will not coincide with zone radii calculated in the approximation of 'thin' conical diffractive optical element,

$$R_K = (b+x)/\tan\alpha + [(b+x)/\tan^2\alpha + x^2 - b^2]^{1/2}$$

$$x = \frac{h\cos\alpha[(1 - \cos^2\alpha/n^2)^{1/2} - \sin\alpha/n]}{(1 - \cos^2\alpha/n^2)^{1/2}\tan\alpha} \qquad (3.18)$$

$$+ \frac{hn}{(1 - \cos^2\alpha/n^2)^{1/2}} - \frac{hn}{\sin\alpha} - b - \frac{k\lambda}{2}$$

where h is the thickness of the technological substrate along the normal to the cone surface and n is the refractive index of the material. As $h \to 0$ the expression (3.18) transforms into (3.17), and as $\alpha \to \pi/2$ it transforms into the familiar expression for Fresnel zone radii of a zone plate. In fact, equation (3.18) is an expression for Fresnel zone radii not of a purely diffractive but of a refractive–diffractive element.

As follows from expressions (3.17) and (3.18), not including the optical depth of the substrate into calculations results in errors in the values of zone radii and reduces the number of Fresnel zones on the surface on a 'conical' diffractive optical element.

If the apex of a 'conical' element is illuminated with a spherical wave generated by a pointlike radiation source sitting on the optical axis of the system at a distance $-a$ from the apex, the Fresnel zone boundaries are calculated using the expression

$$R_n = (-x + (x^2 - 4yz))/2y \qquad (3.19)$$

where

$$x = \frac{(A+B)(c^2 + B^2 - A^2) - 2c^2 B}{\tan \alpha}, \qquad y = \frac{(A+B)^2 - c^2}{\tan^2 \alpha} - c^2$$

$$z = \frac{(c^2 + B^2 - A^2)^2}{4} - c^2 B^2, \qquad c = A + B + n\lambda/2.$$

In the general case the following algorithm is used to calculate Fresnel zone radii for diffractive optical elements fabricated on second-order surfaces of revolution, taking into account the technological substrate. The radii of Fresnel zones are calculated from the in-phase conditions for waves at the geometric focal point on the inner surface of the 'technological' substrate, taking into account the refractive index of the material and the shape of the surface. The phase profile of the diffractive element is constructed from the known boundaries of Fresnel zones.

To simplify the presentation we assume that all diffractive optical elements discussed below are axisymmetric. The equation of the outer surface

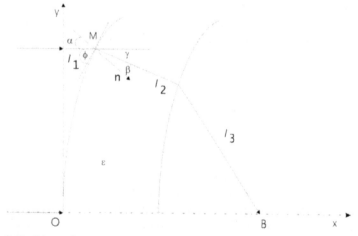

Figure 3.18. Schematic diagram of computation for phase-inversion diffractive optical element.

can be written in the explicit form

$$a_{11}x^2 + a_{12}xy + a_{22}y^2 + 2a_1x + 2a_2y + a = 0.$$

The in-phase condition of waves at a point B (figure 3.18) can be written as

$$l_1 + l_2N + l_3 = HN + (b_0 - H) + i\lambda_0/p, \qquad i = 0, 1, 2, \ldots \qquad (3.20)$$

where l_1 is the optical length of the ray from the outer surface of the diffractive optical element, l_2 is the length of the refracted ray, l_3 is the length of the ray from the inner surface of the diffractive optical element to the focal point, H and N are the thickness of the technological substrate and the refractive index of the substrate material, respectively, b_0 is the focal length of the diffractive element, $b_0 = |OB|$; λ_0 is the radiation wavelength and p is the number of phase quantization levels.

The geometry of the problem implies that $l_1 = x$. Let us find the length l_2. We draw a tangent to an arbitrary point M located on the outer side surface of the diffractive optical element. The slope coefficient is

$$\tan\varphi = \frac{a_{11}x + a_{12}y + a_1}{a_{21}x + a_{22}y + a_2}.$$

The angle between the normal to the surface at a point M and the incident ray is $\alpha = (\pi/2 - \varphi)$. For a low-curvature surface we have

$$l_2 \approx H/\cos\beta, \qquad \beta = \arcsin(\sin\alpha/N).$$

The ray length $l_3 = [(b - x')^2 + y'^2]^{1/2}$, where x' and y' are the coordinates on the inner surface of the diffractive optical element (here b is the projection of the Fresnel zone boundary on to the x axis). The relation between the coordinates of points on the outer and inner surfaces of the substrate are given by the equations

$$y' = y - l_2\sin\gamma$$
$$x' = x + l_2\cos\gamma$$
$$\gamma = \alpha - \beta.$$

The solution of equation (3.20) gives Fresnel zone radii on the inner surface of the diffractive optical elements.

The phase profile of the diffractive element for the technological substrate of thickness $H + HH$ is found in a similar manner:

$$HH = [\lambda_0/(p(N-1))]\,\text{abs}\{i + 1 - p^*\,\text{entier}((i+p)/p)\}.$$

For instance, the phase step for $p = 2$ is

$$\lambda/2(N-1),\ 0,\ \lambda/2(N-1), \ldots.$$

A comparison of expressions obtained to calculate Fresnel zone radii for a flat surface and for second-order surface of revolution shows the following. The number of Fresnel zones that are found on the surface of a diffractive element fabricated on a second-order surface of revolution is greater than for a 'flat' diffractive optical element for the same values of aperture, nominal radiation wavelength, and the front and rear segments.

Another factor must be mentioned. Taking into account the 'refractive layer' of the diffractive element makes it possible to treat this 'layer' as an additional degree of freedom, and by choosing its profile it is possible to correct for the focusing properties of the element as a whole. However, the formulas are very cumbersome and will not be given here. They are readily derivable by the reader in case of need.

Let us evaluate (as an example) the admissible error of manufacturing the phase profile on the surface of a 'conical' diffractive optical element. The admissible phase deviation $\Delta\varphi$ for which the level of signal from each zone reduces by not more than 25%, is $\Delta\varphi \approx \pi/10$ [38]. For a 'conical' diffractive optical element the distance from the focal point to the boundary of each zone (provided the optical paths of all rays within each zone are equal and they change by $\lambda/2$ at zone boundaries) equals

$$l_n = [(b - R_n/\tan\alpha)^2 + R_n^2]^{1/2} + R_n/\tan\alpha - b - n\lambda/2.$$

We differentiate this expression in the approximation $b \gg n\lambda/2$,

$$\delta R_n \approx (\delta b l \sin^2\alpha)/R_n$$

and obtain, using the Rayleigh criterion $\delta \leq \lambda/4$,

$$\delta R_n \approx (0.25\lambda b \sin^2\alpha)/R_n.$$

Typically $2R_{n_{\max}} \approx b$ for a high-aperture diffractive optical element, so that $\delta R \approx (\lambda/8)\sin^2\alpha$.

The field intensity distribution along the optical axis in the focal region of a diffractive element is shown in figure 3.19 for various wavelengths of microwave radiation. The abscissa plots the relative distance along the optical axis in units of longitudinal resolving power for the equivalent zone plate.

An analysis of the results obtained shows the following:

- half-width (at half-height) of field intensity distribution along the optical axis for a 'conical' diffractive optical element, with parameters as shown above, is twice as narrow as that of an equivalent zone plate (when radiation is incident on the side of the apex of the diffractive optics);
- when radiation is incident on the side of the base of the diffractive optical element, the width of field intensity distribution along the optical axis is approximately 2.5 times wider than for the equivalent zone plate;
- the shape of this distribution, plotted in relative units, varies very little (by about 3%) in the range of wavelengths that deviate from the nominal value by less than ±17%;

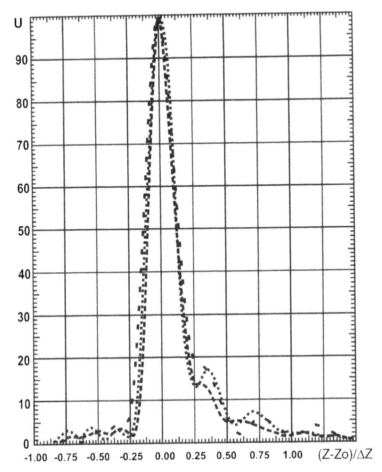

Figure 3.19. Microwave field intensity distribution along the optical axis of a diffractive optical element on a conical surface, for three values of $\Delta\lambda/\lambda_0$ [8]: +, −21.15%; □, 0.0%; ×, −12.59%.

- as wavelength decreases in comparison with the nominal value, the intensity of the first side lobe increases; this is the lobe that is located farther from the zone plate relative to the distribution maximum. The amount of increase of relative intensity approximately coincides with the amount of wavelength detuning.

When a plane electromagnetic wave is incident on a diffractive optical element, the longitudinal resolving power (3.16) can be rewritten in the form

$$\Delta_z^{(\pm)} = \Delta_{zp}^{(\pm)} \pm \langle x_n \rangle \qquad (3.21)$$

where $\Delta_{zp}^{(\pm)}$ is the longitudinal resolving power of the equivalent zone plate, and $\langle x_n \rangle$ is the average flexure of the surface of the diffractive optical element.

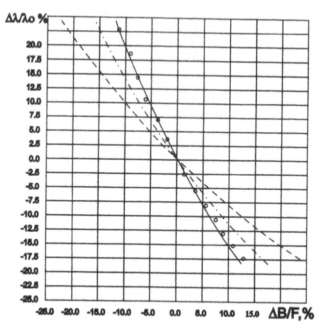

Figure 3.20. Frequency properties of a diffractive optical element [8] fabricated on a conical surface for plane incident wavefront: ——, theory; O, experiment; – – –, equivalent zone plate; – · – · –, zone plate with $B' = B - \langle x \rangle / 2$, where $\langle x \rangle$ is the flexure of the conical element surface.

Therefore, the main conclusion is that the longitudinal resolving power of the diffractive optical element can be controlled by choosing the flexure of the diffractive optical element surface and its spatial orientation; note that the longitudinal resolving power (3.21) remains invariant.

Let us consider another characteristic of the diffractive optical element: its frequency properties. Figure 3.20 gives the dependence of the rear segment on the relative detuning of wavelength of radiation incident on the apex. The ordinate axis is the relative detuning of radiation wavelength, and the abscissa axis, the relative displacement of the focal area along the optical axis.

An analysis of the frequency functions shown for a 'conical' element of diffractive optics and a zone plate shows that

- the rate of displacement of the position of the focal point along the optical axis is faster for a conical objective by a factor of 1.5–1.6 than for the equivalent zone plate;
- the frequency characteristics of 'conical' diffractive optical element co-incide with those of a zone plate with the same length of the rear segment as for the 'conical' objective lens but are reduced by the value of

the flexure of the surface; this last property can also be used for modelling frequency properties of a 'conical' element.

This last statement follows from the frequency invariant of the diffractive element considered, which can be rewritten, taking into account expression (3.16) and the condition $x_n \ll y_n$, as

$$B(\lambda) \approx 0.5(B^+(\lambda) + B^-(\lambda))$$

where $B(\lambda)$ are the frequency properties of the equivalent zone plate, and B^+ and B^- are the frequency properties of the diffractive optical element when radiation is incident on it on the side of the base or apex, respectively.

An analysis of the transverse resolving power of the diffractive element discussed above demonstrated that the resolving power it provides is at least as good as the diffraction limit in the entire working spectral range.

Therefore, the focusing and frequency properties of diffractive elements can be controlled by choosing the surface shape and spatial surface orientation relative to the position of the focusing area. This allows the designer a chance to considerably expand the functionality of diffractive optics and thereby widen the range of its possible applications.

In addition, the introduction of non-flat surfaces for diffractive focusing elements is one of the promising ways of reducing (suppressing) their chromatic properties. An example of such an element that focuses radiation on to a prescribed spatial region within a chosen spectral range is an element that was designed and studied in [39].

The investigation of focusing and frequency properties of diffractive optics both on a flat surface and on a prescribed second-order surface of revolution demonstrated that fixing the shape of the surface and its geometric parameters (front and rear segments, the nominal wavelength, the element diameter) results in prescribed frequency properties that are much stronger than or weaker than in the case of zone plates.

In a number of cases the reverse situation is preferable: to determine the discrete phase function and surface profile from the required frequency properties of the element of diffractive optics within the prescribed spectral range. This is essentially a solution of an ill-posed inverse problem of synthesizing diffractive optics having prescribed frequency properties.

Let us consider an example of such an element. Assume that we need to focus radiation of wavelength λ_1 at a point B_1, and radiation of wavelength λ_2 at a point B_2. The principle of determining zone boundaries and the surface profile of a diffractive optical element is readily understandable by analysing equiphase contours (figure 3.21). Let radiation sources be located at the same point. Then it is not difficult to find zone boundaries and the element profile from the points of intersection of equiphase contours built for two elements, corresponding to (λ_1, B_1) and (λ_2, B_2). In the case of a plane incident wavefront the zone boundaries (x_n, y_n) are found from the set

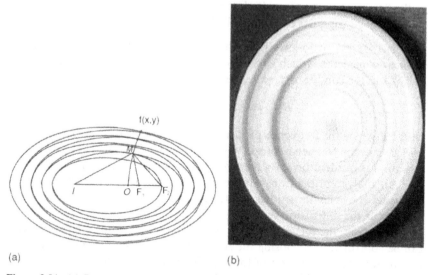

Figure 3.21. (a) Reconstruction of the surface of diffractive element based on equiphase contours [39]. (b) Photograph of 'achromatic' diffractive element for the 54–76 GHz range.

of equations

$$x_n = \frac{K_1 - K_2}{n(\lambda_1 - \lambda_2)}, \qquad y_n = (-n\lambda_1 x_n + K_1)^{1/2}$$

where $K_i = B_i n \lambda_i + (n\lambda_i/2^{m-1})^2$; $i = 1, 2$.

Numerical and experimental studies with this element showed that its focal distance does vary from B_1 to B_2 as wavelength varies from λ_1 to λ_2 [39].

3.9 Correction of aberrations of a given order by choosing a surface profile for the diffractive element

The introduction of an additional degree of freedom of a diffractive element—the shape of its surface—makes it possible to compensate for aberrations of a given order characterizing this element. For instance, it was shown in [40, 41] that comatic aberrations are eliminated in the paraxial approximation when diffractive optical elements are fabricated on a spherical surface.

As an example, we will consider a diffractive element working in the 'reflection' mode. Let a pointlike radiation source be located at a point A at a distance f (figure 3.22). The phase profile of the diffractive optical element is fabricated on the inner surface having the generatrix $\varphi(x, y)$ and the apex at the origin of the coordinates.

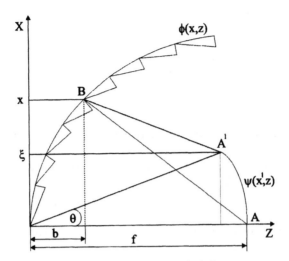

Figure 3.22. Aberration correction for diffractive optical elements.

We move the radiation source in the focal plane to a point A', that is, $\psi(x', z)$ is a straight line. We need to find the phase shift at a point B at the boundary of the nth zone of the diffractive element. Obviously,

$$\Delta\Phi = (2\pi/\lambda)\{|BA'| - |BA|\}.$$

Writing the expressions for eikonals in explicit form we find

$$\Delta\Phi = (2\pi/\lambda)\{[(F - b)^2 + (x - F\tan\theta)^2]^{1/2} - [x^2 + (F - b)^2]^{1/2}\}. \quad (3.22)$$

Assume now that the diffractive optical element has a long focal distance, that is, $x \ll (F - b)$. Expanding then expression (3.22) into a series in a small parameter, we obtain

$$\Phi = (2\pi/\lambda)\{\varepsilon(1 - 1/2(\Delta x/\varepsilon)^2 - 1/8(\Delta x/\varepsilon)^4 + 1/16(\Delta x/\varepsilon)^6 + \cdots)$$
$$- \varepsilon(1 + 1/2(x/\varepsilon)^2 - 1/8(x/\varepsilon)^4 + 1/16(x/\varepsilon)^6 + \cdots)\}$$

where $\varepsilon = (F - b)$.

As $x \ll \varepsilon$, we can assume that condition $\tan\theta \approx \theta$ holds. Then

$$\Delta\Phi \approx (2\pi/\lambda)\{xF\theta/\varepsilon(-1 + x^2/2\varepsilon^2) + F^2\theta^2/2\varepsilon(1 - 3x^2/2\varepsilon^2)$$
$$+ xF^3\theta^3/8\varepsilon^3(4 - 9x^2/\varepsilon^2) + \cdots\}.$$

From this we can find, for instance, the condition for $\Delta\Phi$ to reach its minimum value. Taking into account terms up to second order of smallness, we obtain

$$\Delta\Phi_{min} = F\theta + (F - b)[F^2\theta^2/(F - b)^2 - 2/3]^{1/2}.$$

Assume now that the surface of the diffractive element is slightly concave, that is, $(F - b) \approx F$. Then

$$\Delta\Phi \approx (2\pi)\lambda\{x\theta(x^2/2\varepsilon^2 - 1) + \theta^2/2(1 - 3/2(x/\varepsilon)^2)$$
$$+ x\theta^3/8(4 - 9x^2/\varepsilon^2) + \cdots\}.$$

This expression implies that in the paraxial approximation:

1. For the flat profile of the diffractive optical element surface $(b \to 0)$, comatic aberrations are important [41].
2. For nearly spherical surface of diffractive optical element, comatic aberrations are eliminated:

$$(F - b)^2 = F^2 - x^2, \qquad x^2/[2(F - b)^2] = F^2/[2(F - b)^2] - \tfrac{1}{2} \approx 0.$$

3. To eliminate second-order aberrations, the surface profile of a diffractive element must be parabolic:

$$3x^2 = 2(F - b).$$

Comatic aberrations are not completely eliminated here but they are substantially lower than for a flat diffractive optical element.
4. To eliminate third-order aberrations, the optimal profile of diffractive optical element is nearly conical:

$$x = 2(F - b)/3.$$

In other words, the above elementary analysis of aberrations in non-flat diffractive optical elements demonstrates that in a number of cases it is expedient to choose an optimum surface profile of the element corresponding to prescribed aberration characteristics.

The problem can still be expanded, namely: in some cases of practical importance it is not necessary to consider an image in the plane or to shift the radiation source. Let the image be considered on a surface with a generatrix of the type $\psi(x', z)$ with the radiation source moving along this surface from point A to point A'. By analogy to the previous case, we can write

$$\Delta\Phi \approx (2\pi/\lambda)\{[(F - b - \delta)^2 + (x - (F - \delta)\tan\theta)^2]^{1/2} - \{x^2 - (F - b)^2]^{1/2}\}$$

or, assuming $(F - b - \delta) \gg x$, $\delta \ll 1$ (δ is the projection of the distance between the points A and A' on to axis z)

$$\Delta\Phi \approx (2\pi/\lambda)\{xF\theta/(F - b)[-1 + x^2/[2(F - b - \delta)^2)]$$
$$+ F^2\theta^2/[2(F - b)]\{1 - 3x^2/[2(F - b - \delta)^2]\} + \cdots\}.$$

Comparing this expression for $\Delta\Phi$ with the preceding one, we find that they are equivalent, with the replacement $(F - \delta) \to F$. This leads to the conclusion that by introducing a non-flat image surface (or the surface on which the radiation source moves) it is possible either to eliminate aberrations generated by a flat diffractive optical element or to control the flexure of the non-flat surface of diffractive elements.

3.10   Polarization diffractive elements

The subject of the earlier sections was the properties and ultimate character-istics of diffractive elements based on the scalar theory of diffraction. This means that the vector nature of electromagnetic field was not ignored. However, it is possible to create new types of diffractive optical elements that essentially employ another degree of freedom—the polarization of radiation. An example of this type of element is the amplitude-phase-type zone plate suggested in [42, 43]. As a starting element, the amplitude-type zone plate was chosen (as the simplest and technologically the most flexible element). However, the opaque-for-radiation Fresnel zones were fabricated as phase-rotating diffractive structures, for example, as quasi-one-dimen-sional diffractive grids or sets of holes of prescribed diameter.

Typically, one-dimensional grids of period d much smaller than the radiation wavelength ($d \ll \lambda$, $\chi = d/\lambda \ll 1$) are used for the transformation of polarization in the mm- and sub-mm-wavelength ranges. If the vector E of the wave coincides with the orientation of conductors in the grid, the main part of the incident wave in this configuration is reflected by the grid. If E is perpendicular to the orientation of grid conductors, the grid transmits the main part of the wave. Therefore, as a wave of arbitrary polarization passes through such a grid, it becomes plane polarized and the direction of the vector E in the transmitted wave is perpendicular to the orientation of grid conductors.

The results of studying the focusing and frequency properties of a zone plate of this type are described in [42]; the phase-rotating diffractive structure of this zone plate was a one-dimensional wire grid with relative period about 1 and filling coefficient 0.3–0.5. With such grid parameters, E- and H-type waves have a phase shift of $\pi/2$ and the radiation transmitted through this structure is circularly polarized [44]. This effect appears because the transmis-sion coefficients in the resonance area ($\chi \approx 1$) for the parallel and vertical polarizations of the electromagnetic wave on this grid are comparable and nearly equal unity at certain values of the wire-to-space ratio.

Figure 3.23 plots the ellipticity coefficient P for the wave transmitted through this grid as a function of parameter χ (the ratio of the squared ampli-tudes along the smaller and larger half-axes of the polarization ellipse of the transmitted wave). The maximum value of P corresponds to $\chi \approx 0.885$. The band width at the level $P = 90\%$ is nearly 10 GHz [44].

An experimental study of the shape of the diffractive spot formed by the 'polarization' zone plate in the 4 mm wavelength range showed that for two

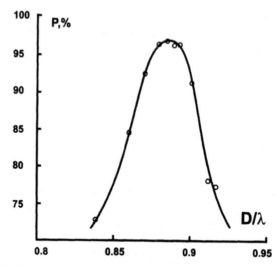

Figure 3.23. Ellipticity of transmitted wave [44].

mutually orthogonal radiation polarizations the intensity distribution at the focal point was the same, to the accuracy of the experiment, as the similar characteristic of the conventional Fresnel zone plate. The zone plate diameter was $D/\lambda \approx 100$, with $D/B = D/A = 1$.

Figure 3.24 plots the experimental dependence of the signal amplitude at the focal point of the zone plate on the angle of orientation of grid conductors (in degrees of arc) for three values of frequency of the radiation used.

Note that frequency f corresponding to the maximum value of P can be changed by varying the grid parameters b and d (b stands for the conductor width): reduce them to increase f or increase them to reduce f, leaving the wire-to-space ratio s of the grid unchanged [44]:

$$s = b/d \approx 0.4, \qquad \chi = d/\lambda \approx 1, \qquad f \approx 12\chi/b$$

where f is in GHz.

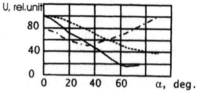

Figure 3.24. Signal amplitude at the focal point of a polarization zone plate as a function of the angle of orientation of grid conductors (experimental data) as a function of relative radiation frequency increment: ——, $\Delta f/f_0 = 0\%$; ---, $\Delta f/f_0 = -12.8\%$; ---·--, $\Delta f/f_0 = +19.3\%$.

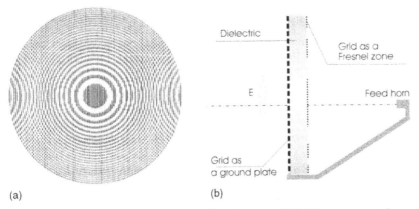

Figure 3.25. Polarization zone plate: (a) phase-inversion [42], (b) transparent for one polarization.

Figure 3.25(a) shows what the amplitude mask of such a zone plate looks like. The zone plate designed in this manner is selective to the polarization of the incident radiation and transforms the polarization of the wave field. Our experiments demonstrated that both the focusing and frequency properties of such a zone plate are invariant and coincide with similar characteristics of the amplitude zone plate. By using diffraction gratings that reflect radiation of one selected polarization, it is possible to design antennas that are 'transparent' for this selected polarization (figure 3.25(b)).

Elements of diffractive optics designed following this principle can be successfully applied in such promising fields as signal reception from satellite television and antennas for automobile locators (where it is necessary to work with only one polarization in order to filter out false signals reflected from the road surface). It is also possible to use them in other fields where an antenna must automatically select the required polarization of radiation. This method of building a planar antenna actually leads to a phase-inversion antenna for the chosen type of polarization.

Another type of planar antennas stemming from zone plates can be created using perforated structures; for instance, holes of specific size and shape can be created in radiation-opaque zones, and the choice of hole parameters can control the phase shift in the corresponding zone and also the transmission coefficient, thus creating the required amplitude-phase distribution within the aperture of the antenna.

3.11 Diffractive elements without axial symmetry

Diffractive elements without axial symmetry or without any symmetry are variations on the same theme. Let us briefly consider some of them.

In order to focus radiation on to a region lying off the line drawn through the centre of the diffractive element and the radiation source, it is possible to use either elements with off-axis position of the focal point or off-axis zone plates [45]. In these cases the diffraction on such an element can be treated using the Fresnel–Kirchhoff integral,

$$U(\theta) = \iint \frac{\exp[ik(r+S)]}{rS} \left[\frac{z_2}{S} - \frac{z_1}{r} \right] R \, dR \, d\varphi$$

where $A = -z_1$, $B = z_2$, $r^2 = (x_1 - R\cos\varphi)^2 + (y_1 - R\sin\varphi)^2 + (z_1 - z)^2$, $S = (x_2 - R\cos\varphi)^2 + (y_2 - R\sin\varphi)^2 + (z_2 - z)^2$; (x_1, y_1, z_1) are the coordinates of the radiation source and (x_2, y_2, z_2) are the coordinates in the focal region.

Let us consider how to construct zone boundaries of a diffractive element that provides off-axis focusing (following [46]). If a kinoform mirror is used, the source of the wave incident on the mirror and the point of focusing of the reflected radiation should not lie on the same axis perpendicular to the mirror plane. In the case of focusing of a parallel beam to a point located on the axis of the focusing element, Fresnel zones are circles with radii that depend on the zone number and are given by the familiar formula for the Fresnel zone plate.

Let us assume that a rectangular reference frame (x, y, z) (figure 3.26) is used to design a flat diffractive element for working in the 'reflection' mode.

A spherical wave that diverges from a point P_α after reflection and diffraction by the element must focus to a point P_β. Without restricting the generality of the argument, we assume that these two points lie in plane $(y, 0, z)$ that we refer to as the 'vertical' plane. The coordinates of this plane will be denoted by $(0, y_\alpha, z_\alpha)$ and $(0, y_\beta, z_\beta)$, respectively, and all linear dimensions will be normalized with respect to wavelength λ.

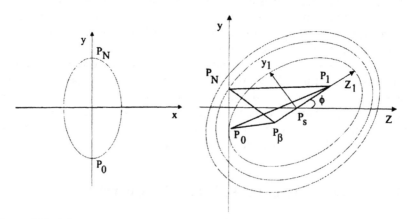

Figure 3.26. Schematic diagram for designing diffractive element to focus radiation on to an off-axis region.

To focus radiation to the point P_β it is necessary for the optical paths of all rays diffracted at zone boundaries of the diffractive element to differ by a constant equal to an integral multiple of the wavelength λ:

$$R_{\alpha,N} + R_{\beta,N} = \text{const} + N, \qquad N = 0, 1, 2, \ldots \qquad (3.23)$$

where $R_{\alpha,N} = \overline{P_\alpha P_N}$; $P_{\beta,N} = \overline{P_\beta P_N}$.

It is not difficult to notice that equality (3.23) defines a family of ellipsoids of revolution whose foci coincide with points P_α and P_β. Obviously, the zone boundaries of the diffractive element are the curves of intersection of these ellipsoids with the plane of the element $(z = 0)$; hence, their shape is elliptical.

Equality (3.23) represents a family of ellipsoids of revolution that can be transformed to a canonical form in a rectangular system of coordinates $(x_1 = x_2, y_1, z_1)$, referring to points P_α and P_β, with the centre at a point $P_s(0, y_s, z_s)$, where $y_s = (y_\alpha + y_\beta)/2$, $z_s = (z_\alpha + z_\beta)/2$:

$$z_1^2/C_{1,N}^2 = 1$$

where the ellipsoid parameter $C_{1,N}$ (defined below) is a function of the number N of the ellipsoid.

There exists a certain 'minimal' number N_0 (not necessarily an integer) for which the ellipsoid of revolution is tangent to the plane of the diffractive element at a single point P_0. This means that the Nth zone of the element contracts to a point. The coordinates of this point are defined as follows:

$$x_0 = 0, \qquad y_0 = \frac{y_\alpha z_\beta + y_\beta z_\alpha}{z_\alpha + z_\beta}, \qquad z_0 = 0.$$

The number of the 'zeroth' ellipse N_0 depends on the parameters that characterize the ellipsoid of revolution:

$$N_0 = 2(z_s^2 + D^2 \cos^2 \varphi)^{1/2} \qquad (3.24)$$

where $2D$ is the distance between the points P_α and P_β, and φ is the angle between the direction $P_\alpha P_\beta$ and the axis Oz.

The parameters of the 'zeroth' ellipsoid of revolution $A_{1,0}$ are $C_{1,0}$ are found from the formulas

$$A_{1,0}^2 = (N_0/2)^2, \qquad C_{1,0}^2 = A_{1,0}^2 - D^2.$$

The parameters of all consecutive ellipsoids of revolution with numbers $N = N_0 + n$ $(n = 1, 2, 3, \ldots)$ are found from

$$A_{1,N}^2 = (N/2)^2, \qquad C_{1,N}^2 = A_{1,N}^2 - D^2.$$

Now we need to find formulas that describe zone boundaries on the diffractive element. They create a family of ellipses displaced relative to one another, with their centres at the points $P_{0,N}(0, y_0, N)$; their equations in the kinoform plane $(x, 0, y)$ can be written in the canonical form

$$x^2/A_N^2 + (y - y_{0,N})^2/B_N^2 = 1 \qquad (3.25)$$

$$y_{0,N} = y_s + \frac{z_S D^2 \sin\varphi \cos\varphi}{A_{1,N}^2 - D^2 \cos^2\varphi} \qquad (3.26)$$

while A_N and B_N depend on $A_{1,N}, D, z, \varphi_s$:

$$A_N^2 = \frac{A_{1,N}^2(A_{1,N}^2 - D^2) - t_S^2 \dfrac{A_{1,N}^2 - D^2 \sin\varphi - D^4 \sin^2\varphi \cos^2\varphi}{A_{1,N}^2 - D^2 \cos^2\varphi}}{A_{1,N}^2 - D^2 \cos^2\varphi} \qquad (3.27)$$

$$B_N^2 = A_N^2(A_{1,N}^2 - D^2 \cos^2\varphi)/A_{1,N}^2.$$

When deriving the formulas describing zones boundaries of the diffractive element, no simplifications or approximations were used. These formulas are exact for any position of a pointlike object and for any point of focusing of the reflected wave.

A particular case of an off-axis **MWDO** element was analysed in [47]. Figure 3.27(a,b) shows the structure of a diffractive element designed to work at an angle θ to the incident radiation with relation to the z axis.

The phase difference in the plane X–Y between a plane wave illuminating the diffractive element at an angle θ and a spherical convergent wave is given by the expression

$$\Phi(x, y) = n'k[(x^2 + y^2 - 2yf \sin\theta + f^2)^{1/2} - f + y \sin\theta].$$

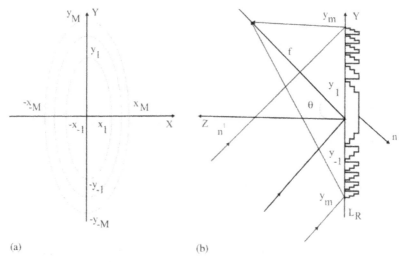

(a)   (b)

Figure 3.27. (a,b) The structure of 'reflection'-type diffractive element for oblique incidence of radiation [47].

The phase function Φ_f of a diffractive element can be obtained from Φ by taking it modulo 2π:

$$\Phi_f(x,y) = \Phi(x,y) - 2m\pi$$

where m is an integer satisfying the condition $0 \leq \Phi_f \leq 2\pi$. The phase function Φ_f is symmetrical with respect to the X axis and asymmetrical with respect to the Y axis. This means that the period of the zones of the diffractive element is smaller in the positive direction of Y than in the negative one.

The zone boundaries of the diffractive element are found from the condition $\Phi_f = 0$:

$$\frac{x^2}{[m\lambda/(n'\cos\theta)]^2 + 2mf'\lambda/n'} + \frac{(y - y_{0N})^2 \cos^2\theta}{[m\lambda/(n'\cos\theta)]^2 + 2mf'\lambda/n'} = 1$$

$$y_1 = -m\lambda \tan\theta \cos\theta/n'.$$

This equation describes an ellipse with a centre at a point $(0, y_1)$. The length of the larger and smaller axes of the ellipsoid are found from the expressions

$$S_{ym} = (2/\cos\theta)[(m\lambda/(n'\cos\theta))^2 + 2mf\lambda/n']^{1/2}$$

$$S_{xm} = 2[(m\lambda/(n'\cos\theta))^2 + 2mf\lambda/n']^{1/2}.$$

The ratio of the larger to smaller axes of the diffractive element depends only on the angle of incidence of radiation on to the element:

$$S_{ym}/S_{xm} = 1/\cos\theta.$$

The depth (or height) of the phase profile is

$$L(x,y) = (\lambda/2n\cos\theta)[1 - \Phi_f(x,y)/2\pi].$$

The maximum diffraction efficiency for this element is achieved near the Bragg angle $\theta_B = \arcsin(\lambda/2n\Lambda)$, where Λ is the period of the diffraction grating.

In polar coordinates, the equation determining the topology of the diffraction zones of the focusing element arranged at an angle α is given by

$$r = \frac{[2kf'\lambda(1 - \sin^2\alpha\cos^2\varphi) + k^2\lambda^2]^{1/2} - k\lambda\sin\alpha\cos\varphi}{1 - \sin\alpha\cos\varphi} \tag{3.28}$$

where r and φ are the radius-vector and polar angle in the plane of the element, respectively [48]. If $\varphi = 0, \pi$ this formula dictates the dimension of the zones in the vertical plane, and if $\varphi = \pi/2, 3\pi/2$ then the dimension in the horizontal plane.

We need to mention that if the radiation wavelength deviates from the nominal value, the spatial position of the focal region in the 'distance-angle' coordinates changes too.

An analysis of aberrations of such zone plates is given in [49].

3.12   Square zone plate

Janicijevic [50] suggested to prepare Fresnel zone plates as sets of consecutively embedded squares. The transmission function of a positive 'square' zone plate (with transparent first zone) is described in the paraxial approximation by the expression $t(x, y) = 1$, if

$$(x, y) \in \mathrm{rect}\left(\frac{x}{a(2j + 1)^{1/2}}\right)\mathrm{rect}\left(\frac{x}{a(2j + 1)^{1/2}}\right)$$

$$- \mathrm{rect}\left(\frac{x}{a(2j)^{1/2}}\right)\mathrm{rect}\left(\frac{x}{a(2j)^{1/2}}\right), \qquad j = 0, 1, 2, \ldots, N.$$

For a negative zone plate (with opaque first zone) we have $t(x, y) = 1$, if

$$(x, y) \in \mathrm{rect}\left(\frac{x}{a(2j)^{1/2}}\right)\mathrm{rect}\left(\frac{x}{a(2j)^{1/2}}\right)$$

$$- \mathrm{rect}\left(\frac{x}{a(2j - 1)^{1/2}}\right)\mathrm{rect}\left(\frac{x}{a(2j - 1)^{1/2}}\right), \qquad j = 0, 1, 2, \ldots, N$$

where a is the length of the side of the square.

In the Fresnel diffraction approximation, when a 'square' zone plate is illuminated with a wave $A \exp(i\delta)$, the field intensity in the focal plane is described by the expression [50]

$$
\begin{aligned}
I(\xi, \eta, z) = \frac{A^2}{4} \Bigg\{ &\left[\sum_m (-1)^m \left\{\left[C\sqrt{\tfrac{2}{\lambda z}}(a\sqrt{m} - \xi)\right] + C\left[\sqrt{\tfrac{2}{\lambda z}}(a\sqrt{m} + \xi)\right]\right] \right.\\
&\times \left[C\left[\sqrt{\tfrac{2}{\lambda z}}(a\sqrt{m} - \eta)\right] + C\left[\sqrt{\tfrac{2}{\lambda z}}(a\sqrt{m} + \eta)\right]\right] \\
&- \left[S\left[\sqrt{\tfrac{2}{\lambda z}}(a\sqrt{m} - \xi)\right] + S\left[\sqrt{\tfrac{2}{\lambda z}}(a\sqrt{m} + \xi)\right]\right] \\
&+ \left.\left[S\left[\sqrt{\tfrac{2}{\lambda z}}(a\sqrt{m} - \eta)\right] + S\left[\sqrt{\tfrac{2}{\lambda z}}(a\sqrt{m} + \xi)\right]\right]\right]^2 \\
&+ \left[\sum_m (-1)^m \left\{\left[C\sqrt{\tfrac{2}{\lambda z}}(a\sqrt{m} - \xi)\right] + C\left[\sqrt{\tfrac{2}{\lambda z}}(a\sqrt{m} + \xi)\right]\right]\right.\\
&\times \left[S\left[\sqrt{\tfrac{2}{\lambda z}}(a\sqrt{m} + \eta) + S\left[\sqrt{\tfrac{2}{\lambda z}}(a\sqrt{m} + \eta)\right]\right]\right] \\
&+ \left[S\left[\sqrt{\tfrac{2}{\lambda z}}(a\sqrt{m} - \xi)\right] + S\left[\sqrt{\tfrac{2}{\lambda z}}(a\sqrt{m} + \xi)\right]\right] \\
&+ \left.\left[C\left[\sqrt{\tfrac{2}{\lambda z}}(a\sqrt{m} - \eta)\right] + C\left[\sqrt{\tfrac{2}{\lambda z}}(a\sqrt{m} + \eta)\right]\right]\right]^2 \Bigg\}
\end{aligned}
$$

where

$$C(u) = \int_0^a \cos[(\pi/2)^2 u^2] \, du \qquad S(u) = \int_0^a \sin[(\pi/2)^2 u^2] \, du;$$

(x, y, z) are the Cartesian coordinates in the plane of the zone plate, and (ε, η, z) are the Cartesian coordinates in the observation plane. For a positive plate, the summation is run from 0 to $2N - 1$, and for a negative plate from 1 to $2N$.

Another approach to constructing diffracting focusing elements with rectangular structure of Fresnel zones was suggested in [51]. The proposal is essentially as follows. If we take two one-dimensional (cylindrical) zone plates and juxtapose them, and if the angle between their axes equals $\pi/2$, we obtain an analogue of a conventional axisymmetric zone plate. Obviously, the scattering function of the resulting diffractive element will be dictated by the convolution of the scattering functions of the original elements. Furthermore, this implies that the method can generate diffractive elements capable of focusing on to sufficiently arbitrary areas. Thus, depending on the angle between the axes of two one-dimensional zone plates, it is possible to obtain diffractive elements with the focusing region that changes from a line to an ellipse to a circle. A diffractive element which is an analogue of a zone plate is shown in figure 3.28.

An analysis of the properties of such an element showed [51] that

- this design retains the frequency properties of the element, and these in turn coincide with similar properties of the zone plate;

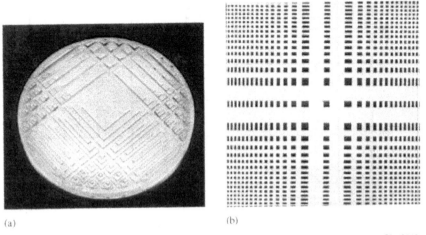

(a)   (b)

Figure 3.28. (a) Phase-inversion four-level zone plate with rectangular zone profile [51]. (b) Amplitude zone plate with rectangular zone profile [51] (the size of the first zone was chosen in a special way—see figure 3.1(a,b)).

- mathematically, its shape is dictated by the convolution of the corresponding functions of the initial elements.

This is a familiar fact that when a cylindrical zone plate is illuminated with a plane wavefront, a focal region is a line situated transversely to the optical axis, parallel to the plane of the zone plate. Therefore, in contrast to the Fresnel zone plate, a 'square' zone plate composed of two cylindrical zone plates produces a focusing region of an asymmetrical shape.

The focal area has the shape of a 'cross' with a maximum in the intersection area. Along the diagonal of the cross, the field intensity distribution is similar to that produced by the Fresnel zone plate. The intensity of the side lobe is independent of the type of the zone plate and does not exceed, for the Rayleigh–Wood-type zone plate, 10% of the maximum intensity of the field at the focal point (figure 3.29).

The following zone plate parameters were used: $A = B = D = 45.65\lambda$, $\lambda = 4.6\,\text{mm}$.

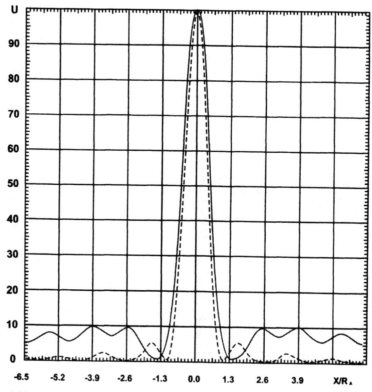

Figure 3.29. Experimentally obtained shape of the diffraction spot for four-level zone plate with rectangular zone profiles [51]. – –: □, orientation of the zone plate; ——: ◇, orientation of the zone plate.

Thus, if the uncertainty function of each one-dimensional element is described by the Airy function (i.e. each one of them is a diffractionally constrained system), it can be shown that the new synthesized element formed of the two by 'superposing' them at a certain angle α (with some offset) has the uncertainty function of the type

$$U(x) = \left(\frac{2J_1(x)}{x} \frac{2J_1(y \sin \alpha)}{y \sin \alpha} \right)^2.$$

A two-component radio lens for the millimetre wavelength range, similar to the one described earlier in chapter 2, was fabricated on the basis of two four-level phase-inversion zone plates with rectangular profile; its properties were then investigated. The main results of the study carried out by the authors in 1987–1988 are shown in figures 3.30 and 3.31.

Figure 3.30 plots the shape of the diffraction spot for a two-component diffractive radio lens based on zone plates with square zone profiles, for three main orientations of the components relative to each other, as shown in the figures.

The results shown here demonstrate, among other things, that if phase-inversion zones of the two components of the objective lens are oriented normally to each other, the field intensity distribution along the optical axis has a single maximum and the signal level at the focal point is approximately twice as high as in the case of 'parallel' orientations of phase-inversion zones. The presence of the second maximum in the field intensity distribution along the optical axis is caused by the discrepancy between the boundaries of the rectangular Fresnel zone and the exact value corresponding to the

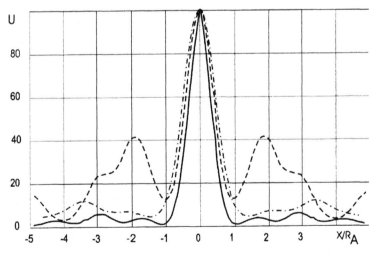

Figure 3.30. Diffraction spot for a two-component objective lens built of phase-inversion zone plates with rectangular zone profiles. ——: orientation ◇◇; – – –: orientation □□; –·–·–, orientation ◇□.

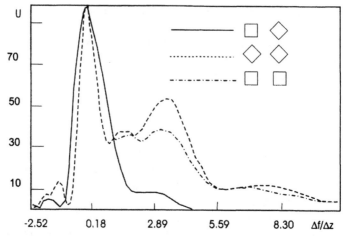

Figure 3.31. Normalized field intensity distribution along the optical axis for a two-component objective lens built of phase-inversion zone plates with rectangular zone profiles. Shown on the right is the mutual orientation of zone profiles for each component of the objective lens.

diagonal length and can be optimized by choosing the zone structure on the surface of the element.

3.13   Field intensity distribution function along the optical axis

In the first approximation for multi-component focusing systems, field intensity distribution along the optical axis can be described in analytical form as follows. Verzel [52] derived the formula describing the axial distribution of field intensity in the focal region of optical systems with high Fresnel number valid for one particular case. In a more general case, the following expression can describe this distribution assuming the thin-lens approximation and the Huygens–Fresnel approximation with unit illumination level on the pupil [53]:

$$I(z) = \left[2\frac{R}{z} \sin\left(\frac{u}{4}\right) \right]^2, \qquad R = \frac{rf}{r+f-d}, \qquad u = 2\pi N \frac{z/R}{1+(1-d/f)z/R}$$

where d is the distance from the aperture diaphragm to the lens that is equivalent to that part of the system that lies downstream of the diaphragm, f is the front focal distance of this lens, r is the curvature radius of the wave incident on the diaphragm, and N is the Fresnel number characterizing the diffraction quality of the quasioptical system,

$$N = \frac{a^2}{\lambda R}$$

where R is the distance from the exit pupil to the paraxial image of the axial source that lies in the plane of the object, and a is the radius of the exit diffraction pupil. Note that both in microwave and in millimetre wavelength ranges, the Fresnel number is less than that in the optical range by one to three orders of magnitude, which is of principal importance for a number of features of the elements under discussion.

In the case of a single lens, the formulas transform into familiar ones [54]:

$$I = I_0\left(1 - \frac{u}{2\pi N}\right)\left[\frac{\sin(u/4)}{(u/4)}\right]^2, \qquad u = 2\pi N\,\frac{(z/R)}{1 + (z/R)}.$$

In these expressions, the factor (R/z) is the field in the absence of diffraction—the 'geometrical' component.

The following expression can describe the axial intensity distribution in the aberration-free case of one-component radio lens, normalized to the intensity at the point of the paraxial image [55]:

$$I \approx 2\left(\frac{f}{\pi N z}\right)^2\left[1 - \cos\left(\frac{\pi N z}{f + z}\right)\right]$$

where N is the Fresnel number. This distribution clearly demonstrates certain specifics of the field in the focal zone of quasioptical systems: the shift of the 'focal point', that is, the shift of the maximum in the field intensity distribution along the optical axis from the paraxial image point towards the focusing element, by the amount

$$z_m \approx -f\left[1 + \frac{\pi^2 N^2}{24} - \sqrt{\left(1 + \frac{\pi^2 N^2}{24}\right)^2 - 1}\right].$$

Furthermore, the expression for field intensity distribution along the optical axis also implies another characteristic: the asymmetry of this distribution—the number of maxima in the area of positive Z is constrained by the quantity $N/2$, and the distribution in the main diffractive order is also asymmetric.

It is easily seen that the coordinates of the minima of the axial distribution are [56]

$$z_n = \frac{2nf}{N - 2n}, \qquad z_{-n} = -\frac{2nf}{N + 2n}, \qquad n = 1, 2, \ldots$$

In their turn, the data for relative values for the minima make it possible to determine the linear dimensions of diffraction orders in the neighbourhood of the main maximum:

• the size of the main zeroth order

$$\Delta Z_0 = \frac{4fN}{N^2 - 4}$$

- the size of the first positive order

$$\Delta Z_1 = \frac{2fN}{(N-2)(N-4)}$$

- the size of the first negative order

$$\Delta Z_{-1} = \frac{2fN}{(N+2)(N+4)}.$$

For instance, estimates of the displacement of the 'focal point' from its geometrical position in the microwave and the millimetre wave ranges show that its value is $(0.1–0.01)f$ and is easily detectable. In the optical range of wavelengths this 'displacement' is smaller by two to three orders of magnitude than in the preceding case and measuring it is extremely difficult. This property is a principal difference between elements for the optical range and those for the microwave range.

It must be noted that relations given above for finding the parameters of field intensity distribution along the optical axis can be used not only to evaluate the focusing properties in microwave optics but also to find from the distribution the geometrical-optics parameters of the optical system as such.

The most acceptable from the practical standpoint is the following technique [57]. The field of the wave formed by a pointlike source and transmitted through the quasioptical system is scanned along the optical axis, the energy distribution in the zone of the main diffraction order is recorded, and the asymmetry of this distribution is estimated; this asymmetry can be used as a measure of difference between optical and quasioptical systems:

$$\delta = \frac{\Delta Z_1 + \Delta Z_{-1}}{\Delta Z_1 - \Delta Z_{-1}}.$$

Geometric-optics parameters and the Fresnel number of the quasioptic system are calculated from the measured value of asymmetry; in the case of a spherical lens these values are

$$N = \frac{2(\pi^3 - 3)}{\delta\pi^2}\left[1 + \frac{2}{3}\cosh\left(\frac{\varphi}{3}\right)\right], \qquad F = \frac{L_0(\pi^2 - 4)}{4N},$$

$$D = \frac{\lambda L_0(N^2 - 4)}{4N}$$

where

$$\varphi = a\cosh\left[\frac{6}{\pi^2 - 3}\left(\frac{2(\pi^2 - 3)^3}{2r\pi^4 8^2} + 3\right)\right].$$

3.14   Field intensity distribution in the focal area of zone plates with low number of zones

Let us consider the behaviour of the field in the neighbourhood of the focal point of phase-type zone plates as a function of the number of working zones and quantization levels P [65].

The field intensity $I(0, z)$ at an arbitrary point on the optical axis can be calculated analytically:

$$I(0, z) = \left| \sum_{l=1}^{L} U_l(0, z) \right|^2 = \left| \sum_{l=1}^{L} U_l^1 \right|^2 + \left| \sum_{l=1}^{L} U_l^2 \right|^2.$$

The corresponding formulas for U_l^1 and U_l^2 have the form

$$U_l^1 = \cos(\varphi_l + \sigma \rho_{l+1}^2) - \cos(\varphi_l + \sigma \rho_l)$$
$$U_l^2 = \sin(\varphi_l + \sigma \rho_{l+1}^2) - \sin(\varphi_l + \sigma \rho_l), \qquad |U_0|^2 = 1.$$

Here the axisymmetric zone plate with the complex transmission function $\tau(\rho) = \exp[j\widehat{\varphi}(\rho)]$ has diameter $2a$, $\widehat{\varphi}(\rho)$ is the value of phase in each zone, $\sigma = \pi a^2 / \lambda z$ and z determines the position of the observation plane.

When a refractive lens is illuminated with a uniform plane wave, the intensity $I_R(0, z)$ at the points on the optical axis is calculated using the following formula:

$$I_R(0, z) = \frac{f_R}{(f_R - z)^2} [\{\cos[\sigma(1 - z/f_R)] - 1\}^2 + \{\sin[\sigma(1 - z/f_R)]^2\}].$$

Such calculations [65] yielded the dependence of the focal displacement on the number of steps of the zone plate relief (figure 3.32(a,b)). Dots in these figures indicate the calculated values $|\Delta f|/f_0$ and $\Delta I/I_0$ for the zone plate, and straight lines mark the values for the classic lens; L stands for the number of full Fresnel zones. Therefore, zone plates produce smaller focal displacements. The focal displacement for a lens is less than 1% of the focal distance only on the condition that $L > 6$; note that the same displacement is achieved with a quarter-wave zone plate already for $L = 5$.

The intensity distributions in diffraction focal planes of phase-type zone plates and refractive lenses are not very different. Small differences appear in the position and height of minima and secondary maxima. Figure 3.33 shows the intensity distribution in the scattering circle within the focal diffraction plane of a zone plate with diameter equal to $L = 1$ and the number of phase quantization levels $P = 2$ (solid line) and $P = 4$ (dashed line); the dash-dot curve corresponds to the equivalent classic lens. The scale of plots in the upper part of figure 3.33 was magnified by a factor of 10 ($P = 8$, dotted line).

As the number of steps on phase-inversion profile increases, the diameter of the scattering surface of the zone plate rapidly approaches the scattering

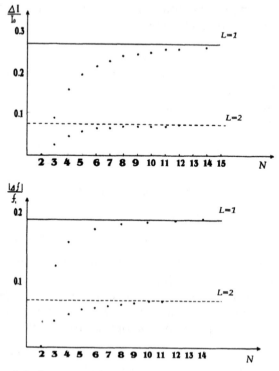

Figure 3.32. Focal displacement of the zone plate as a function of a number of steps of the phase-inversion profile [65].

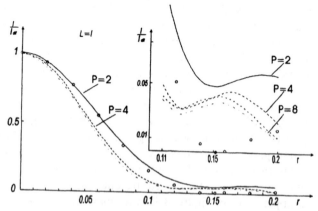

Figure 3.33. Field intensity distribution within the scattering circle in the focal plane of the zone plate with only one full Fresnel zone, as a function of the number of phase quantization levels [65].

diameter of the refractive lens. Thus, for $L = 1$ the difference is 20.9% for $P = 2$, reduces to 2.4% for $P = 4$, and practically vanishes for $P = 8$. At the same time the difference between relative intensities of the secondary minimum and secondary maximum is invariant. For instance, it is 67.7% and 58.6% for $P = 2$ (for min and max, respectively), 12.4% and 20.8% for $P = 4$, and 14.2% and 12.3% for $P = 8$. The increase in the intensity of side lobes in the diffraction circle of the zone plate is caused by scattering of energy by the stepped profile.

3.15 Zone plates with dynamically variable focal area

The approach, presented above, to designing diffractive elements on the basis of several one-dimensional elements can also be used to generate dynamically variable forms of focusing zones, such as areas of focusing radiation for scanning specific areas along specific prescribed arbitrary curves, or similar tasks, if it is possible to change the angle between the two initial elements dynamically, for instance, by rotating one of them. Figure 3.34 shows the shape of focusing area for three values of the angle α: 0, $\pi/4$, $\pi/2$.

It must be taken into account, however, that the focusing properties of, for example, axisymmetric and cylindrical zone plates are not quite the same. For example, as the number of phase quantization levels diminishes, the intensity of the main maximum diminishes for a cylindrical zone plate, and

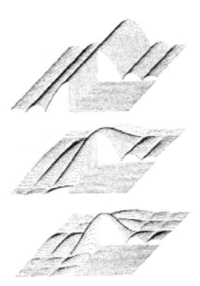

Figure 3.34. Shape of the focal area for a MWDO element formed by juxtaposition of two one-dimensional zone plates at an angle α equal to 0, $\pi/4$ and $\pi/2$.

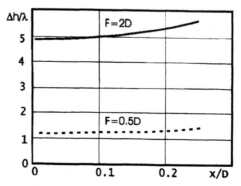

Figure 3.35. Diffraction width of the focusing region of a cylindrical zone plate as a function of distance from the optical axis, for two values of aperture ratio.

the intensities of side lobes of scattering increase. Furthermore, the field intensity at the positions of the first and second minima ceases to tend to zero, that is, there are no zeros in the corresponding scattering function. For instance, if $m = 2$, the intensity at the first minimum is higher than the maximum of intensity at the first side lobe for $m = 12$.

Another decisive difference between a cylindrical zone plate and an axisymmetric one is that the former can be said to smoothen the intensity distribution in the focal plane at low values of m and to reduce the size of the main scattering lobe. It is possible to evaluate the dimensions of the focusing area of a cylindrical zone plate (from the first minima) using the Kirchhoff integral which describes the diffraction of a cylindrical wave on a circular aperture: $\Delta h \approx 1.2\lambda F/(D^2/4 - x^2)^{1/2}$, where D is the aperture diameter (figure 3.35). Experiments confirmed this relation. As was shown above, the Airy formula holds for axisymmetric zone plates: $R_A \approx 1.22\lambda F/D$.

In contrast, as the value of m of an axisymmetric zone plate increases, the width of the main lobe and the absolute values of maxima in the field intensity distribution remain practically unchanged. In addition, as m diminishes, the intensities of side lobes grow in proportion to the diffraction efficiency.

3.16 Diffractive elements in off-design modes

Choosing the principle of design of diffractive elements while taking into account specific degrees of freedom strongly influences both the quality of focusing and the properties of these elements. These factors were briefly discussed above. Regardless of the principles of designing such elements, it was assumed that they were meant for working under the conditions for which they were designed, that is, prescribed shape of the incident wavefront, prescribed wavelength of incident radiation, the angle of incidence of this radiation on to the element (if it works in the 'reflection' mode), and so forth. Typically, it is necessary to synthesize two different diffractive elements in order to generate two close but different focusing regions, for instance, a ring and an ellipse [58].

In some cases a modification of the focusing region of a diffractive element is better produced by distorting the wavefront of the incident radiation. As an example, we consider two very simple problems: focusing of radiation on to a ring and on to an ellipse. Assume that a diffractive element was synthesized originally by one of the familiar methods and produces focusing of radiation on to a ring. Then, if we change, say, the angle of incidence of this radiation on to the element, the shape of the focusing region will be modified (distorted) as a result of the violation of tautochronism. Obviously, the new shape of the focusing region in this case will be an ellipse.

The corresponding numerical studies were conducted in the approximation of scalar waves. The field intensity distribution in the focal area was calculated using the Fresnel–Kirchhoff integral (without expanding the integrand phase function into a series). The parameters of the high-aperture diffractive element that provides focusing of radiation from a pointlike source on to a ring of a prescribed radius R_{ring}, were: the distance A from the pointlike radiation source to the diffractive element plane and the distance B from it to the focal plane were $2D$, the diameter $D = 200\lambda$, where λ is the radiation wavelength, and the focusing ring radius was $R_{ring} = 0.1D$.

Figure 3.36 is the isometric projection of the field intensity distribution in the focal area of the diffractive element. These results show that in the designed mode ($\Delta x = 0$, figure 3.36(a)) the diffractive element provides focusing of the radiation on to the focusing zone of prescribed shape,

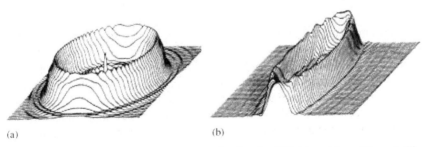

(a)   (b)

Figure 3.36. Field intensity distribution in the focal region [59]: (a) position of the pointlike radiation source according to the original design, (b) displacement of the radiation source transversely to the optical axis by $\Delta x = 20R_A$.

annular in this particular case. If the shape of the incident wavefront is distorted (the pointlike radiation source is shifted transversely to the symmetry axis of the diffractive element), the shape of the focusing region transforms into an ellipse (figure 3.36(b)). The admissible displacement of a pointlike source of radiation from the optical axis is determined by the information properties of the appropriate diffractive element [59]. In this case the ratio of ellipse's half-axes with the indicated value of Δx is 3:1. The quality of radiation focusing must be considered as good.

Therefore in certain cases it is expedient to use focusing diffractive elements in essentially off-design operation modes, that is, when the parameters of the incident wavefront are deliberately made to deviate from the designated mode of operation. This method makes possible the use of the same diffractive element for focusing the radiation on to flat curves of different shapes.

3.17 The method of constructing diffractive elements for focusing radiation on to an arbitrary focal curve

We have discussed the design principles and the basic properties of diffractive elements that focus radiation on to a ring of a prescribed radius. It is easy to see that there exists a certain relationship between such a diffractive element and a zone plate, namely: each narrow sector of such an element is a one-dimensional section of the zone plate, displaced by one radius of the ring from its optical axis. Therefore the wave incident on the diffractive element is focused in the geometrical optics approximation by each such one-dimensional zone plate (a sector) to a point, and since each sector is shifted from the previous one by some small angle, on aggregate this element provides focusing of radiation on to an annular area.

It is quite clear that by applying this principle it is possible to design a diffractive optical element for focusing the incident radiation on to an arbitrary curve located in the focal plane [60]. It is also not difficult to derive an analytical expression that would describe the zone structure on the surface of such an element.

To simplify the interpretation, we consider a flat diffractive element that focuses a plane wavefront on to a prescribed planar curve. A well-known expression for Fresnel zone radii holds for each sector of such an element (that focuses radiation to a point). To ensure focusing of radiation on to a prescribed curve, it is necessary to introduce the shape of this curve, in other words, to replace the radius of the ring in the familiar expression with the equation of the focusing curve, which may be given in polar coordinates:

$$R_n(\varphi) = \{\rho(\varphi') + (C^2 + F^2)^{1/2}\}$$
$$C = n\lambda/2^{m-1} + (F^2 + \rho^2(\varphi'))^{1/2}, \qquad \varphi = \psi(\varphi')$$

where $R_n(\varphi)$ is the equation of Fresnel zone boundaries in the plane of the diffractive element, $\rho(\varphi')$ is the equation of the focusing curve in the focal plane in polar coordinates, and $\varphi = \psi(\varphi')$ is the law of transformation of angular coordinates. A more detailed explanation is needed for the function $\psi(\varphi')$. If the focusing curve is closed, then $\varphi' = \varphi$. If, however, the focusing curve is not closed (a segment, an arc, and so forth), it is necessary to fix the correspondence of the interval of change of the polar angle φ in the plane of the element and of φ' in the focal plane. Physically this means that the 'filling' of the focusing curve with points of focusing from each elementary sector of the diffractive element will be denser (will have a different step) than in the case of a closed curve.

To illustrate this explanation, figure 3.37 shows a three-dimensional view of the 'rectangular' region of focusing. Taking into account that a binary amplitude-type diffractive element was used for modelling, the quality of focusing must be considered good. Using phase-inversion multi-level elements will make it possible to eliminate spurious field intensity maxima at singular points, that is, at the break points of the focusing curve.

In some cases the method of designing diffractive elements may be based on other physical principles. For instance, the so-called 'deformational' and

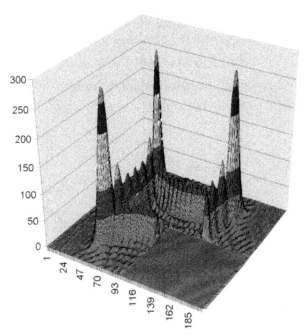

Figure 3.37. An example of a rectangular focusing area of an amplitude-type binary diffractive element.

'conformal' principles of designing such elements were suggested in [61]. The former of these two is based on the similarity theorem for the intensities of the diffracted waves on screens of different shapes, and the latter principle is based on the Riemann and Kristoffel–Schwartz formulas of the theory of conformal mappings.

3.18 Principle of design of diffractive elements for spatial focusing of radiation

The principle of design of diffractive elements that focus monochromatic radiation on to an arbitrary prescribed surface is based on the generalized principle of tautochronism and the functioning of axicon.

An MWDO element is a combination of a lens (zone plate) and a cone that deflects the rays at a certain angle α to the normal to the surface of the element. This principle can be easily extended to MWDO elements fabricated on arbitrary-shape surfaces. It is also necessary to add the phase condition for rays at a certain point M on the surface of focusing of the radiation. The angle of deflection α is found from the condition of focusing the radiation at the point B—the apex of the surface of focusing (figure 3.38).

Let us write the basic relations. To simplify the exposition, we consider the case of axial symmetry: the surface of focusing is described by a second-order curve with the generatrix of the form $\varphi(z, x) = 0$.

For a binary element with the boundary condition at the angle α

$$|AO'| + |O'B| = |AG| + |GB| + \lambda/2$$

the in-phase condition for the radiation has the form

$$|AO| + |OM| = |AG| + |GM| + n\lambda/2, \qquad \varphi(z, x) = 0.$$

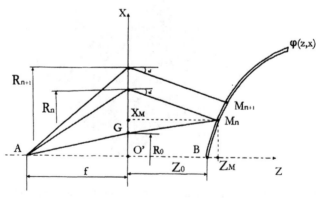

Figure 3.38. Schematic diagram of designing a MWDO element.

The terms in detail are

$$|AO| = (F^2 + r_n^2)^{1/2}, \qquad |AG| = (F^2 + R_0^2)^{1/2}, \qquad |O'B| = (z_0^2 + r_1^2)^{1/2}$$

$$|AO'| = (F^2 + r_1^2)^{1/2}, \qquad |GB| = (R_0^2 + z_0^2)^{1/2}$$

where R_0 is the reference radius [8, 11].

The distances $|OM|$ and $|GM|$ are found from the specific type of the curve $\varphi(z, x)$:

$$|OM| = ((r_n - x_M)^2 + z_M^2)^{1/2} = (r_n - x_M)/\sin \alpha$$

$$|GM| = (z_M^2 + (x_M - R_0)^2)^{1/2}.$$

Therefore,

$$(F^2 + r_n^2)^{1/2} + (r_n - x_M)/\sin \alpha$$

$$= (F^2 + R_0^2)^{1/2} + (z_M^2 + (x_M - R_0)^2)^{1/2} + n\lambda_0/2^{m-1} \qquad (3.29)$$

$$\varphi(z, x) = 0, \qquad m = 2, 3, 4, \ldots, k$$

where m is the number of phase quantization levels of the diffractive element.

The set of equations (3.29) shows that two equations are given:

$$(F^2 + r_n^2)^{1/2} + (r_n - x_M)/\sin \alpha = C + (z_M^2 + (x_M - R_0)^2)^{1/2} \qquad (3.30)$$

$$\varphi(z, x) = 0.$$

Let us find the requirements to the form of the function $\varphi(z, x)$ that provides for the existence of a solution to set (3.30). We rewrite the first equation of set (3.30) as

$$(F^2 + r_n^2)^{1/2} = \Lambda - \mu(r_n - x_M) \qquad (3.31)$$

where

$$\Lambda = (F^2 + R^2)^{1/2} + n\lambda/2^{m-1} + z_M^2 + (x_M - R_0)^2)^{1/2} > 0$$

$$\mu = 1/\sin \alpha, \qquad C = (F^2 + R_0^2)^{1/2} + n\lambda_0/2^{m-1}.$$

We rewrite equation (3.31) by gathering similar terms:

$$(\mu^2 - 1)r_n^2 - 2(\mu\Lambda + \mu^2 x_M)r_n + \mu^2 x_M^2 + \Lambda^2 + 2\Lambda\mu x_M - F^2 = 0. \qquad (3.32)$$

Equation (3.32) is always of second degree because the following condition is met:

$$\mu^2 - 1 = \cotan \alpha > 0.$$

We can find the discriminant of the equation (3.32):

$$D = (\Lambda\mu + \mu^2 x_M)^2 - (\mu^2 - 1)[(\mu x_M + \Lambda)^2 - F^2]$$

$$= (\mu x_M - \Lambda)^2 + F^2(\mu - 1) \geq 0.$$

Therefore, equation (3.32) always has a positive root:

$$r_{1,2} = \{\mu(\mu x_M + \Lambda) \pm [(\mu x_M + \Lambda)^2 + F^2(\mu - 1)^2]^{1/2}\}/(\mu^2 - 1). \quad (3.33)$$

It is obvious from equation (3.31) that

$$\Lambda - \mu r_n + \mu x_M > 0.$$

Let us find what sign must be taken in front of the root (3.33). It follows from (3.31) that

$$\Lambda + \mu x_M > \mu r_n. \quad (3.34)$$

Taking into account (3.34), we obtain from (3.33)

$$(\Lambda + \mu x_M)(\mu^2 - 1) - \mu^2(\mu x_M + \Lambda) - \mu[(\mu x_M + \Lambda)^2 + F^2(\mu^2 - 1)]^{1/2}$$

$$= -(\mu x_M + \Lambda) - \Lambda[(\Lambda x_M + \Lambda)^2 + F^2(\mu^2 - 1)]^{1/2} < 0.$$

Therefore, the sign of the square root in expression (3.33) must be negative. Finally, remembering that $\mu = (\sin \alpha)^{-1} > 1$, we have

$$r_n = \{\mu(\mu x_M + \Lambda) - [(\mu x_M + \Lambda)^2 + F^2(\mu^2 - 1)]^{1/2}\}/(\mu^2 - 1). \quad (3.35)$$

Let us find what requirements must be satisfied by $\varphi(z, x)$. Expression (3.35) can be rewritten as a condition

$$\mu^2(\mu x_M + \Lambda) > (\mu x_M + \Lambda)^2 + F^2(\mu^2 - 1) \quad (3.36)$$

or

$$F < \mu x_M + \Lambda. \quad (3.37)$$

Taking into account (3.31), we find for μ and Λ

$$x_M/\sin \alpha + C - F + [z_M^2 + (x_M - R_0)^2]^{1/2} > 0. \quad (3.38)$$

As $C = (f^2 + R_0^2)^{1/2} + n\lambda_0/2^{m-1} > F$, condition (3.38) always holds. In other words, if $x = \eta(z)$-type 1:1 mapping can be defined by a function $\varphi(z, x)$, set (3.29) always has a solution.

Therefore, only one restriction is imposed on the type of function representing the focusing surface $\varphi(z, x)$: for the solution of the set of equations (3.29) to exist, it is sufficient for the first derivative of the function $\eta'(z)$ to have one and the same sign in its domain, that is, it is sufficient for the function $\eta(z)$ to be piecewise-monotonous.

We will now give some very simple examples of focusing the radiation on to most frequently encountered curves that the authors of this book obtained in 1982 for diffractive elements designed for the microwave wavelength range [8].

3.19   Focusing on to a linear segment

A diffractive element that focuses radiation on to a linear segment has independent importance. The formation of conical wavefronts generates the so-called Bessel fronts which are free of diffraction constraints in a certain range of parameters. In other words, the field intensity distribution across the optical axis is described along the focusing zone (the length of the segment) by the Bessel function $I \approx J_0^2(kr \sin \alpha)$. Since the coordinate Z does not enter the argument of the Bessel function, the cross section of such a beam does not change along the length of the segment, while the width of this distribution is close to the radiation wavelength. A peculiar feature of diffraction-free beams (rather, diffractionally compensated beams) is the lateral influx of energy at a constant angle to the axis along the length of the segment (figure 3.39). This type of wave propagation compensates for the decrease of intensity in the neighbourhood of the axis that results from the diffraction-caused divergence. The field intensity may vary as a function of the longitudinal coordinate but the dimensions of the central cylindrical zone of energy concentration and of 'pipe-like' secondary maxima remain unchanged and dictated by the properties of the Bessel function J_0.

In our opinion, the following fields are promising for the application of such elements: adjustment of optical systems using a ray of millimetre wavelength range radiation, monitoring of mirror surface quality, three-dimensional quasioptics vision systems in the millimetre range, tomographic systems, security systems in the millimetre range, and so forth. A number of novel applications are possible if one-dimensional analogues of diffractive focusing devices with Bessel wavefronts are used.

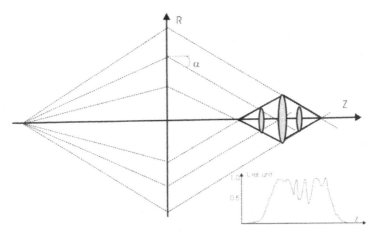

Figure 3.39. Schematic diagram of formation of Bessel beams and field intensity distribution along the optical axis for a binary diffractive element.

For simplification, we assume the reference radius to be zero. For a linear segment $(z = 0)$ we have

$$\varphi(z, x) = \begin{cases} x \equiv 0 \\ z_n = r_n \cot \alpha. \end{cases} \tag{3.39}$$

We then have from expression (3.29) that

$$r = \frac{1}{v^2 - 1}\left\{ -v\left(F + \frac{n\lambda_0}{2^{m-1}}\right) + \left[F^2 v^2 + 2F_n\frac{\lambda}{2^{m-1}} + \left(\frac{n\lambda_0}{2^{m-1}}\right)^2\right]^{1/2}\right\} \tag{3.40}$$

where $v = -\tan(\alpha/2)$. If $\alpha \ll 1$ expression (3.40) for binary type $(m = 1)$ element of diffractive quasioptics transforms into a familiar formula [61]:

$$r = \frac{(F \sin \alpha + 2Fm\lambda_0 + m^2\lambda_0^2)^{1/2} - (F + m\lambda_0)\sin\alpha}{\cos^2 \alpha}.$$

Figure 3.39 gives a schematic diagram of a diffractive element designed to form Bessel beams and the field intensity distribution along the focusing segment in the millimetre wavelength range [8].

3.20 Focusing on to a circular segment (tube)

By analogy to the preceding case, the in-phase condition for radiation along a circular segment of radius a can be written as $(R_0 = 0)$

$$(F^2 + r_n^2)^{1/2} + (r - a)/\sin\alpha = (F^2 + a^2)^{1/2} + n\lambda_0/2^{m-1} + (r_n - a)\cot\alpha.$$

Hence, using the notation $C = (F^2 + a^2)^{1/2} + n\lambda_0/2^{m-1}$, we readily obtain that zone radii on the surface of an MWDO element are described by the equation

$$r_n = \frac{v(av - c)}{v^2 - 1} - \frac{[v^2(c - av)^2 - (v^2 - 1)((c - av)^2 - 2a^2v^2 - F^2)]^{1/2}}{v^2 - 1}. \tag{3.41}$$

Obviously, as $a \to 0$, expression (3.41) transforms into (3.40) for the linear segment.

The length of a circular segment is determined, as it was for the linear segment, by the quantities a, r_1, r_N, α. The initial point of the corresponding focusing curve is at a point z_N with the coordinate

$$Z_{\text{ini}} = \begin{cases} r_1/\sin\alpha & \text{for linear segment} \\ (r_1 - a)\sin\alpha & \text{for annular segment}. \end{cases}$$

The coordinate of the final point of the curve is

$$Z_f = \begin{cases} r_1/\sin \alpha & \text{for linear segment} \\ (r_1 - a) \sin \alpha & \text{for annular segment.} \end{cases}$$

3.21   Focusing on to a cone

The set of equations (3.29) is written in this case in the form (assuming for simplicity that $z_0 = R_0 = 0$):

$$(F^2 + r_n^2)^{1/2} + (r_n - x)/\sin \alpha = (F + n\lambda_0/2) + (z^2 + x^2)^{1/2}$$

$$x = z \tan \beta.$$

(3.42)

Simple geometric arguments lead to the expression

$$x = \rho r_n, \qquad \rho = \cotan \alpha/(\cotan \alpha + \cotan \beta). \tag{3.43}$$

By solving the first equation of (3.42) in view of (3.43) we obtain

$$r_n = \frac{1}{\psi^2 - 1}[-\psi c - (c^2 + F^2(\psi^2 - 1))^{1/2}] \tag{3.44}$$

where $\psi = \rho/\sin \beta + \varepsilon(\rho - 1)$, $\varepsilon = 1/\sin \alpha$ and $c = F + n\lambda_0/2^{m-1}$.

Note that the shape of the focusing region—from cone to truncated cone—can be controlled by choosing the angle α.

If the reference radius R_0 is non-zero, the expression for r_n becomes more complicated. Thus the corresponding equation in the case we consider here is

$$(F^2 + r_n^2)^{1/2} + \gamma r_n = c + [(a - \rho r_n)^2 + r_n^2\rho^2 \cotan^2 \beta]^{1/2}$$

and reduces to an equation of fourth degree with respect to r_n (here $\gamma = \varepsilon(1 - \rho)$).

In the general case radiation is focused on to a cone, the equation for zone radii r_n has the following form:

$$(F^2 + r_n^2)^{1/2} + \gamma r_n + \varepsilon b = (F^2 + R_0^2)^{1/2} + n\lambda_0/2^{m-1}$$

$$+ [((\rho r_n - b) \cotan \beta + z_0)^2 + (R_0 - \rho r_n + b)^2]^{1/2}$$

where $b = z_0/(\cotan \alpha + \cotan \beta)$.

It is readily seen that if $\beta > \pi/2$, focusing will occcur on to an 'inverted' cone.

Note also that if radiation is focused on to a spatial curve (a surface), the problem has several solutions. Axisymmetric shapes (if the reference radius is ignored) lead to two such solutions [4]. For instance, if focusing is done to a cone, the second solution has the form of equation (3.44) with the coefficients (for $R_0 = z_0 = 0$)

$$\psi = \rho/\sin\beta - \varepsilon(1 + \rho), \qquad \rho = \cotan\alpha/(\cotan\beta - \cotan\alpha).$$

By choosing a specific type of solution, it is possible to modify the energy distribution function in the radiation focusing zone; this is especially important for high-aperture diffractive elements with low lens aperture D/λ [63].

3.22   Focusing to a disk

In this case we easily obtain

$$r_n = Ad/(d - c) + [(A^2 d^2/(d - c)^2 + (F^2 - A^2)]^{1/2}$$

where $A = (z_0^2 + d^2 - F^2 - c^2)/2c$, $d = R_0 + z_0 \tan\alpha$ and $c = z_0/\cos\alpha - (F^2 + R_0^2)^{1/2} - n\lambda_0/2^{m-1}$.

3.23   Frequency and formatting properties of MWDO elements

The simplest version of a diffractive element that focuses radiation to a linear segment was discussed in [62] (in paraxial approximation). To increase the length of the curve of focusing S it was suggested in this paper to move the radiation source closer to the element's plane. In this case the estimate given in [49] holds in the paraxial approximation for long-focal elements:

$$S \approx D/[2(\alpha - \Delta\alpha)] \approx D/[2\alpha - Dx\cos\varphi_x/(D^2 + F(F - x))]$$

where φ_x is the angle between the optical axis and the beam drawn from the pointlike radiation source to the edge of the diffractive element. As a result of this displacement of the radiation source the angles at which rays intersect the optical axis do not remain constant but decrease from α to $D/2S$. The diameter of the focusing line varies correspondingly.

At the same time, frequency properties of MWDO elements can be used to change (correct) the shape of the area on to which radiation is focused, by varying the diffraction angle α. For instance, it follows from (3.29) that

$$\alpha = \arcsin\left(\frac{r_n - x_m}{(F^2 + R_0^2)^{1/2} - (F^2 + r_n^2)^{1/2} + [z_m^2 + (x_m - R_0)^2]^{1/2} + n\lambda/2^{m-1}}\right)$$

$$(3.45)$$

that is, $\alpha = \alpha(\lambda)$. Thus for focusing on to a linear segment, we have from (3.40)

$$\alpha = 2\arctan\left(\frac{F + n\lambda/2^{m-1} - (F^2 + r_n^2)^{1/2}}{r_n}\right).$$

Taking into account that $(\sin \psi)^{-1} = (1 + \cotan^2 \psi)^{1/2}$, expression (3.45) can in the general case be rewritten in the form

$$a(\lambda) = \arctan\{(d/[c\lambda - (a - b)]^2 - 1)\}^{1/2} \qquad (3.46)$$

where

$$d = (r_n - x) > 0, \quad c = n/2^{m-1} > 0, \quad a = (F^2 + r_n^2)^{1/2} - (F^2 + R_0^2)^{1/2} > 0$$

and

$$b = (z^2 + (x - R_0)^2)^{1/2} > 0.$$

Formula (3.46) above describes how the angle varies when wavelength deviates from λ.

We shall list the basic properties of MWDO elements in the off-design frequency mode (i.e. when $\lambda \neq \lambda_0$) [8]:

- the wave diffraction angle α depends on the number $\alpha = \alpha(r_n, n)$ of the zone on the surface of the diffractive element. Therefore, in this mode of focusing, the diffractive width of the focusing curve, measured transversely to the generatrix surface of focusing $\varphi(z, x)$, will vary in a monotone fashion, increasing with increasing distance from the element plane;
- the form of the function $\alpha(\lambda)$ will depend on the shape of the surface of focusing.

Let us find a relation between the position of a pointlike radiation source y on the optical axis and the shape of the surface of focusing. We have from (3.29)

$$[(F - y)^2 + r^2 n]^{1/2} - [(F - y)^2 + R_0^2]^{1/2} - n\lambda/2^{m-1}$$
$$= [z_1^2 + (x_1 - R_0)^2]^{1/2} - (r_n - x_1)/\sin(\alpha - \Delta\alpha)$$
$$[F^2 + r_n^2]^{1/2} - [F^2 + R_0^2]^{1/2} - n\lambda/2^{m-1}$$
$$= [z_1^2 + (x_n - R_0)^2]^{1/2} - (r_n - x_n)/\sin\alpha. \qquad (3.47)$$

The analysis will be conducted in the paraxial approximation, that is, $(z, z_1) \gg (x, x_1)$ and $F \gg r_n$, $\sin\alpha \approx \alpha$. We consider two cases. In the first of them we assume that the generatrix of the focusing curve is quasiparallel to the optical axis, that is, $x_n \approx x_1$. Expanding then (3.47) into a series in powers of a small parameter, subtracting the first equation from the second and retaining only the terms of first degree of smallness, we obtain

$$\frac{(r_n^2 - R_0^2)y_T}{2F(F - y_T)} \approx (z_1 - z_n) - \frac{(r_n - x_n)\Delta\alpha}{\alpha(\alpha - \Delta\alpha)}.$$

Likewise, in the second case where $(z_1 \approx z_n)$, we obtain

$$\frac{(r_n^2 - R_0^2)y_T}{2F(F - y_T)} \approx (x_1 - x_n)\left(\frac{1}{\alpha(\alpha - \Delta\alpha)} - \frac{R_0}{z_n}\right) - \frac{(r_n - x_n)\Delta\alpha}{\alpha(\alpha - \Delta\alpha)}. \qquad (3.48)$$

Among the expressions obtained above and characterizing the 'stability' of radiation focusing by MWDO elements with respect to small changes in the position of a pointlike radiation source on the optical axis, only one will be mentioned: the introduction of the reference radius R_0 in the second of the cases discussed, (3.48), makes it possible to control this property. Similar behaviour occurs in the case of the displacement of a point-like radiation source by y_v in the vertical plane.

3.24 Numerical modelling of radiation focusing to a conical area

The study of the structure of radiation focused by MWDO elements used numerical experiment as a tool. The diffraction field was found by calculating the Fresnel–Kirchhoff diffraction integral.

The simplest configuration of a diffractive element was chosen for the study: a binary amplitude-type element. Amplitude elements can be justly considered thin since they only attenuate the radiation they transmit but leave the phase unchanged. Phase elements that do not affect the amplitude of the incident radiation change only the eikonal. Rigorously speaking, phase elements are never thin because phase is usually changed by varying the geometry of their surface. The zone structure on the surface of a diffractive element was found from equation (3.42). The parameters of the element studied were: $D/\lambda_0 = 100$, $D/F = 1$, $\alpha = 10°$, $\beta = 12°$. Taking into account the axial symmetry of the problem (Oz axis is at the same time the symmetry axis), numerical experiments determined the structure of the energy density distribution $W = UU^*$ in the $(z - r)$ plane, where r is the radius normal to the axis Oz.

Figure 3.40 shows the energy density distribution in the calculated field of dimensions $650\lambda_0(z) \times 63\lambda_0(r)$. The initial value of the coordinate z_{ini} in figure 3.40 corresponds to $D/z_{ini} \approx 1$. The computation domain was divided into 24 layers (in z), equidistant from one another. In each of these layers, the distribution U_p was found and, from it, the curve $W(r)$ was

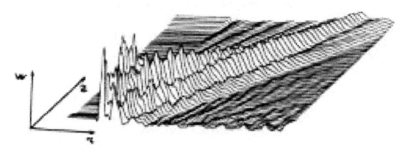

Figure 3.40. Focusing of radiation to a conical region [8].

plotted. Intermediate values of energy density between layers were found by bicubic two-dimensional interpolation.

Figure 3.40 shows the sought isometric distribution where the values W_j $(j = 1,\ldots,24)$ were normalized to the maximum within the jth layer. The distribution shown makes it possible to visualize the field structure in each section and to evaluate the relative contribution to interference of the spurious maximum on the optical axis and on the generatrix surface of focusing: dividing (3.54) by (3.51) and assuming that $\omega \approx \lambda$ we obtain that the gain in the maximum zone number is nearly 2.

When radiation is focused, for instance, on to a conical target fabricated of a material that is opaque to this radiation, the central spurious radiation maximum (along the optical axis) is absent; hence, to have an idea of a realistic energy density distribution along the generatrix of the target, the corresponding areas were not considered. Normalization was then run to the common maximum of W in the entire domain of computation.

The results shown above demonstrate that enhanced energy density is produced along the conical surface even in the case of a binary element.

Further improvement in the 'quality' of radiation focusing can be achieved by employing various familiar techniques, for instance, by resorting to two phase-inversion elements with a large number of phase quantization levels, or by using additional degrees of freedom of the diffractive element, such as reference radius, non-flat shape of surface and so on.

Furthermore, it is possible to combine the principles of design of MWDO elements with those of elements that focus radiation on to an arbitrary curve. This opens a possibility of creating diffractive elements that focus radiation to arbitrary three-dimensional configurations [61, 63].

3.25 Zone screening by diffractive elements on curvilinear surfaces

Black and Wiltse [17] considered screening of Fresnel zones in a high-aperture zone plate. The problem was later treated in detail by Petosa and Ittipiboon [67]. An experimental study was run on a high-aperture-ratio $(F/D = 0.25)$ two-level phase-inversion Plexiglas zone plate $(\varepsilon \approx 3)$ at a frequency of 30 MHz. Measurements showed that the directivity of the zone plate varies by approximately 0.3 dB, depending on the direction of incidence on a smooth surface or on the side of phase-inversion zones. The main conclusion of this paper was that shading (blocking) effect must be taken into account for high-aperture-ratio zone plates $(F/D < 0.5)$.

It will be shown below that this effect can be reduced for diffractive elements fabricated on an arbitrary curvilinear surface.

A ray drawn from the focal point of the zone plate to the boundary of the ith Fresnel zone screens part of the $(i+1)$th zone. It is easy to write, using

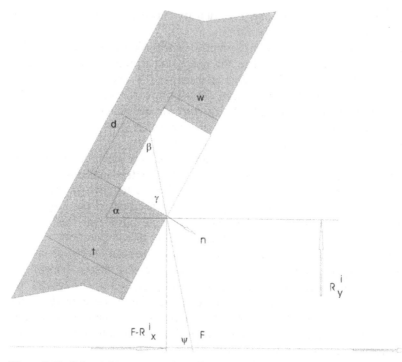

Figure 3.41. Schematic representation of zone blocking on a high-aperture zone plate.

geometric relations (figure 3.41), that

$$\tan\psi = R_i/F, \qquad d = w\tan\psi \qquad (3.49)$$

where w is the depth of the phase step, d is the width of the shadowed part of the zone and F is the focal length.

Let us evaluate the maximum number of that zone beginning with which the entire subsequent zone is completely in the shadow. It is easy to derive from the geometry of the problem that

$$d_{n_{\max}} = R_{i+1} - R_i = \frac{w}{F} R_i. \qquad (3.50)$$

Let $R_i = \sqrt{i\lambda F}$ where λ is radiation wavelength. It is then easy to show that

$$n_{\max} = \frac{F^2}{w^2 + 2wF}. \qquad (3.51)$$

Formula (3.51) allows us to evaluate the limiting number of 'working' zones of a flat binary phase-type zone plate, and to evaluate, in combination with (3.49), the coefficient of utilization of zone plate surface from the standpoint of the blocking action of its stepwise phase profile.

One of the methods of reducing this blocking effect of a phase profile on a zone plate is to use curvilinear surfaces for zone plate fabrication.

For simplification, let us consider a zone plate fabricated on a conical surface [8].

Fresnel zone boundaries on the surface of a conical zone plate are given by the expressions [8]

$$R_y^i = \sqrt{R_{if}^2 + \left(\frac{i\lambda}{2\tan\alpha}\right)^2} - \frac{i\lambda}{2\tan\alpha}, \qquad R_x^i = \frac{R_y^i}{\tan\alpha} \tag{3.52}$$

where R_{if} is the radius of a flat zone plate, $R_{if} = \sqrt{i\lambda F + (i\lambda/2)^2}$, R_x, R_y are the boundaries of the *i*th zone on the abscissa and ordinate axes, respectively, and α is one half of the cone opening angle. It is easily seen that

$$d = w \tan\gamma, \qquad \gamma = (\alpha + \psi) - \frac{\pi}{2}. \tag{3.53}$$

Let us evaluate the maximum value of $n_{\max}$. For clarity, we assume $\alpha = \pi/4$. We also assume in the first approximation that $R_f \approx \sqrt{n\lambda F} \approx R_x \approx R_y$. Then, taking into account that

$$R_{(n+1)f}^2 = R_{nf}^2\left(1 + \frac{1}{n}\right) + \left(\frac{\lambda}{2}\right)^2 \approx R_{nf}^2\left(\frac{1}{n} + 1\right)$$

and $\tan\psi = R_n/(F - R_n)$, we easily obtain the following estimate from (3.52) and (3.53):

$$n_{\max} \approx F\left(\frac{w + 4\lambda}{4w\lambda}\right). \tag{3.54}$$

By comparing (3.54) and (3.51) we immediately conclude that the fact of Fresnel zone blocking for diffractive elements fabricated on a curvilinear surface becomes appreciable at higher zone numbers than on a flat surface; dividing (3.54) by (3.51) and assuming $w \approx \lambda$ we conclude that the maximum gain in zone number is about 2.

3.26 Interrelation between zone plate parameters on curvilinear surfaces

Let us find interrelations between such main parameters of a diffractive element as its normalized diameter (D/λ), aperture ratio (D/F), measure of deviation from flatness of the surface $((F - X_N)/D)$ and the number of Fresnel zones N within the aperture. To do this, we take into account that the diameter of a diffractive element on an arbitrary surface is limited by a certain number N of Fresnel zones, that is, $D = 2Y_N$. Now we solve the quadratic equation for the zone number N, taking into account the formulas

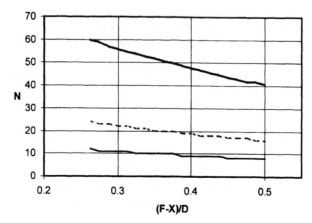

Figure 3.42. The number N of Fresnel zones within the aperture of a diffractive element of arbitrary cuvilinear shape as a function of relative deviation from flat surface for three values of D/λ (from bottom up): 10, 20 and 100 wavelengths.

for projections of Fresnel zone radii (3.12) and readily obtain

$$N = \frac{1}{2}\left(\frac{D}{\lambda}\right) \frac{1}{\left((F - x_N)/D + \sqrt{((F - x_N)/D)^2 + 1/4}\right)} \qquad (3.55)$$

where N is the entier (maximum integral part) of the real value of the right-hand side of equation (3.55).

The relation thus derived for the diffractive element fabricated on an arbitrary surface thus implies that aperture ratio and the number of Fresnel zones are linear functions of relative diameter.

It is not difficult to generalize equation (3.55) to the case of an arbitrary number of phase quantization levels,

$$N = \frac{p}{2}\left(\frac{D}{\lambda}\right) \frac{1}{\left((F - x_N)/D + \sqrt{((F - x_N)/D)^2 + 1}\right)}.$$

That is, the number of Fresnel zones is directly proportional to the number p of phase quantization levels.

As an example, figure 3.42 plots the number N of Fresnel zones within the aperture of a diffractive element of arbitrary curvilinear shape as a function of relative surface flexure for several values of aperture ratio.

Bibliography

[1] Engel A, Steffen J and Herziger G 1974 'Laser machining with modulated zone plates' *Appl. Opt.* **13**(12) 269–273

[2] Fedotowsky A and Lehovec K 1974 'Far field diffraction patterns of circular gratings' *Appl. Opt.* **13**(11) 2638–2642

[3] Fedotowsky A and Lehovec K 1974 'Optimal filter design for annular imaging' *Appl. Opt.* **13**(12) 2919–2923

[4] Koronkevich V P, Palchikova I G, Poleschuk A G and Yurlov Yu I 1985 'Kinoform optical elements with annular momentum response' Preprint No 256 IAiE CO AN SSSR Novosibirsk, p 19

[5] Kazansky N L 1992 'Numerical experimental study of diffraction characteristics of a focusator onto a ring' *Kompyuternaya Optika* **10/11** 128–144

[6] Golub M A, Kazansky N L, Sisakyan I N and Soifer V A 1988 'Numerical experiment with flat optics elements' *Avtometriya* **1** 70–82

[7] S T, Greisukh G I and Turkevich Yu G 1986 *Optics of Diffractive Elements and Systems* (Leningrad: Mashinostroenie) 223

[8] Minin I V and Minin O V 1992 *Diffraction Quasioptics* (Moscow: Research and Production Association 'InformTEI') (in Russian)

[9] Minin I V and Minin O V 1990 'Control of focusing properties of diffractive elements' *Kvantovaya Elektronika* **17**(2) 249–251

[10] Minin I V and Minin O V 1989 'Focusing elements of diffractive quasioptics and their applications' in *3rd USSR School on Propagation of Millimeter and Sub-millimeter Waves in the Atmosphere, Kharkov*, pp 251–252

[11] Minin I V and Minin O V 1989 'Optimization of focusing properties of diffractive elements' *Pisma v ZTF* **15**(23) 29–33

[12] Born M and Wolf E 1999 *Principles of Optics* 7th edn (Cambridge University Press)

[13] Steward E G 1983 *Fourier Optics: An Introduction* (Chichester: Ellis Horwood; New York: Halsted Press, a division of Wiley,)

[14] Volostnokov V G 1993 'Diffractive optics elements and the phase problem in optics' *Proc. P N Lebedev Phys. Inst. (FIAN)* **217** 151–158

[15] Hidaka T 1991 'Generation of a diffraction free laser beam using a specific Fresnel zone plate' *Jap. J. Appl. Phys.* **30**(8) 1738–1739

[16] Black D N and Wiltse J C 1987 'Millimeter wave characteristics of phase-correcting Fresnel zone plates' *IEEE Trans. Microwave Theory Tech.* **MTT-35**(12) 1122–1129

[17] Wiltse J C 1985 'The Fresnel zone-plate lens' *Proc. Soc. Photo-Opt. Instrum. Eng.* **544** 41–47

[18] Wiltse J C 1985 'The phase correcting zone plate' in *10th Int. Conf. Infrared and Millimeter Waves, Lake Buena Vista, FL*, 9–13 December, No. 4, pp 345–347

[19] Minin I V and Minin O V 1989 'Selection of harmonics of coherent radiation using computer optics' in *Abstracts of All-Union Seminar on 'Methods and Technology of Processing Two-Dimensional Signals'*, Part II (Moscow: TsNIINTIKPK) pp 36–37

[20] Minin I V and Minin O V 1989 Inventor's Certificate (USSR) No 176651 Device for radiation focusing

[21] Minin I V and Minin O V 1989 'Elements of optical computer based on diffractive quasioptics' in *Abstracts of All-Union Seminar on 'Methods and Technology of Processing Two-Dimensional Signals'*, Part II (Moscow: TsNIINTIKPK) pp 38–39

[22] Markin A S, Studenov V B and Ioltuchovsky A A 1980 'Filtering of harmonics of coherent radiation source by zone plates' *ZTF* **50** 2482

[23] Markin A S and Studenov V B 1987 'Spatial filtration of harmonics of coherent radiation' *Kompyuternaya Optika* **1** 147–151

[24] Maluzhinetz G D 1946 'Diffraction near the optical axis of a zone plate' *Doklady AN SSSR* **54**(5) 403–406

[25] Futhey J and Fleming M 1992 'Superzone diffractive lenses' in *Diffractive Optics: Design, Fabrication and Application Technical Digest 1992* (Washington, DC: Optical Society of America) vol 9, pp 4/MA2-1, 6/MA2-3

[26] Slusarev G G 1957 'Optical systems with phase layers' *Doklady AN SSSR* **113**(4) 780–783

[27] Raysky S M 1952 'Zone plate' *UFN* **47**(4) 516–536

[28] Minin O V and Minin I V 1989 'Information potential of diffractive quasioptics on second-order surfaces of revolution' in *Abstracts of All-Union Seminar on 'Methods and technology of processing two-dimensional signals'*, Part II (Moscow: TsNIINTIKPK) pp 29–30

[29] Golub M A, Doskolovich L L, Kazanskiy N L, Kharitonov S I and Soifer V A 1992 'Computer generated diffractive multi-focal lens' *J. Modern Optics* **39**(6) 1245–1251

[30] Gabor D 1972 (Edwards Memorial Lecture) *Holography: 1948–1971* City University, London

[31] Dey K K and Khastgir P 1973 'Comparative focusing properties of spherical and plane microwave zone plate antennas' *Int. J. Electronics* **35**(4) 497–506

[32] Dey K K and Khastgir P 1977 'Design and focusing characteristics of a microwave paraboloidal zone plate' *Indian J. Radio Space Phys.* **6** 202–204

[33] Khastgir P, Chakravorty I N and Dey K K 1973 'Microwave paraboloidal, spherical and plane zone plate antennas: comparative study' *Indian J. Radio Space Phys.* **2** 47–50

[34] Minin I V and Minin O V 1998 'Diffractive objective lenses on parabolic surfaces' *Kompyuternaya Optika* **3** 8–15

[35] Baybulatov F Kh, Minin I V and Minin O V 1985 'Study of focusing properties of a Fresnel zone plate' *Radiotekhnika i Elektronika* **30**(9) 1681–1688

[36] Minin I V and Minin O V 1986 'Wide-angle multicomponent diffractive microwave lens' *Radiotekhnika i Elektronika* **31**(4) 800–806

[37] Minin I V and Minin O V 1989 'Invariant properties in elements of diffractive quasioptics' *Kompyuternaya Optika* **6** 89–97

[38] Bazarsky O V, Kadashov N G and Khlyavich Ya L 1979 'On designing binary Fresnel zone plates' *Radiotekhnika i Elektronika* **24**(4) 834

[39] Minin I V and Minin O V 1989 'Diffraction-based radio optics systems: achievements and prospects' in *Abstracts of All-Union Conference on 'Optical, Radiowave and Thermal Radiation Methods of Nondestructive Control'*, Mogiliov, pp 204, 205

[40] Bolostozky A L and Leonov A S 1992 'Aplanatic waveguide Fresnel lenses' *Kvantovaya Elektronika* **19**(8) 804–806

[41] Kock W E 1975 *Engineering Application of Lasers and Holography* (New York and London: Plenum Press) p 400

[42] Minin I V and Minin O V 1989 'Amplitude-phase-type polarization zone plate' in *Abstracts of All-Union Seminar on 'Methods and Technology of Processing Two-Dimensional Signals'*, Part II (Moscow: TsNIINTIKPK) p 33

[43] Minin I V and Minin O V 1990 Inventor's Certificate 1596417, SSSR, MKI Í01Q 15/12. 'Zone plate' Published in Bulletin No. 36

[44] Korshunov B P, Lebedev S P and Masalov S A 1984 'Using metal grids as phase plates in sub-mm wavelength range' *ZTF* **54**(2) 825–827

[45] Minin I V and Minin O V 1987 'Diffractive radio optics systems for the microwave range' in *Abstracts of All-Union Conference on 'Methods and Technology of Measuring Electromagnetic Characteristics of Materials in HF and Microwave Ranges'*, Novosibirsk, pp 169–170

[46] Dubik B, Zayonz M and Novakh E 1990 'Focusing kinoform mirror' *Avtometriya* **2** 85–88

[47] Shiono T and Ogawa H 1991 'Diffraction-limited blazed reflection diffractive microlens for oblique incidence fabricated by electron-beam lithography' *Applied Optics* **30**(25) 3643–3649

[48] Lenkova G A 1985 'Rotatable focusing kinoform' *Avtometriya* **6** 7–12

[49] Greisukh G I and Stepanov S A 1989 'Diffractive elements based optical systems with multi-sectional fields of view' *Kompyuternaya Optika* **4** 49–52

[50] Janicijevic L J 1982 'Diffraction characteristics of square zone plates' *J. Optics (Paris)* **13**(4) 199–206

[51] Minin I V and Minin O V 1989 'A novel type of computer optics element' in *Abstracts of All-Union Seminar on 'Methods and Technology of Processing Two-Dimensional Signals'*, Part II (Moscow: TsNIINTIKPK) pp 34–35

[52] Verzel R G 1987 'Effect of the aperture-lens separation on the focal shift in large *F*-number systems' *J. Opt. Soc. Amer. A* **4**(2) 340–345

[53] Kazemirchuk S S and Smirnov S A 1989 'Axial intensity distribution in the focal zone of diffraction-constrained systems' in *Abstracts of All-Union Seminar on 'Methods and Technology of Processing Two-Dimensional Signals'*, Part II (Moscow: TsNIINTIKPK) pp 19–20

[54] Li Y and Platrer H 1983 'An experimental investigation of the diffraction pattern in low Fresnel-number systems' *Optica Acta* **30** 1621–1643

[55] Li Y and Wolf E 1984 'Three-dimensional intensity distribution near the focus in systems of different Fresnel numbers' *J. Opt. Soc. Amer. A* **1** 801–808

[56] Kazemirchuk S S and Smirnov S A 1989 'Parameters of axial distribution and problems of controlling design characteristics of radiowave lens systems' in *Abstracts of All-Union Seminar on 'Methods and Technology of Processing Two-Dimensional Signals'*, Part II (Moscow: TsNIINTIKPK) pp 21–22

[57] Kazemirchuk S S, Krilov K I and Smirnov S A 1989 'Finding the Fresnel number and geometric-optics parameters from the asymmetry of the main diffraction peak' in *Abstracts of All-Union Seminar on 'Methods and Technology of Processing Two-Dimensional Signals'*, Part II (Moscow: TsNIINTIKPK) pp 23–24

[58] Danilov V A, Ionov V V, Prokhorov A M *et al* 1982 'Synthesis of optical elements generating a focal line of arbitrary shape' *Pisma v ZTF* **8**(13) 810–815

[59] Minin I V and Minin O V 1988 'Information properties of the zone plate' *Computer Optics* **3** 15–22

[60] Minin I V and Minin O V 1991 'Principle of designing diffraction quasioptics elements for focusing coherent radiation onto an arbitrary spatial configuration' in *Abstracts of 1st Ukrainian Symposium on 'Physics and Technologies in the mm and Sub-mm Wavelength Ranges'*, Kharkov, 15–18 October, Part 1, pp 14–15

[61] Minin I V and Minin O V 1989 'Focusing diffractive quasioptics elements and their application' in *Abstracts of 3rd USSR School on propagation of mm- and sub-mm waves through the atmosphere*, Kharkov, 12–17 October, pp 251–252

[62] Korolkov V P, Koronkevich V P, Poluschuk A G *et al* 1989 'Kinoforms: technologies, novel elements and optical systems' *Avtometriya* **4** 47–64

[63] Minin I V and Minin O V 1989 'Diffractive optics—achievements and prospects' in *Abstracts of All-Union Seminar on 'Methods and Technology of Processing Two-Dimensional Signals'*, Part II (Moscow: TsNIINTIKPK) pp 25–26

[64] Sussman M 1960 'Elementary diffraction theory of zone plates' *Am. J. Phys.* **28**(4) 394–398

[65] Palchukova I G 1989 'Mathematical modeling of diffraction of light on phase-type zone plates' Preprint, Siberian Branch of AN SSSR, Institut Avtomatiki i Electrometrii, No. 433

[66] Minin I V and Minin O V 1988 Inventor's Certificate (CCCP) No. 1617398, G 02 B 27/42 'Zone plate'

[67] Petosa A and Ittipiboon A 2000 'Shadow blockage effects on the aperture efficiency of dielectric Fresnel lenses. 'Microwaves, antennas and propagation' *IEE Proc.* **147**(6) 451–454

Selected bibliography

Minin I V and Minin O V 2004 'Shadow effects of curvilinear zone plate lens' *Autometria* **40**(3) 106–117

Webb G W 2003 'New variable for Fresnel zone plate antennas' in Proc. 2003 Antenna Applications Symposium, Allerton Park, Monticello, IL, 15–17 September

Chapter 4

Alternative methods of synthesizing diffractive elements

4.1 Synthesis of diffractive elements that create a flat focal line with prescribed intensity distribution along it

This chapter gives a brief review of alternative methods of designing of MWDO elements and discusses interrelations between all algorithms outlined earlier.

As we indicated earlier, the problem of synthesizing a diffractive element that focuses electromagnetic radiation on to a flat curve does not have a unique solution. Furthermore, different approaches are possible to obtaining a solution to this problem. One of such approaches is considered when we discuss the synthesis of a diffractive element known as focusator [1, 2].

In the general sense, the direct aim of diffractive optics is to solve the problem of wave diffraction on a diffractive element and to obtain the field distribution in the area of interest. The inverted problem in this case is to find the transmission function or reflection function of the diffractive element that creates the prescribed wave field.

The inverse focusing problem seems to have been formulated and solved for the first time in the geometric optics approximation [3]. Then it was given rigorous mathematical formulation and a number of mathematical foundations were built [4–13]. From the mathematics standpoint, the inverse problem is ill posed [14]. First, a solution may not exist at all. Typically, the absence of solution may be caused by restrictions imposed on the area of focusing by fundamental laws of physics. For instance, radiation cannot be focused to an area of size smaller than the size imposed by the diffraction limit. Second, the solution may not be unique, that is, there may be a set of various solutions for the phase function that would result in the same focusing zone. Third, the solution may be unstable. The instability manifests itself in that small changes in the shape of the focusing zone or in intensity distribution within the zone may correspond to large changes in the phase function of the diffractive element. The instability results in substantial changes of the shape of the focal line and in reducing the energy efficiency of the element.

We shall consider the synthesis of focusators that create a flat focal line with prescribed intensity distribution along it. Let $U(x, y, z)$ be a scalar wave field satisfying the Helmholtz equation in free space:

$$\Delta U + k^2 U = 0$$

where $k = 2\pi/\lambda$ is the wave number.

Assume now that a scalar field $U_0(x, y, z)$ incident on a flat element in the plane $z = 0$ is defined in the half-space $z < 0$. This element transforms the incident field in such a way that

$$U(x, y, +0) = U_0(x, y, -0)\tau(x, y). \qquad (4.1)$$

The field in the half-space $z \geq 0$ is found from the corresponding boundary problem with boundary conditions given by (4.1). Therefore, an arbitrary flat element is characterized by the transmission function $\tau(x, y)$ [7]. In general, the function $\tau(x, y)$ is complex and defines the type of the focusing element. If $|\tau(x, y)| = 1$, the element does not absorb the incident radiation. Such elements are known as phase elements. In this case

$$\tau(x, y) = \exp[ikh(x, y)]$$

where $h(x, y)$ is a real function. Amplitude-type elements are characterized by a real transmission function that assumes values from 0 to 1.

The solution (4.1) is an approximation. The quality of approximation is determined by the wavelength λ: the smaller λ is in comparison with the characteristic dimensions of the flat element, the better the approximation.

In most problems the Kirchhoff approximation [15] is a sufficiently good one; the field at a point $z = f$ is then calculated using the expression

$$U(x, y, f) = \text{const} \iint \frac{\exp\{ik[(x-u)^2 + (y-v)^2 + f^2]^{1/2}\}}{[(x-u)^2 + (y-v)^2 + f^2]^{1/2}} U(u, v, +0) \, du \, dv.$$

$$(4.2)$$

Integration in (4.2) is carried out over the entire plane $z = 0$. After radiation passes through the focusing element with characteristics $h(u, v)$, the intensity of radiation in the plane $z = f$ is

$$I(x, y) = \text{const} \left| \iint \frac{\exp[-ikR + ikh(u, v)]}{R} U(u, v, -0) \, du \, dv \right|^2 \qquad (4.3)$$

where $R = [(x-u)^2 + (y-v)^2 + f^2]^{1/2}$.

The inverse problem of synthesizing a phase function of the diffractive element that we discuss here consists in solving the nonlinear operator equation (4.3) with respect to the unknown function $h(u, v)$. It is expedient in problems of synthesis of optical elements to transform equations to the geometric optics approximation. The field $U(x, y, z)$ in the half-space $z > 0$ is expressed in the geometric optics approximation $(k \rightarrow 0)$ as

$$U(x, y, z) = A(x, y, z) \exp[ik\Phi(x, y, z)].$$

Note that $A(x, y, z)$ and $\Phi(x, y, z)$ are slowly varying functions. The function $\Phi(x, y, z)$ satisfies the eikonal equation

$$|\nabla\Phi(x, y, z)| = 1. \qquad (4.4)$$

As an example, we consider, in the geometric optics approximation, the solution of the problem of focusing a uniform beam on to a linear segment with constant intensity along it [12]. The phase function for this element is described by the expression

$$\varphi(u, v) = -\frac{U^2}{2f/(1 - l_0/R)} - \frac{V^2}{2f}.$$

Here l_0 is the length of the segment, f is the focal distance of the focusator, and R is the aperture radius of the element. The surface defined by this expression is a paraboloid. On reflection at a point with coordinates (u, v) the phase changes by

$$\psi(u, v) = (2\pi/\lambda)\varphi(u, v) = -\frac{\pi U^2}{\lambda f/(1 - l_0/R)} - \frac{\pi V^2}{\lambda f}.$$

Assuming $\psi = n\pi$, $n = 1, 2, \dots$ we obtain

$$\frac{U^2}{n\lambda f/(1 - l_0/R)} + \frac{V^2}{n\lambda f} = 1. \qquad (4.5)$$

Equation (4.5) defines the contour lines for the function $\psi(u, v)$. On the surface of the diffractive element, these contours lines are the zone boundaries for which the phases of oscillations differ by π, that is, they are analogues of conventional annular Fresnel zones.

We will give another solution of the problem of focusing a plane monochromatic wave by a diffractive element on to a line. A segment of a straight line on which radiation is focused is $2L$ long, and is situated along the axis x, symmetrically with respect to the origin of coordinates. The plane of the diffractive element and the focal plane are parallel and lie at a distance f from each other, and (x, y) and (u, v) are the coordinates in the focal plane and in the plane of the diffractive element, respectively. The zone boundaries on the surface of the diffractive element are found from the equation

$$\tan(GV^2) = -\frac{S[\chi(u + L)] - S[\chi(u - L)]}{C[\chi(u + L)] - C[\chi(u - L)]}$$

where $G = \pi/\lambda f$, $\chi = (2/\lambda f)^{1/2}$, $S[\chi(u \pm L)]$ and $C[\chi(u \pm L)]$ are the Fresnel integrals. As $L \rightarrow 0$ we obtain the boundaries of the classic Fresnel zones, and for $L \to \infty$ we make use of the properties of Fresnel integrals and arrive at the following expression for zone boundaries:

$$V_n = [\lambda f(n + 3/4)]^{1/2}, \qquad n = 1, 2, 3, \dots .$$

When the direct problem is solved in the Fresnel approximation [16], an analysis of the image generated by the element (designed in the geometric optics approximation) shows that the width of the focal segment is not the same along this segment and increases towards its end. Therefore, the geometric optics approximation is not adequate to reality in the problem of focusing to a line, and for further improvement we need to use the wave approximation.

4.2   Design of diffractive elements using the diffraction method of computations

Calculating diffractive elements by geometric optics techniques results in a number of drawbacks. First, the diffraction of waves in free space is ignored. Second, as a result of diffraction, the radiation intensity distribution in the plane in which the required wavefront is formed is not constant while the geometric-optics approach assumes that it is.

Let us consider the diffraction technique [17] of calculating a phase-type element which produces a required phase distribution in a certain plane perpendicular to the optical axis and located at a given distance from the diffractive element. The method is based on modernizing the interactive Gerhberg–Saxton algorithm [18]; this means that it is necessary to solve by successive approximations the nonlinear integral equation of scalar diffraction in the Fresnel approximation:

$$\varphi_0(\varepsilon, \eta) = \arg\left[\iint_\infty A(x, y) \exp[i\varphi(x, y)]\right.$$

$$\left. \times \exp\{ik[(x - \varepsilon)^2 + (y - \eta)^2]/2z\}/2z \, dx \, dy\right] \quad (4.6)$$

where $A_0(x, y)$ is the amplitude of the wave incident on the element, $\varphi(x, y)$ is the phase we seek, $\varphi_0(\varepsilon, \eta)$ is the required phase distribution in the plane at a distance z from the element, and $\arg(a + ib) = \arctan(b/a)$.

The algorithm for solving equation (4.6) is as follows. The initial approximation of the sought phase $\varphi_0(x, y)$ is chosen randomly. The Fourier transform is calculated from the function $A_0 \exp(i\varphi_0)$; the resulting function $F_n(\varepsilon, \eta)$, where n is the number of the iteration, is replaced by a function $F'(\varepsilon, \eta)$ using the rule

$$F'_n(\varepsilon, \eta) = |F_a(\varepsilon, \eta)| \exp[i\psi_0(\varepsilon, \eta)].$$

The inverse Fourier transform is then calculated for the function $F'(\varepsilon, \eta)$ and the function $f_n(x, y)$ obtained in the plane of the element is replaced with $f'_n(x, y)$ using the rule

$$f'_n(x, y) = \begin{cases} A_0(x, y)f_n(x, y)|f_n(x, y)|^{-1}, & (x, y) \in \Omega \\ 0, & (x, y) \notin \Omega \end{cases}$$

where Ω is the shape of the element's aperture.

The process is repeated until the error of the approximation δ does not reach the predetermined value

$$\delta = \left(\frac{\iint_{-\infty}^{\infty} [\psi_0(\varepsilon, \eta) - \psi_0(\varepsilon, \eta) - c]^2 \, d\varepsilon \, d\eta}{\iint_{-\infty}^{\infty} [\psi_0^2(\varepsilon, \eta) \, d\varepsilon \, d\eta} \right)^{1/2}. \tag{4.7}$$

Here $\psi_n(\varepsilon, \eta)$ is the phase calculated at the nth step of iteration, and $\psi_0(\varepsilon, \eta)$ is the required phase. Both phases are chosen in the range $[0, 2\pi]$. The constant $C = \psi_0(\varepsilon_0, \eta_0) - \psi_n(\varepsilon_0, \eta_0)$ is added to the expression (4.7) because the approximate phases may converge to a phase that may differ from the predetermined one by a constant value.

4.3 Physical restrictions on the possibilities of focusators and their relation to other diffractive elements

The previous section discussed the principle of designing diffractive elements known as focusators. In fact, however, such elements can also be treated as generalized zone plates [19]. In what follows we briefly discuss the relation of focusators to other types of focusing diffractive elements. To simplify the presentation, we limit the scope to the paraxial approximation $D \gg F$.

Let us consider diffraction of a plane wave U on an element of MWDO of diameter $2d$ with transmission function $t(u, v)$ placed in a plane (u, v) perpendicular to the axis Oz [20]. The point of intersection of this plane with the axis Oz is chosen as the origin of coordinates. The complex amplitude $I(x, y, z)$ of electromagnetic field in the plane z is described by the Fresnel–Kirchhoff integral and is given by a two-dimensional convolution

$$I(x, y, z) = t(x, y) \otimes h(x, y, z) \tag{4.8}$$

where

$$h(x, y, z) = z \frac{\exp\{i2/\lambda[z^2 + (x - y)^2 + (y - v)^2]^{1/2}\}}{[z^2 + (x - y)^2 + (y - v)^2]^{1/2}}.$$

In the paraxial approximation (that is, if the condition $z \gg (x - u), (y - v)$ is satisfied) we obtain, instead of (4.8),

$$h(x, y, z) = i/(\lambda z) \exp[i\pi(x^2 + y^2)/\lambda z + i2\pi z/\lambda].$$

Assume now that the function $h(x,y,z)$ has a restricted spectrum (the Fourier transform in the variable z) of width Ω_z and consider it on the axial segment $f_1 \leq z \leq f_n$ that is sufficiently remote from the origin of coordinates, and assume also that the condition $(f_n - f_1)/z \ll 1$ is satisfied on this segment.

The function $h(x,y,z)$ can be approximately represented on the segment $f_1 \ll z \ll f_n$ as a series

$$h(x,y,z) \approx \sum_{l=1}^{L} h_l(x,y) \frac{\sin \Omega_z(z - lz_0)}{\Omega_z(z - lz_0)} \tag{4.9}$$

where $z_0 = \pi/\Omega_z$ is the discretization step along the axis Z.

The function $h_l(x,y)$ describes the diffraction of the field in the plane that is separated from the diffractive element by a distance $l - z_0$. In accordance with formula (4.9), the medium is treated as a set of two-dimensional spatial filters that have transit characteristics of identical shape, depending on a variable parameter.

The field amplitude $I(x,y,z)$ is formed by a weighted summation of the outputs of different filters. Starting with the physics of the diffraction process, we assume that the first spatial filter selectively affects the corresponding segment of the diffractive element with its characteristic $t_l(u,v)$, that is, one of the non-overlapping segments is responsible for field formation in the plane $z = lz_0$. However, it is quite clear that if no special measures are taken to select the regions $t_l(u,v)$, the neighbouring regions will also affect one another in the formation of $I(x,y,z)$. Let us consider the conditions under which this mutual influence is small, and evaluate the number L in the expansion (4.9) using the analysis of Fresnel zone.

Consider a set of hypothetical zone plates having focal distances in the range from f_1 to f_n. A zone plate focusing radiation to a point f_1, in paraxial approximation has a ring of radius $r_m = (f_1 m \lambda)^{1/2}$, and that focusing radiation to a point f_n, a ring of radius $r_k = (f_n k \lambda)^{1/2}$.

The maximum radius of the ring must satisfy the condition $r_{max} \leq d$, whence we find the restriction on the maximum number of the ring, $M \leq d/f_n \lambda$.

We consider now two neighbouring points on the axis Oz: f and $f - \Delta f$, corresponding to two zone plates P and P' with the radii

$$r_m = (fm\lambda)^{1/2} \tag{4.10}$$

$$r'_m = [(f - \Delta f)m\lambda]^{1/2}. \tag{4.11}$$

We superpose the zone plates P and P' on to one another. For visual clarity, we consider them as amplitude-type plates with alternating transparent and opaque rings. If the condition $\Delta f/f_0 \ll 1$ is satisfied, their central parts are hardly different. However, as we move from the centre to the periphery, opaque rings of the zone plate P begin blocking the transparent rings of the plate P' more and more and finally, at a critical value $m = \mu$, when the condition

$$r_\mu \geq r_{\mu+1} \qquad (4.12)$$

is satisfied, a transparent ring of the zone plate P' will be completely covered with an opaque ring of the zone plate P.

We consider now a diffractive element composed of the two zone plates described above. The central part of the diffractive element will have μ rings with radii (4.10) corresponding to the zone plate P. The peripheral part has rings with radii (4.11) corresponding to the zone plate P'. In view of the arguments given above, we find the minimal achievable difference between the focal distances f and f' of the zone plates P and P' combined in a single element. From (4.10)–(4.12) we find

$$\Delta f \geq f/\mu. \qquad (4.13)$$

The parameter μ is determined for a fixed wavelength by the geometric parameters of the quasioptic system.

This analysis allows us to suggest a method of calculation of a diffractive element to focus radiation on to the axial segment $f_1 \leq z \leq f_n$ with uniformly distributed intensity. As we wish to generate uniform intensity distribution on the segment

$$I(x, y, z) = \begin{cases} 1, & f_1 \leq z \leq f_n \\ 0, & z < f_1, \quad z > f_n \end{cases}$$

it is necessary to choose the number of rings identical for all segments of the diffractive element. The central segment of the diffractive element will be a circle while the peripheral segments will be concentric rings of variable widths. With the problem formulated as above, the synthesis of the diffractive element reduces to finding the number of segments L (or the displacement of the focal distances Δf). Let us calculate them.

The maximum number M of the ring on a diffractive element will be found, according to (4.13), from the relation

$$M = d^2/f_n\lambda.$$

The number of segments on the diffractive element is then

$$L = \text{entier}[M/\mu]. \qquad (4.14)$$

where $\text{entier}(x)$ is the largest integer not greater than x. On the other hand, the value L can be found as a number of 'nodes' in the focusing zone using the formula

$$L = \text{entier}[(f_n - f_1)/\Delta f] \qquad (4.15)$$

where we set, according to (4.13), that $\Delta f = f/\mu$. Equating the right-hand sides of formula (4.14) and (4.15) and substituting into them the expressions for the parameters M and Δf, we obtain the formulas for finding the sought quantities:

$$\Delta f = f_n[\lambda(f_n - f_1)]^{1/2}/d \qquad (4.16)$$

$$L = \text{entier}[d(f_n - f_1)/\lambda]^{1/2}/f_n. \qquad (4.17)$$

Let us evaluate the mean cross dimension of the focusing area (in this case this is the thin cylinder or rather a cone). Applying the Rayleigh criterion, we find

$$\bar{\rho} = 1.22\bar{f}\lambda/2d_{\text{eff}}$$

where

$$\bar{f} = (f_n - f_1)/2, \qquad d_{\text{eff}} = d/L.$$

We can suggest a procedure for a discrete control of intensity on the axial segment (at L points). The number of segments is found from formula (4.17), and the number of rings on the lth segment μ_l is chosen to be proportional to a required value of intensity on the corresponding focal point.

4.4   Diffractive elements for analysing transversal radiation modes

Diffractive elements that made it possible to analyse the transversal mode composition of radiation are known as *modans* [21, 22]. The amplitude and phase modans are holograms with the transmission function that describes an irregular sinusoidal diffraction grating. A modan separates distinct transversal modes to different angles.

The transmission function of a multichannel modan, in which the splitting of modes is achieved by holographically recording each mode with its specific carrier spatial frequency, is in the general case an amplitude-phase function; to record it on to a purely phase-type medium, a method of coding is used that generates additional diffraction orders that reduce the resulting energy efficiency.

An iterative method of designing multichannel phase modans was considered in [22]; these modans form Hermitian beams in prescribed diffraction orders, with the predetermined energy distribution over diffraction orders.

4.5   Logarithmic axicons

We will consider the design of diffractive elements meant to focus radiation on to a linear segment [2, 23].

Let us consider a diffraction Fresnel–Kirchhoff integral on the surface of a diffractive element placed in plane G and focusing a wavefront $U_0(\rho, v)$ on to a linear segment along the optical axis (a focal segment):

$$U(r, \varphi) = e^{ikh}/(i\lambda h) \iint_G U_0(\rho, v) \exp ik[S + f(\rho)]\} \rho \, d\rho \, dv \qquad (4.18)$$

where (ρ, v) are the coordinates in the plane G; S is the distance between the arbitrary point P_0 in the plane G and a point P', $\exp[ikf(\rho)]$ is the transmission coefficient and $f(\rho)$ is the phase transmission function of the diffractive element.

For any point on the optical axis we have $r = 0$, $S = (\rho^2 + h^2)^{1/2}$, therefore

$$U(0, \varphi) = e^{ikh}/(i\lambda h) \iint_G U_0(\rho, v) \exp\{ik[(\rho^2 + h^2)^{1/2} + f(\rho)]\} \rho \, d\rho \, dv.$$

$$(4.19)$$

In contrast to the case of diffraction on apertures of a different shape and on lenses, the field amplitude at each point of a focal segment of a family of axicons is defined by a neighbourhood of the corresponding critical points located on a circle in plane G. This property allows us to use the stationary phase technique for integrating (4.19). Using this method, we obtain the following expressions for field amplitude (4.19) in the paraxial approximation:

$$S \approx h[1 + 0.5(\rho/h)^2]$$

$$U(0, \varphi) = ikU_0(\rho) \exp\{ik[f(\rho) + (\rho^2 + h^2)^{1/2}] + i\pi/4\} f'[\lambda/(f'' - f'/\rho)]^{1/2}.$$

$$(4.20)$$

In the general case, assuming $S = (\rho^2 + h^2)^{1/2}$, we find

$$U(0, \varphi) = ikU_0(\rho) \exp\{ik[f(\rho) + S] + i\pi/4\} f'[\lambda/(f'' - f'/\rho)(1 - f'^2)]^{1/2}.$$

$$(4.21)$$

Expression (4.21) was written taking into account the expansion

$$\rho = -f'h/(1 - f'^2)^{1/2} \approx -f'h(1 + f'^2/2 + \cdots).$$

Here and further on we assume that the incident wave has plane wavefront and its amplitude $U_0(\rho)$ depends on the radial coordinate only. A more general case is very easily analysed.

4.5.1 Axicons of the first type

We will find the transmission function of the A1 axicon that generates constant intensity along the entire length of a focal segment of the optical axis:

$$|U(0,\varphi)|^2 = \text{const} = a. \tag{4.22}$$

In the paraxial approximation we obtain the following differential equation

$$f''_1(\rho) - f'_1(\rho)/\rho - f'_1(\rho)U_0(\rho)/a = 0. \tag{4.23}$$

In the general case, (4.23) does not have an analytic solution; it must be solved numerically, taking into account the specific form of $U_0(\rho)$. In the particular case of $U_0(\rho) = \text{const}$ equation (4.23) defines the transmission function $f_1(\rho)$ up to constants a, b, c:

$$f_1(\rho) = -a\ln(c - \rho^2) + b.$$

The integration constants are determined by the geometry of the problem and by the following axicon parameters: the length of the focal segment and its position relative to the axicon (which also defines the internal and external diameters of the axicon), and the admissible changes in the diameter of the focal spot along the focal segment.

In the axicon plane G the transmission function $f(\rho)$ is the phase function of the wavefront and hence is related in a single-valued manner to the direction of rays: $f'(\rho) = -\sin\alpha$. The diameter d_0 of the focal spot is determined by the convergence angle α of the angles at the chosen point of the focal segment, $d_0 = 5/(k\sin\alpha) = 5/(kf')$ and, in the final count, by the derivative of the transmission function $f'(\rho)$.

Therefore, integration constants are found from the following two conditions:

$$f''_1(\rho_0) = 0 \quad \longrightarrow \quad a = -f'_{1,0}\rho \tag{4.24}$$

$$f'_1(\rho_0) = f_{1,0} \quad \longrightarrow \quad c = -\rho^2. \tag{4.25}$$

Finally, we obtain

$$f'_1 = f'_{1,0}\rho_0 \ln(\rho_0^2 + \rho^2) + b. \tag{4.26}$$

We will discuss below the physical meaning of the condition (4.24) and (4.25), and also of the constant b.

Let us find the transmission function of the A1 axicon in the general case, using equation (4.21) and condition (4.22)

$$f''_1(\rho) - f'_1(\rho)/\rho - f'_1(\rho)U_0(\rho)/a + (f'_1(\rho)/\rho)^3 = 0. \tag{4.27}$$

This equation allows only a numerical solution.

4.5.2 Axicons of the second type

The approach shown above allows the determination of transmission function of the A2 axicon which generates constant 'energy filling' in the transverse section of the focal segment.

Let intensity along the focal segment of the axis and the focal spot diameter d vary but in a way that maintains the power delivered by radiation to the focal spot constant:

$$\int_0^{d/2} I(r)r\,dr = \text{const for any } h. \tag{4.28}$$

Integral (4.28) can always be replaced with an expression of the type $I_0 S_{\text{eff}} = \text{const}$, where I_0 is the intensity on the axis, $S_{\text{eff}} \approx d^2 \text{const}/f' = \text{const}/f'^2$ is the effective area, and

$$I_0 = \text{const}\, f'^2. \tag{4.29}$$

In the paraxial approximation, we combine (4.29) and (4.20), and again arrive at a differential equation:

$$f_2'' - f_2'/\rho - U_0(\rho)/\alpha = 0.$$

An analytical solution exists in the case of $U_0(\rho) = \text{const}$,

$$f_2(\rho) = \rho^2/\alpha[\ln(\gamma\rho)/2 - 1/4] + \beta$$

where α, β, γ are integration constants determined by the conditions (4.24) and (4.25):

$$f_2''(\rho_0) = 0 \;\longrightarrow\; \alpha = \rho_0/f_{2,0}', \qquad f_2'(\rho_0) = f_{2,0}' \;\longrightarrow\; \gamma = 1/\rho_0.$$

The axicon transmission function in the paraxial approximation, provided $U_0(\rho) = \text{const}$, has the form

$$f_2(\rho) = -\rho^2 f_{2,0}'/4\rho_0[2\ln(\rho/\rho_0) - 3] + \beta. \tag{4.30}$$

In the general case, the differential equation of the transmission function $f_2(\rho)$ of the A2 axicon follows from equation (4.21) and the condition (4.29):

$$f_2''(\rho) - f_2'(\rho)/\rho + [f_2'(\rho)]^3/\rho - U_0(\rho)/\alpha = 0.$$

Let us now consider the field intensity distribution on the axicon's caustic. As follows from expression (4.18), the field in the focusing zone of an arbitrary axicon with the phase transmission function $f(\rho)$ is described by the expression

$$U(r,h) = 1/(\text{i}\lambda h) \int_0^{2\pi} \int_0^a U_0(\rho) \exp\{\text{i}k[h + (r^2 + \rho^2)/2h$$
$$- \rho r \cos \upsilon'/h + f(\rho)]\}\rho\,d\rho\,d\upsilon'$$

where $\upsilon' = \upsilon - \varphi$. It is easy to see that the main contribution to the integral υ' is given by two diametrically opposed critical points for which $\sin\varphi = 0$ ($\varphi = 0, \varphi = \pi$). We rewrite the integral over the angle in term of the Bessel function of order zero, J_0:

$$U(r,h) = \frac{k}{ih}\int_0^a U_0(\rho)\exp\{ik[h+(r^2+\rho^2)/2h+f(\rho)]\}J_0(k\rho r/h)\rho\,d\rho.$$

The radius ρ_s of the circle on which critical points are located is given by

$$\rho_s/h+f'(\rho_s)=0, \qquad \rho_s=-f'(\rho_s)h. \qquad (4.31)$$

Using the stationary phase method, we obtain

$$U(r,h) = -ikU_0(\rho)f'(\rho_s)J_0[krf'(\rho_s)]\{\lambda h/[1+hf''(\rho_s)]\}^{1/2}$$
$$\times \exp\{ik[h+(r^2+\rho_s^2)/2h+f(\rho_s)]+i\pi/4\}.$$

The intensity distribution on the caustic is described by the formula

$$I(r,h) = |U(r,h)|^2 = 2\pi k|U_0(\rho_s)|^2\{f'(\rho_s)J_0[krf(\rho_s)]\}^2h/[1+hf''(\rho_s)].$$

Let us analyse the expressions obtained for different types of axicons.

The transverse intensity distribution on the caustic is completely defined by the Bessel function of zero order $J_0(x)$ that has a maximum at the point $x=0$ and the first zero at a point $x=2.5$.

The focal spot diameter is found in an ordinary fashion:

$$kdf'(\rho_s)=5, \qquad d=5/[kf'(\rho_s)]. \qquad (4.32)$$

Taking into account (4.31) and (4.32), we readily see that the transversal size of the focusing zone varies along the optical axis. The nature of this variation follows from the specific shape of the function $f'(\rho)$. Therefore, the condition (4.24) fixes the position of the bottleneck on the focal segment of the optical axis while the condition (4.25) defines the transversal dimension of the bottleneck.

The traditional conical axicon possesses the transmission function

$$f_a = \varepsilon\rho \qquad (\varepsilon = \text{const})$$

and the diameter of the focal segment

$$d = 5/(k\varepsilon) = 5H/(kR)$$

where $2R$ is the axicon diameter and H is the length of the focal segment. The focusing zone is an H-long cylinder with diameter d_a, since d_a is independent of the coordinate h.

The focal spot diameter d of the axicons A1 and A2 is a function of h, since $f_1'(\rho) \neq \text{const}$, $f_2' \neq \text{const}$ and $\rho = -f'(\rho)h$. The focusing zone is a body of revolution, which is a paraboloid in the neighbourhood of the bottleneck.

4.5.3 Longitudinal distribution

In the case of a conical axicon and constant intensity of the initial beam $U_0(\rho) = \text{const}$, the intensity on the axis grows linearly along the focal segment and reaches maximum for $h = H$. If, however, the intensity of the original beam falls off from the centre to the periphery, for instance, following Gaussian shape, the intensity on the axis near the focal segment has a maximum whose position is dictated by the shape of the function $U_0(\rho)$.

The A2 axicon provides constant intensity on the axis along the entire focal segment. Note that the transmission function f_1 can be calculated for any prescribed intensity distribution $U_0(\rho)$ using equation (4.23) and the corresponding initial and boundary conditions. If, however, the transmission function and the field distribution $U_0(\rho)$ are not correlated, the intensity on the axis has a maximum inside the focal segment.

The A2 axicon is of interest because both the diameter of the focal segment and the intensity of radiation on the axis vary along the focusing line but the 'energy filling' (the intensity) of the focal spot remains constant. The intensity on the axis is proportional to squared derivative of the transmission function $I_0 = \mathrm{const} f_2'^2$, the maximum being located on the bottleneck of the focal segment.

If the transmission function f_2 and the original distribution $U_0(\rho)$ are not correlated, the condition of constant 'intensity' is violated.

The A1 and A2 axicons with logarithmic transmission functions can be fabricated as kinoforms. Let us look at how axicons are implemented in practice. First of all, it is necessary to determine the geometric parameters of the axicons on the basis of the prescribed parameters. Typically, the following parameters are given: the length of the focal segment L, the focal distance F_1, the maximum admissible kinoform radius R_{max}, and the focal spot radius $d_0/2$ at the bottleneck. From these data, the constant f_0' is calculated: $f_0' = 2.5\lambda/(\pi d)$. It is also necessary to fix the acceptable change in the diameter Δd of the focal spot along the axis. Knowing the admissible value of the derivatives f'_{max} and f'_{min} at the ends of the focal segment, we calculate the inner ρ_{min} and outer ρ_{max} radii of the working zone of the axicon:

$$\rho_{min} = F_1 \tan \alpha_{min}, \qquad \sin \alpha_{min} = f'_{min}$$

$$\rho_{max} = (F_1 + L) \tan \alpha_{max}, \qquad \sin \alpha_{max} = f'_{max}.$$

Therefore, all constants in the expression of the phase function, with one exception of B (or β), are defined. The constant B (β) does not affect the form of the phase function—this is an additional phase summand, a 'prop' that allows shifting of the origin of measuring the phase; this is necessary to create convenient enumeration of kinoform zones. As it is sufficient to define the phase of electromagnetic wave up to 2π, the continuous phase transmission function f can be modulated with the function Φ, which has discontinuities at the points $f(r) = 2\pi n$ ($n = 0, 1, 2, \ldots$); however, this function must equal the function $f - 2\pi n$ in each interval $2\pi n < f < 2\pi(n + 1)$, that is, $\Phi = f - 2\pi n$. By virtue of technological requirements, the continuous function Φ within each interval has to be replaced with a stepped function. The boundaries of the steps are given by the equation

$$kf(\rho) = 2\pi(m/N + n) \tag{4.33}$$

where $m = 0, 1, 2, \ldots, N - 1$ is the number of the step, N is the number of steps and $n = 0, \pm 1, \pm 2, \ldots$ is the number of the kinoform zone in which the step is located.

Zone boundaries are given by the equation (4.33) by using a sample in m, n. It is readily shown using equations (4.26) and (4.33) that the radii of the steps for the A1 axicon are given by the equation

$$\rho_{n,m} = \left((\rho_0^2 + \rho_{min}^2) \exp \left[\frac{2\pi(m + N_n)}{kNf'_0\rho} \right] - \rho_0^2 \right)^{1/2}$$

where $n = 0, 1, 2, \ldots, N$, $m = 0, 1, 2, \ldots, N - 1$ and N is an integer.

The constant B was chosen in view of $f(\rho_{min}) = 0$. The calculation of the boundaries of profile steps is fairly straightforward if the transmission function is described by differential equations. For instance, equation (4.27) for the case of $U_0(\rho) = 1$ is transformed to the form

$$\rho'' + \rho'^2/\rho - 1/\rho + \rho'/a^2 = 0. \tag{4.34}$$

The numerical solution of (4.34) for the values of f that follow from (4.34) with the initial conditions $\rho(0) = \rho_0$; $\rho'(0) = 1/f'_{1,0}$; $\rho''(0) = 0$ allows us to specify the required array of boundaries.

The phase transmissivity function of the A2 axicon is described by equation (4.30) and the boundaries of the profile steps—by the expression

$$\rho^2 \frac{f'_{2,0}}{4\rho_0} \left[3 - 2\ln\left(\frac{\rho}{\rho_0}\right) \right] - \rho_{min}^2 \frac{f'_{2,0}}{4\rho_0} \left[3 - 2\ln\left(\frac{\rho_{min}}{\rho_0}\right) \right] = \frac{2\pi}{k}\left(\frac{m}{N} + n\right).$$

Kotlyar *et al* [24] suggested a diffraction-based method of calculating the phase function of the focusator that focuses on to a longitudinal segment, with average deviations of 1–10% from the required distribution.

The method is based on the fact that the complex amplitude of the electromagnetic radiation behind the focusator is related to the complex amplitude of radiation on the optical axis by the Fourier transformation. To calculate the phase function, the interactive Gerhberg–Saxton algorithm [18] was suggested to solve the problem by consecutive approximation.

4.6   Two-orders diffractive elements

Golub *et al* [25] suggested a method of designing diffractive elements that concentrate radiation at a constant intensity on to a set of arbitrarily arranged straight segments and semicircles. The method is based on a consistent

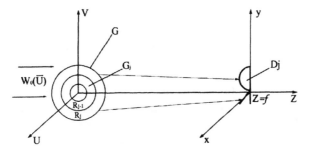

Figure 4.1. Geometry of the problem.

splitting of the element into concentric rings and takes into account the symmetry of the illuminating beam even if the eikonal function is asymmetric.

The image consists of N partial segments and L semicircles. Note that the diffractive element is split into segments in the form of concentric rings G_j of radii R_{0-1}, R_j that reflect the symmetry of the illuminating beam.

The presentation of an element as a set of rings G_j (figure 4.1), each of which focuses radiation on to an arbitrarily situated element D_j (a rectilinear segment or a semicircle), makes it possible to propose a universal method of computing the eikonal of the focusator element $\psi(u, v)$.

In focusing on to an element of the focal area, the eikonal $\psi_j(u, v)$ within an annular segment G_j is described by a set of parameters $Q = \{d_j, (x_j, y_j), \alpha_j\}$ and can be derived from the eikonal $\psi'_j(u, v)$ of the annular segment calculated for focusing with 'reference geometry', that is, on to an element with the parameters $Q'_j = \{d_j, (0,0), 0\}$.

Indeed, by rotating the coordinate system (u, v) by an angle α_j and adding functions linear in variables u and v that describe the displacement of the element of focusing relative to the origin of coordinates, we can obtain, in the paraxial approximation, that

$$\psi_j(u, v) = \psi'_j(u', v') + x_j u/f + y_j v/f$$

where

$$\begin{pmatrix} u' \\ v' \end{pmatrix} = \begin{pmatrix} \cos \alpha_j & \sin \alpha_j \\ -\sin \alpha_j & \cos \alpha_j \end{pmatrix} \begin{pmatrix} u' \\ v' \end{pmatrix}.$$

Therefore, splitting into rings makes the problem of calculating the eikonal of an annular segment invariant with respect to the geometry of the position of the corresponding focusing element.

Owing to the complexity of solving inverse focusing problems, the focusator phase functions were obtained only in the case of focusing on to simple focal elements, such as a rectilinear segment, an annulus, a semi-annulus, a ring, a rectangle etc. To focus on to more complicate shapes, segmented elements are used. The aperture of the focusing element is then split by one of the available methods into segments each of which serves to focus to a chosen focal element. Doskulovich *et al* [26] analysed diffractive elements that implement focusing to two focal elements without segmenting the aperture, through simultaneous formation of elements of the focal area in different orders of diffraction.

Let us see how this computation is carried out. It is assumed that the focusing area consists of two elements 1 and 2. The initial data for computing a two-orders diffractive element are the phase functions φ_a, φ_b that provide focusing of the illuminating beam to the focal elements 1 and 2, respectively. The phase function of such an element is defined as

$$F(\mathbf{u}) = 0.5[\varphi_a(\mathbf{u}) + \varphi_b(\mathbf{u})] + \Phi[h(\mathbf{u})], \qquad \mathbf{u} \in G \qquad (4.35)$$

where $\mathbf{u} = (u, v)$ are the Cartesian coordinates in the plane of the element, G is the aperture of the element, and

$$h(\mathbf{u}) = \mathrm{mod}_{2\pi}(0.5[\varphi_a(\mathbf{u}) - \varphi_b(\mathbf{u})]).$$

A two-order diffractive element (4.35) is given as a mathematical superposition of two elements: an element with the phase function $0.5[\varphi_a(\mathbf{u}) + \varphi_b(\mathbf{u})]$ and a zone plate $\Phi[h(\mathbf{u})]$ that corresponds to the results of nonlinear transformation $h(\mathbf{u})$, normalized to the interval $[0, 2\pi)$.

The height of the phase profile is proportional to the phase function (4.35), normalized to the interval $[0, 2\pi)$.

Let us find the form of nonlinearity of $\Phi[h]$. We expand the function $\exp(i\Phi[\varepsilon])$ into a Fourier series $[0, 2\pi)$,

$$\exp(i\Phi[\varepsilon]) = \sum_{n=-\infty}^{\infty} c_n \exp(in\varepsilon) \qquad (4.36)$$

where

$$c_n = 1/2 \int \exp(i\Phi[\varepsilon] - in\varepsilon)\, d\varepsilon \qquad (4.37)$$

are the Fourier coefficients.

Assuming $\varepsilon = h(\mathbf{u})$ in (4.36), we write the function of complex transmission of the element (4.35), taking into account the 2π-periodicity, in the form

$$\exp[iF(\mathbf{u})] = \sum_{n=-\infty}^{\infty} c_n \exp[0.5i(\varphi_a(\mathbf{u})(1+n) + \varphi_b(\mathbf{u})(1-n))]. \qquad (4.38)$$

According to (4.38), as the illuminating beam is diffracted by the element (4.35), many images (orders of diffraction) are formed. Each image can be interpreted as a result of transformation of the illuminating beam by an element with a phase function

$$\varphi_n(\mathbf{u}) = 0.5[\varphi_a(\mathbf{u})(1+n) + \varphi_b(\mathbf{u})(1-n)]. \qquad (4.39)$$

The required process of focusing is realized in the orders $+1$ and -1: $\varphi_1(\mathbf{u}) = \varphi_a(\mathbf{u})$, $\varphi_{-1}(\mathbf{u}) = \varphi_b(\mathbf{u})$. The fraction of the illuminating beam energy focused into each order of diffraction is proportional to the squared modulus of the corresponding Fourier coefficient (4.37). Therefore, the non-linearity $\Phi[\varepsilon]$ must be chosen to satisfy the condition of vanishing Fourier coefficients (4.37) for $n \neq \pm 1$.

This nonlinearity corresponds to the phase profile of the diffraction grating that concentrates radiation into the orders $+1$ and -1.

When focusing on to a shape with uniform intensity, the quotient $\gamma = |c_1|^2/|c_{-1}|^2$ must be proportional to the quotient of dimensions of the focal elements 1, 2. If $\gamma = 1$, $\Phi[\varepsilon]$ takes the form

$$\Phi[\varepsilon] = \begin{cases} 0, & \varepsilon \in [0, \pi) \\ \pi, & \varepsilon \in [\pi, 2\pi). \end{cases} \tag{4.40}$$

The Fourier coefficients are then given by the relation

$$C_n = \begin{cases} \varepsilon[1 - (-1)^n]/\pi i n, & n \neq 0 \\ 0, & n = 0. \end{cases} \tag{4.41}$$

The energy efficiency of the grating (4.40) is 81% ($|c_1|^2 + |c_{-1}|^2 = 0.81$). It is impossible to synthesize a two-order phase grating with 100% efficiency. Therefore, in addition to the required focusing, the element (4.35) also creates spurious images that correspond to nonzero coefficients C_n in expression (4.38) for $n \neq \pm 1$.

As an example, we consider a two-orders diffractive element designed to focus radiation on to a cross-shaped zone [26]. The phase function of a two-orders element for focusing on to a cross is described by a general equation (4.35) where a function $\Phi[\varepsilon]$ has the form (4.40), and the phase functions $\varphi_a(\mathbf{u})$ and $\varphi_b(\mathbf{u})$ are found from the condition of focusing on to the straight segments that form the cross.

Thus, with a square aperture G and a plane-wavefront illuminating beam, $\varphi_a(\mathbf{u})$ and $\varphi_b(\mathbf{u})$ have the form

$$\begin{aligned} \varphi_a(u, v) &= -k(u^2 + v^2)/2f + k\,du^2/c2f \\ \varphi_b(u, v) &= -k(u^2 + v^2)/2f + k\,dv^2/c2f \end{aligned} \tag{4.42}$$

where f is the focal distance from the element to the plane of focusing, d is the length of straight segments that form the cross, and c is the side length of the aperture.

Comparing the two-orders cross-forming element with segmented geometric optics focusators, we note that the segmentation of the element into two parts leads to unequal diffraction widths of the segments of the cross. For a two-orders element, the total aperture contributes to both segments of the cross and, therefore, the diffraction width of the cross is constant and equals $\lambda f/c$. The segmentation of the element into four parts results in the diffractive width of the cross being constant and twice as large as in the two-orders element. A two-orders element loses 19% of the energy of the illuminating beam to spurious orders of diffraction. However, as a result of the smaller diffraction width, the field intensity along the cross is on average greater by a factor of 1.6 than in a segmented element.

4.7   Diffractive elements for focusing radiation on to flat zones

In the preceding section we considered diffractive elements designed to focus radiation on to a focal curve. Golub *et al* [27] suggested a numerical method for computing elements designed for focusing radiation to complicated-shape planar objects.

A laser radiation beam is incident on to an element Φ with aperture G in plane $\mathbf{u} = (u, v)$; it has intensity $I_0(\mathbf{u})$ and eikonal $\psi_0(\mathbf{u})$, that is, its complex amplitude is

$$W_0(\mathbf{u}) = [I_0(\mathbf{u})]^{1/2} \exp[ik\psi_0(\mathbf{u})].$$

In what follows we assume that the area G corresponds in shape to the cross-section of the incident beam. The aim is to form a wave field in the zone D on the plane (x, y) with a prescribed intensity distribution I. The solution of the focusing problem reduces to finding the phase function of the element $\varphi(u)$ that would ensure the formation of the required wave field from the illuminating beam.

The geometric optics phase function of the element $\varphi(\mathbf{u})$ can be obtained from the paraxial approximation from the solution of the following set of equations:

$$\varphi(\mathbf{u}) = k[\psi(\mathbf{u}) - \psi_0(\mathbf{u})]$$
$$x = u + f\psi'_u(\mathbf{u})$$
$$y = v + f\psi'_v(\mathbf{u}) \tag{4.43}$$
$$I_0(\mathbf{u})/I(\mathbf{x}) = |x'_u y'_v - x'_v y'_u|$$
$$x'_u = y'_v.$$

Note that the geometric optics approach is valid for focusing on to a zone whose size exceeds that of the diffraction spot by a factor of 10–100. In the general case, the solution in the set (4.43) is a very difficult problem. Finding a solution to a focusing problem becomes appreciably simpler if the aperture G of the element and the focusing zone D are both rectangular and the functions are factorizable:

$$W_0(\mathbf{u}) = W_1(u)W_2(v), \qquad I(x) = I_1(x)I_2(y)$$

that is

$$I_0(\mathbf{u}) = I_{01}(u)I_{02}(v), \qquad \psi_0(\mathbf{u}) = \psi_{01}(u) + \psi_{02}(v).$$

In this case the solution for two one-dimensional problems makes it possible to find a two-dimensional phase function of the element, $\varphi(\mathbf{u}) = \varphi_1(u) + \varphi_2(v)$. The phase function $\varphi_1(u)$ of a one-dimensional (cylindrical) element that carries out the prescribed transformation of the wave beam is described by the following set of equations:

$$\varphi_1(u) = k[\psi_1(u) - \psi_{01}(u)]$$

$$x = u + f\psi_1/\mathrm{d}u \tag{4.44}$$

$$I_0(u)/I_1(x) = \mathrm{d}x/\mathrm{d}u$$

where $u_0 \le u \le u_1$ and $x_0 \le x \le x_1$.

Solving the set of equations (4.44) is considerably simpler than solving (4.43). For instance, in the case of

$$I_1(x) = \begin{cases} I_1, & \text{if } x_0 \le x \le x_1 \\ 0, & \text{otherwise} \end{cases}$$

the solution of the set (4.44) is

$$\varphi_1(u) = k\left(-u^2/2f + 1/f \int_{u_0}^{u} \left[1/I_1 \int_{u_0}^{\varepsilon} I_{01}(\eta)\,\mathrm{d}\eta - x_0\right]\mathrm{d}\varepsilon - \psi_{01}(u)\right).$$

Therefore, the two-dimensional phase function of an element with rectangular aperture, designed to focus radiation to a rectangular zone with constant intensity over the zone, has the form

$$\varphi(\mathbf{u}) = k(-u^2 + v^2)/2f + 1/f \int_{u_0}^{u} \left[1/I_1 \int_{u_0}^{\varepsilon} I_{01}(\eta)\,\mathrm{d}\eta - x_0\right]\mathrm{d}\varepsilon$$

$$+ 1/f \int_{u_0}^{u} \left[1/I_1 \int_{u_0}^{\varepsilon} I_{02}(\eta)\,\mathrm{d}\eta - y_0\right]\mathrm{d}\varepsilon + \psi_0(\mathbf{u}) \tag{4.45}$$

where (x_0, y_0) are the coordinates of the lower left corner of the focusing rectangle.

To design elements for arbitrary-shape zones D and G with factorizable illuminating beam intensity distribution function, the 'matched rectangles' method is used [27].

In this method, the regions G and D are approximated by sets of rectangles G_i and D_i, respectively, for $i = 1, N$ and the problem of focusing from G_i to D_i is then solved.

As an example, we consider the synthesis of an element with circular aperture that focuses a circular cross section illuminating beam with plane incident wavefront to a rectangle with constant intensity.

The intensity of the illuminating beam has the form:

$$I_0(\mathbf{u}) = \begin{cases} I_0, & \mathbf{u} \in G \\ 0, & \mathbf{u} \notin G. \end{cases}$$

Here $G = \{(u, v), |u^2 + v^2| \leq R^2\}$ is the aperture of the element and R is the radius of the illuminating beam.

The intensity in the focal region is

$$I(\mathbf{x}) = \begin{cases} I, & \mathbf{x} \in D \\ 0, & \text{otherwise} \end{cases}$$

where $D = [-b, b] \times [-a, a]$ is the area where the focal rectangle is located.

In this particular case the area G is approximated by a set of rectangles $G_i = [u_{i-1}, u_i] \times [(R^2 - u_{i-1}^2)^{1/2}, -(R^2 - u_{i-1}^2)^{1/2}]$, $i = 1, N$, $u_0 = -R$, $u_N = R$, and the area D by a set of rectangles $D_i = [x_{i-1}, x_i] \times [-a, a]$, $i = 1, N$, $x_0 = -b$, $x_N = b$.

On the basis of conservation of the energy flux in the propagation of radiation from the aperture rectangle G_i to the corresponding focal rectangle D_i, assuming that the aperture of the element G is constrained by the curves $V = g_1(u)$, $V = g_2(u)$ and straight lines $u = u_{min}$, $u = u_{max}$ (figure 4.2(a)), and the focusing zone D is constrained by the curves $y = f_1(x)$, $y = f_2(x)$ and straight lines $x = x_{min}$, $x = x_{max}$ (figure 4.2(b)), the coordinates of the segment $[x_{min}, x_{max}]$ are found from the solution of the following nonlinear recurrent equation:

$$\int_{u_{i-1}}^{u_i} \int_{g_1 u_{i-1}}^{g_2 u_{i-1}} I_0(\mathbf{u}) \, d^2\mathbf{u} = \int_{x_{i-1}}^{x_i} \int_{f_1 u_{i-1}}^{f_2 u_{i-1}} I(\mathbf{x}) \, d^2\mathbf{x}. \qquad (4.46)$$

If the beam is of circular cross-section, the solution of equation (4.46) that defines the splitting of the focal region is found in the form

$$x_i = x_{i-1} + [I_0(u_i - u_{i-1})(R^2 - u_{i-1}^2)^{1/2}]/I_0, \qquad i = 1, N - 1.$$

(a)   (b)

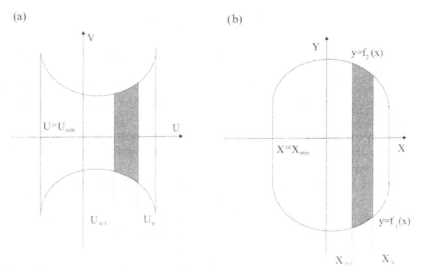

Figure 4.2. (a) Aperture of the element, (b) focusing zone.

The phase function $\varphi_i(u)$, which is a geometric-optics solution of the problem of focusing a rectangle G_i on to a rectangle D_i, can be easily obtained from (4.45) and contains terms quadratic and linear in (u, v).

Goncharsky *et al* [28] proposed a device for focusing monochromatic optical radiation on to a (uniformly illuminated) rectangle. The optical phase element is fabricated as a reflection-type zone plate; the profiles of its zones is given by the formula

$$z(u, v) = 0.5\left\{(u^2 + v^2)/2f\lambda - 1/f\lambda \int_0^u x(\varepsilon)\,d\varepsilon \right.$$

$$\left. - \frac{a\{(2\pi\sigma^2)^{1/2} v\,\mathrm{erf}[v/(2\sigma^2)^{1/2}] + 2\sigma^2 \exp(-v^2/2\sigma^2) - 2\sigma^2\}}{f(2\pi\sigma^2)^{1/2}\,\mathrm{erf}[(R^2 - u^2)/2\gamma^2]^{1/2}\lambda}\right\}\lambda$$

where $z(u, v)$ is the profile height at a point (u, v) of the optical element and (u, v) are the coordinates of a point in the OUV reference frame (the OZ axis is directed oppositely to the direction of incidence of the radiation); f is the focal length of the element and $x(\varepsilon)$ is the function given by the relation

$$x(\varepsilon) = \frac{2b}{(2\pi\sigma^2)^{1/2}[1 - \exp(-R^2/2\sigma^2)]}$$

$$\times \int_0^\varepsilon \exp(-u^2/2\sigma^2)\,\mathrm{erf}[(R^2 - u^2)/2\sigma^2]^{1/2}\,du.$$

Here σ is a parameter of the Gaussian distribution of intensity of the incident radiation in which the intensity at a distance a from the centre is proportional to $\exp(-r^2/2\sigma^2)$, R is the radius of the beam being focused, $2a$ and $2b$ are the sides of the rectangle $(b > a)$; the rectangle lies in the plane

$$z = f, \qquad \mathrm{erf}(x) = 2/(\pi)^{1/2} \int_0^x \exp(-t^2)\,dt.$$

The maximum height of the profile is $\lambda/2$.

4.8 Optimization algorithm for diffraction antennas

To correct for phase errors that manifest themselves as non-uniform phase front in the emitting equipment, an algorithm was developed that optimizes the beampattern of antennas both in the far and in the near zones. The algorithm can be used for both lens-type and mirror antennas, and also for lenses. We will consider it using lens-type antennas as an example [29].

Consider two surfaces S_1 (the surface of the diffractive antenna) and S_2 (arbitrarily chosen surface on the unfolded antenna). The Cartesian system of coordinates is at a point 0 (figure 4.3). The curves by which the plane $x0y$ intersects the surfaces S_1 and S_2 will be described as power-law

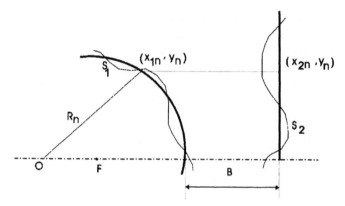

Figure 4.3. Clarification of the optimization algorithm.

polynomials

$$x_1 = a + a_1 y + a_2 y^2 + \cdots, \qquad x_2 = b + b_1 y + b_2 y^2 + \cdots .$$

We assume that the required phase distribution on the section of the surface S_2 can also be described by a polynomial of the type $\psi(y) = \psi_0 + \psi_1 y + \psi_2 y^2 + \cdots$. The source of the monochromatic radiation is placed on the axis x at a point F; its phase characteristic $\Phi(y)$ with respect to F is

$$\Phi(y) = \varphi_0 + \varphi_1 y + \varphi_2 y^2 + \cdots .$$

The optimization reduces to solving the inverse problem—finding the corrected values of the coordinates y_n of Fresnel zone boundaries on the surface of the diffractive antenna. The synchronism condition for the radiation can be written in the form

$$R_n + (x_{2n} - x_{1n}) + \frac{1}{k}(\Psi_n + \hat{O}_n) = R_0 + B + \frac{\lambda}{M(\sqrt{\varepsilon} - 1)} .$$

We readily see from the geometry of the problem that

$$R_n^2 = y_n^2 + x_{1n}^2 .$$

By combining these two equations it is not difficult to find that

$$y_n^2 = R_n^2 - \left[R_n - R_0 - B - \frac{\lambda}{M(\sqrt{\varepsilon} - 1)} + \frac{1}{k}(\Psi_n + \hat{O}_n) + x_{2n} \right]^2 . \qquad (4.47)$$

The following notation is used here: k is the wave number, $k = 2\pi/\lambda$, R_n is the distance from the origin of coordinates to the point with coordinates (x_{1n}, y_n) on the section of the surface S_1, M is the number of phase quantization levels, ε is the dielectric constant of the antenna material, B is the distance along the x axis from the apex of the antenna to the plane of the antenna aperture, and R_0 is the distance from the origin of coordinates to the apex on the surface of the antenna along the x axis.

Therefore, by solving equation (4.47) we find corrected positions of Fresnel zone boundaries y_n for each zone. This takes into account both the phase characteristics of the irradiator and the non-uniformity in the radiation field phase distribution within the aperture of the diffractive antenna.

We must emphasize that it would not be difficult, using a similar algorithm, to design focusing elements with the prescribed wavefront phase distribution in the aperture of the antenna.

4.9   Optimization of parameters of the 'classical' zone plate

Analysis of the properties of diffractive antennas based on zone plates always followed the recipe of geometric-optics calculation of Fresnel zone boundaries, and the optimality of the basic principle (the length of rays from zone boundaries being an integral multiple of half-wavelength (for a binary element)) was never put in doubt.

However, is the Fresnel zone plate optimal as far as its characteristics are concerned? In connection with this question we will discuss the results of some numerical experiments that the authors conducted.

4.9.1   'Compression' of zone plate [30]

Consider a Fresnel–Soret lens-type zone plate (for simplicity we choose a two-level phase inversion zone plate that transforms a plane wavefront into a spherical one). The Fresnel zone boundaries are found using the classical method. Let us now run the following experiment: we 'compress' the zone plate radially, that is, the new Fresnel zone boundaries are found from the relation

$$R_n^{\text{new}} = kR_n, \qquad R_n = \sqrt{n\lambda F + \frac{n^2\lambda^2}{4}}$$

where k is a scale coefficient, $k < 1$. How will this modify the focusing properties of the 'new' zone plate?

In what follows we give the results of a numerical experiment with a zone plate with the following parameters: the lens aperture $D/\lambda = 30$, the aperture ratio $D/F = 1.1$, the number of the Fresnel zones within the aperture $N = 8$, the material of the zone plate—a dielectric with the refractive index $n = 2$ and the loss tangent 10^{-3}. The plate was irradiated through a horn with a radiation intensity decreasing by $-12\,\text{dB}$ at the edge of the aperture.

First the coefficient k was varied and the gain of the zone plate was calculated. It was found that for the parameters given above the optimal value k is not unity (as in the classical case) but equals $k = 0.985$. All the

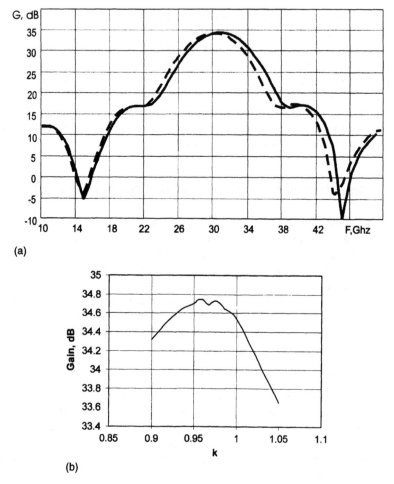

(a)

(b)

Figure 4.4. (a) Directivity of the zone plate as a function of frequency. (b) Directivity of the zone plate as a function of the coefficient k. © 2002 IEEE. Reprinted, with permission, from [30].

basic properties of the 'deformed' zone plate were then investigated for this value of the scale coefficient.

Figure 4.4(a) plots the directivity of the zone plate in the frequency domain. In this and the subsequent figures the dashed curve presents a similar curve for the 'classical' zone plate. The directivity maximum shifts by 1 GHz and grows in magnitude by 0.15 dB. Figure 4.4(b) plots directivity as a function of the coefficient k.

Figure 4.5(a) plots the efficiency of using the area of the zone plate aperture as a function of frequency. We see that the gain is about 1.5% at the reference frequency (30 GHz). The maximum of efficiency also suffers a

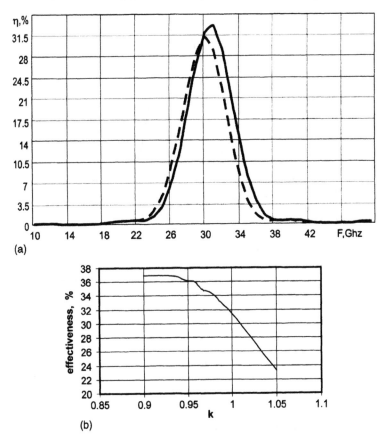

Figure 4.5. (a) The efficiency of using the aperture. (b) The efficiency of using the aperture as a function of the coefficient k. © 2002 IEEE. Reprinted, with permission, from [30].

frequency shift. This shift is easily interpreted: compression of Fresnel zones changes the spatial frequency of zone boundaries, thereby creating the corresponding changes in the frequency domain. This fact is confirmed by figure 4.5(b) which shows the efficiency of using the aperture as a function of the coefficient k.

Figure 4.6 compares the results of modelling of the beampattern. In the case of the 'deformed' zone plate, the level of the first scattering side lobes decreases and the first minima of the beampattern 'degenerate'.

Similar results were obtained for the binary (amplitude-type) Fresnel zone plate. The results are summarized in table 4.1 (showing a comparison with the corresponding 'classical' zone plate).

Summarizing the results above we can suggest an optimization algorithm to improve the properties of antennas and lenses based on zone plates: via optimization of the parameter k for each Fresnel zone.

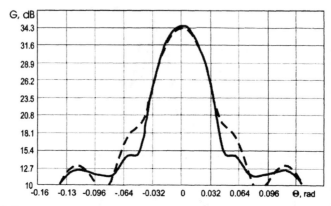

Figure 4.6. Beampattern.

Table 4.1.

Zone plate type	Coefficient k	Gain (dB)	Level of the first side-lobe maximum (dB)	Coefficient of aperture utilization (%)
Amplitude	0.98	+0.2	−1	1.2
Phase-inversion	0.985	+0.15	−1	1.5

4.9.2 Effect of distance to the screen

The simplest two-level mirror-type zone plate is an amplitude-type Fresnel zone plate with a reflecting screen placed at a distance of one-quarter of wavelength from the plane of the zone plate. We will consider the gain by such a zone plate depending on the distance to the screen (figure 4.7(a)). It is found that the distance to the screen of one-quarter of wavelength is not optimal for sufficiently short-focal zone plates.

Figure 4.7(b) plots the signal amplitude at the focal point (in relative units) as a function of distance to the screen (in units of wavelength) for a zone plate with $F = D$, for three values of the ratio $D/\lambda = 10$, 20, 70.

As the aperture ratio increases, the optimal distance between the zone plate and the screen tends asymptotically to the classical value. At the same time, however, this effect is fairly small from the standpoint of optimization of the parameters of a mirror-type zone plate: the gain increases by not more than 0.35 dB and insignificant broadening of the main scattering lobe is observed, plus a small (about 0.8 dB) decrease in the first side lobe of scattering.

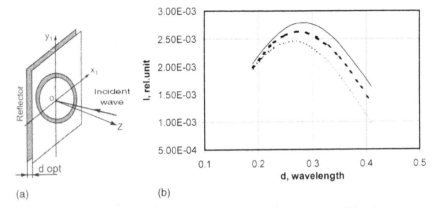

Figure 4.7. (a) Schematic diagram of optimization for a mirror-type diffractive antenna. (b) The signal amplitude at the focal point as a function of distance to the screen for a zone plate with $D/\lambda = 10$ (solid curve), 20 (bold dashes), 70 (thin dashes).

4.10   Optimal zone plate in the Fresnel approximation

It was shown above that an introduction of the scale coefficient for Fresnel zone radii results in improved focusing properties of a zone plate. This behaviour is conveniently interpreted using the Huygens–Fresnel principle. In the arguments below, we follow the results published by Zhang [32, 33]. The following factor has to be taken into account when analysing the results given below: the condition of far field has not been met in the papers quoted above, so that the results obtained are incorrect as far as quantitative estimates are concerned. These results are nevertheless of interest because they provide a qualitative indication that a binary zone plate, built according to laws of geometrical optics, is not an optical one.

Let us consider the diffraction of a plane wave reaching the screen in the Fresnel approximation. In this case it will be convenient to use the concept of Cornu's spiral to analyse the synchronization conditions for waves at the focal point:

$$FI = \int_0^w e^{-j\pi t^2/2}\,dt = C(w) - jS(w)$$

$$C(w) = \int_0^w \cos(\pi t^2)\,dt, \qquad S(w) = \int_0^w \sin(\pi t^2)\,dt.$$

Here $C(w)$, $S(w)$ are the Fresnel integrals and the argument (w) is related to the zone plate parameters in the following manner:

$$w = 2\sqrt{(r-F)/\lambda} = 2\sqrt{(F/\lambda)\left[\sqrt{1 + (\rho/F)^2} - 1\right]}$$

$$(\rho/\lambda) = (w/2)\sqrt{(w/2)^2 + 2(F/\lambda)}.$$

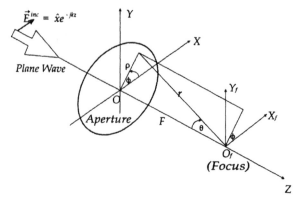

Figure 4.8. Geometry of the problem (with permission of Professor Wen-Xun Zhang).

In these expressions ρ is the vector radius at the plane of the screen, and r is the distance from the current point in the plane of the screen to the observation point (to the focal point) (see figure 4.8).

In the complex plane, the Fresnel integrals $C(w) + jS(w)$ are graphically presented as a Cornu's spiral (figure 4.9).

If Fresnel zone boundaries are found in the classical approximation, the boundary of the first zone corresponds to the coordinates $w = \sqrt{2} = 1.4142$. It is obvious at the same time that the field amplitude is a function of first zone size. The maximum field amplitude is reached at the point of intersection of the Cornu's spiral with the straight line drawn from the origin of coordinates at an angle of 45 degrees of arc (for a two-level zone plate) which corresponds to the value of the parameter $w = 1.2653$. Therefore, the optimal size of the first zone is smaller than the classical one. Similar argument can be applied to every other zone.

Zhang [33] gives the Fresnel zone radii values for zone plates designed for the classical approach and for the approach suggested in [32]. Using these data, the ratio of Fresnel zone radii for the optimal and classical zone plates were plotted as a function of the zone number (figure 4.10).

An analysis of the focusing properties of the optimal zone plate showed [33] that the gain in gain for an amplitude-type binary zone plate is 0.97 dB, and increases to 2.11 dB for a two-level plate. As the number of phase quantization levels increases further to 4, the gain in the total gain decreases to about 0.19 dB.

4.11   The general principle of designing optimal diffraction lenses and antennas

The principle of designing optimal (from the standpoint of Fresnel zone boundaries) zone plates as presented above is a result of Fresnel

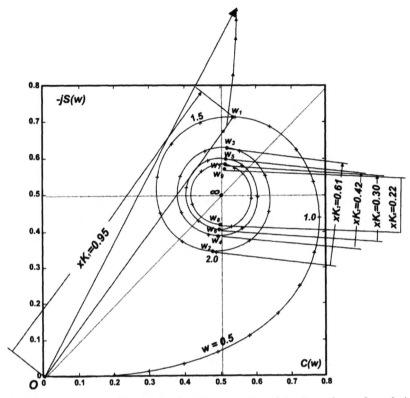

Figure 4.9. Cornu's spiral in the classical determination of the Fresnel zone boundaries (with permission of Professor Wen-Xun Zhang).

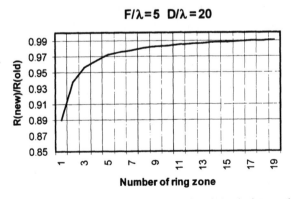

Figure 4.10. Ratio of Fresnel zone radii of the optimal and classical zone plates as a function of the zone number. The parameters of the zone plates are identical and indicated above the plot.

approximation calculations of parameters of Cornu's spiral. There is, however, a different approach to designing arbitrary diffractive elements (lenses and antennas fabricated on an arbitrary surface) which is more general.

We know that for a given geometry of a problem (e.g. the position of the source and receiver of radiation, known shape of the surface of the diffractive element, prescribed pattern of aperture irradiation etc.), the behaviour of the diffraction integral along the generatrix of the element's surface is described by a rapidly oscillated function. Hence, by calculating the diffraction integral along the generatrix of the surface of the diffractive element, the Fresnel zone boundaries will be placed at points that correspond to extremums of the diffraction integral.

This approach permits optimal determination of the structure of Fresnel zone boundaries on the surface of a diffractive element. We consider the implementation of this approach later in chapter 5 using as an example omnidirectional antennas of the millimetre wave range.

4.11.1 Zone-plate-based conformal antenna

Various approaches are possible to receiving signals simultaneously from several sources (e.g. in Spain, two different antennas are used to receive signals from ASTRA and HISPASAT satellites separated in the orbit by an angle of 52 degrees of arc). One of the approaches is to use the scanning properties of diffractive antennas—to use the same antennas with several receivers placed at different points in space. Note that when an antenna is based on an ordinary zone plate, radiation receivers must be protected from the elements of environment (figure 4.11(a)) while if the receivers are located on an arbitrary surface, the antenna itself provides a protective cover [31].

Another approach to this problem is the procedure of segmentation of the antenna's aperture: the total area of the antenna is divided into several subzones, each of which is an independent antenna. This approach was used, for example, in [34] to create two- and three-beam antennas for receiving satellite signals. Figure 4.11(b) shows a segmented diffractive antenna that formats two beams separated by 6 degrees of arc. This figure clearly shows that the aperture of the antenna is divided into three parts: the central part is a complete zone plate while the two symmetric elements on the sides are two fragments of another zone plates. Obviously, since the area of each zone plate that forms the same beam is smaller than the area of the entire antenna, the total area of the antenna is not used efficiently for each of the beams.

Another approach to designing a flat reflective antenna capable of forming simultaneously two beams at a high angular separation uses the concept of conformal Fresnel zones [34, 35].

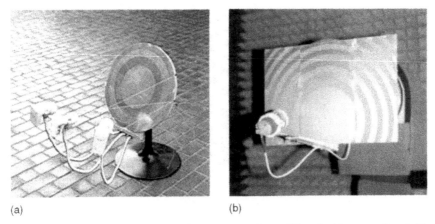

(a)   (b)

Figure 4.11 (a) An example of a three-beam antenna for satellite television reception (with kind permission of Takafumi Hoashi). (b) Two-beam diffractive antenna with segmented aperture (with kind permission of J Vassal'lo).

We know that the conventional axisymmetric zone plate with circular zones forms a beampattern perpendicular to the reflecting surface. A zone plate with elliptic zones forms a beampattern at an angle to the reflecting surface. If we now 'juxtapose' the amplitude masks of two elliptic zone plates, new areas of jointly formed or conformal zones will appear in the area of intersection of transparent and opaque zones. Figure 4.12 shows

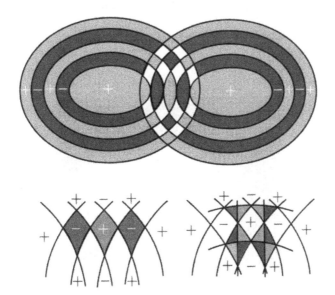

Figure 4.12. The structure of conformal zones (with permission of J Vassal'lo).

Figure 4.13. Amplitude mask of conformal zones (with permission of J Vassal'lo).

such a composite zone plate that joins two zone plates with elliptic zones. Obviously, the new zone plate obtained by this operation generates simultaneously two different beampatterns.

Mathematically, we have two equations for Fresnel zone boundaries [35]. The first equation describes circles:

$$x^2 + y^2 = h(\lambda/2)n\cos\alpha + (n\lambda/4)^2 - [h\sin(\alpha)]^2.$$

The second equation describes parallel straight lines:

$$x = n\lambda/[4\sin(\alpha)].$$

Here λ is the reference wavelength of the radiation, n is the number of Fresnel zones, h is the focal length of the flat reflector and α is the separation angle between beampatterns.

As a result, the amplitude mask of the conformal zone plate is formed; it is shown in figure 4.13.

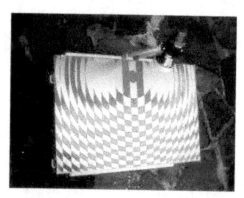

Figure 4.14. A conformal zone plate that forms two beampatterns separated in space by an angle of $\alpha = \pm25°$ [34] (reprinted with permission of J Vassal'lo).

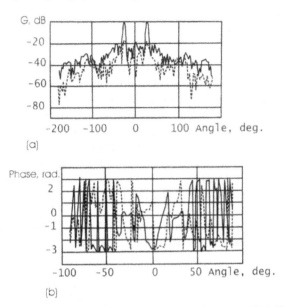

Figure 4.15. (a) Beampattern in the radial plane at a frequency 12.0 GHz [34]. (b) A fragment of the phase characteristic of the 12.0 GHz antenna in the radial plane. The dashed curves show cross-polarization characteristics. (Reprinted with kind permission of J Vassal'lo.)

Pedreira and Vassal'lo [34] published results of their investigation of the focusing properties of a two-beam conformal diffraction antenna in the frequency range around 12 GHz; the antenna was designed using the method described above. The appearance of the antenna that forms two beams at an angle $\alpha = \pm 25°$ is shown in figure 4.14.

Figure 4.15(a) shows the beampattern in the radial plane of a 12.5 GHz antenna, and figure 4.15(b) gives the corresponding wave phase distribution.

The level of the measured side lobes was about -18 dB, the level of cross-polarization was -14 dB. Measurements of the gain of the antenna in 1.5 GHz frequency band for the beam of the antenna tilted at 25 degrees of arc showed that gain varied between 19.19 and 22.16 dB.

Table 4.2 shows the level of the central side lobe (corresponding to zero angle) as a function of radiation frequency.

Table 4.2.

Frequency (GHz)	10.5	11.5	12.5
G (dB)	-13	-18	-21

4.12   Application of holographic principles of diffractive antenna design

The holographic approach can also be used to design diffractive antennas. In this case the simplest zone plate can be interpreted as a hologram of a point-like source. Sazonov *et al* [36] considered the feasibility of designing antennas for receiving satellite TV signals using holographic technique.

The first version of the antenna was an axisymmetric lens-type zone plate tilted at a certain angle to the ground surface; the antenna dimensions were 1.5 × 1.5 m, and frequency was 12.5 GHz. The remaining dimensions are shown in figure 4.16(a).

The results of modelling the focusing properties of such an antenna are plotted in figure 4.16(b).

The level of side lobes did not exceed −15 dB; if the beam deviated by as much as 10 degrees of arc, the level of side lobes changed insignificantly, and

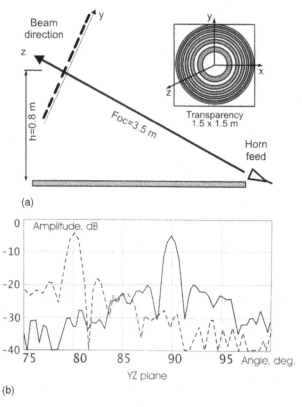

Figure 4.16. (a) Schematic diagram of an axisymmetric holographic antenna [36]. (b) The beampattern of axisymmetric holographic antenna [36] (reprinted with permission of D Sazonov).

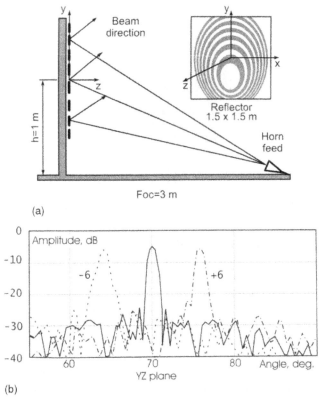

Figure 4.17. (a) Schematic diagram of a reflection-type holographic antenna [36]. (b) The beampattern of the holographic direction-type antenna [36] (reprinted with permission of D Sazonov).

so did gain. Such an antenna is an equivalent of a parabolic reflector of about 1 m in diameter.

It is also possible to fabricate reflective holographic antennas with off-axis position of the focal point (figure 4.17(a)).

The level of side lobes of this antenna is −18 dB for the same geometric dimensions of the antenna (figure 4.17(b)).

Although the level of side maxima grows insignificantly in angular scanning, the width of the beampattern varies appreciably. Efficiency-wise, this antenna is equivalent to a parabolic reflector 1.2 m in diameter.

4.13 Microwave antenna with flat diffractive reflector

In a number of cases, using for a reflector a multilayer diffraction lattice instead of a parabolic mirror may reduce the mass of the antenna and its

Figure 4.18. Diffractive antenna (reprinted with permission of Prof. V I Yudin [37]).

overall dimensions. However, the process of manufacturing such antennas is rather complicated (their irradiating aperture is fabricated of multilayer dielectrics on to which conducting elements of complex shapes are deposited) and the cost is quite high.

Merkulov *et al* [37] suggested using a reflector with waveguide grooves of rectangular cross section, filled with a homogeneous dielectric. The design of the antenna with 13×13 grooves for $10.9\,\text{GHz}$ is illustrated in figure 4.18.

The groove depth is varied in such a way that the phase front of electromagnetic waves reflected by the diffraction grating is as planar as technically possible. Since the electromagnetic field is characterized by a considerable number of oscillation modes that behave as standing waves, the amplitude of the lowest waveguide mode H_{10} is the largest; the input impedance of the H_{10} wave for each groove of depth h can be found using the formula

$$Z_{\text{input}} = \mathrm{j} W_{H_{10}} \tan\left(\frac{2\pi}{\lambda_{H_{10}}} h\right)$$

where

$$W_{H_{10}} = \frac{W_0}{\sqrt{\varepsilon_r - [(\lambda_0/2)a_v]^2}}$$

is the impedance for the H_{10} wave in a $a_v \times a_v$ wide square waveguide filled with dielectric having dielectric permittivity ε_r, and

$$\lambda_{H_{10}} = \frac{\lambda_0}{\sqrt{\varepsilon_r - [(\lambda_0/2)a_v]^2}}$$

is the wavelength of the H_{10} wave in this specific waveguide.

The directivity of the antenna in the $X0Z$ and $Y0Z$ planes is found from the product of the directivity pattern $D_{H_{10}}(\varphi, \theta)$ of the short-circuited wave-guide cell and the grating factor $D_{AP}(\varphi, \theta)$:

$$D_{\Sigma}(\varphi, \theta) = D_{H_{10}}(\varphi, \theta) D_{AP}(\varphi, \theta)$$

where

$$D_{H_{10}}(\varphi, \theta) = \left[\left(\cos(\varphi) + \frac{k_v}{k_0} \frac{1 - |\rho|}{1 + |\rho|} \right) \frac{\cos\left(\frac{k_0 a}{2} \sin(\varphi) \right)}{1 - \left(\frac{2a}{\lambda_0} \sin(\varphi) \right)^2} \right]$$

$$\times \left[\left(1 + \frac{k_v}{k_0} \frac{1 - |\rho|}{1 + |\rho|} \cos(\theta) \right) \frac{\sin\left(\frac{k_0 a}{2} \sin(\theta) \right)}{\frac{k_0 a}{2} \sin(\theta)} \right]$$

$$D_{AP}(\varphi, \theta) = \sum_{i=1}^{N_i} \sum_{n=1}^{N_n} E(x_i, y_n)$$

$$\times \exp[j(\Delta\Phi_{i,n} + k_0 a((i - 1) \sin(\varphi) + (n - 1) \sin(\theta)))]$$

where k_v and k_0 are the wave numbers for the groove and free space, respectively; ρ is the reflection coefficient at the open end of the waveguide; a is the diffraction grating period; N_i and N_n are the number of periods in the diffraction gratings along the axes x and y, respectively; $\Delta\Phi$ is the value of the phase error that depends on the frequency detuning relative to the central frequency.

4.14 The possibility of dispersion distortion correction of femtosecond pulses by choosing surface shape of diffractive optic element

The difference between the phase and group velocities in propagation of femtosecond pulses through a focusing element (lens) results in a time delay between the pulse and phase wavefronts [38]. We will note that while considering the problems concerned with propagation of these pulses, one should take into account their short duration and finiteness of the light velocity value. The pulse front means [38] a surface coinciding with the pulse peak at a fixed instant of time. The time delay in lenses is caused by material dispersion [39] whereas in diffractive optical elements (DOEs) it is a result of their chromatic properties. One of the methods for partial correction of dispersion distortion of femtosecond pulses will be a 'diffractive optical element + dielectric lens' combination [40, 41]. In the present paragraph, we consider only a focusing DOE.

In paper [38], using Fermat's principle in the paraxial approximation it is shown that the delay between the phase and pulse wavefronts for DOE (amplitude zone plate) can be calculated as

$$\Delta T(r) = \frac{r_0^2 - r^2}{2cf^2} \lambda \frac{df}{d\lambda} \tag{4.48}$$

where r is the radial coordinate in the plane of the zone plate, r_0 is the zone plate radius, f' is the focal length, and $df/d\lambda$ is the longitudinal chromatic aberration of the zone plate. In (4.48), the delay for the outer beam was equal to zero $(r = r_0)$.

In the present paper, it is also shown that for typical values of the longitudinal chromatic aberration in optical systems the pulse front leads spatially the phase front by 2–20% of the distance passed in the dispersion medium. These results in distortion of the pulse form in the focusing region and for, femtosecond pulses, one cannot neglect these great delays [38].

Let us consider propagation of femtosecond pulses through a plane DOE (zone plate).

The geometry of beam transmission through a diffractive optical element, which is shown in Figure 19(a), is such that for the axial beam that passes the minimal length (equal to the focal length) the delay between the pulse and phase wavefronts is maximal whereas for the outer beam passing the maximal length the time delay is zero.

The spatial dependence of the time delay between the phase and pulse wavefronts is the ratio of path difference between the outer and arbitrary beams passing through the zone plate to light velocity. This follows from the geometry of the problem. Proceeding from this, the delay between the pulse and phase wavefronts will be defined as

$$\Delta T(r) = -\frac{L(r_0) - L(r)}{c} = -\frac{\sqrt{f^2 + r_0^2} - \sqrt{f^2 + r^2}}{c}. \tag{4.49}$$

Here $L(r_0)$, $L(r)$ are the optical lengths of beams from the point of geometrical focus to the edge of DOE aperture and the point with the coordinate r, respectively.

In a paraxial approximation, for a plane incident wave and for $r_0 \ll f$ expression (4.49) is transformed to the known relationship [38]

$$\Delta T(r) = -\frac{r_0^2 - r^2}{c\left(\sqrt{f^2 + r_0^2} + \sqrt{f^2 + r^2}\right)} \approx -\frac{r_0^2 - r^2}{2fc}. \tag{4.50}$$

Here it was taken into account that the Fresnel zone radii p_n is determined from the formula

$$\rho_n \approx \sqrt{n\lambda f} \qquad \left(\text{i.e. } \frac{df}{d\lambda} \approx -\frac{f}{\lambda}\right).$$

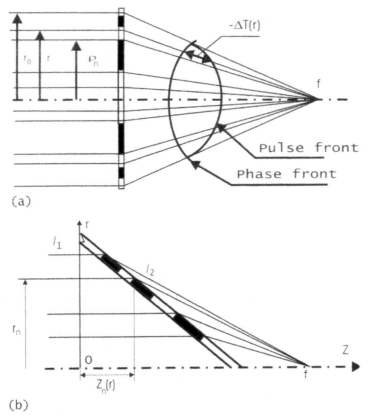

Figure 4.19. Pulse transmission through an amplitude focusing diffractive optical element: (a) on a plane surface and (b) on a conic surface (the reference radius is equal to zero).

One of the methods for delay correction will be the use of diffractive optics on nonplanar surfaces [41].

Really, following the geometrical calculation of DOE fabricated on an arbitrary surface of rotation [42], we may write (figure 4.19(b))

$$l_1 + l_2 = l_3 + n\lambda/2 \tag{4.51}$$

where l_1 is the optical path of a beam parallel to the optic axis and passing through the boundary of the nth Fresnel zone, l_2 is the optical path of the same beam refracted on the DOE surface and directed to the focus, l_3 is the optical path of the reference beam (usually $l_3 = f$), n is the ordinal number of Fresnel zone, and λ is the radiation wavelength. Then

$$z_n(r) + \sqrt{r_n^2 + [f - z_n(r)]^2} = f + n\frac{\lambda}{2}. \tag{4.52}$$

It follows from (4.52) that the dependence of the focal length of the non-planar DOE on the wavelength is [42]

$$f(\lambda) = z_n(r) + \frac{r_n^2 - [n(\lambda/2)]^2}{n\lambda} \tag{4.53}$$

where $z_n(r)$ is the z-axis of the edge of the nth Fresnel zone on the DOE surface, and r_n is the projection of $z_n(r)$ on to the plane XOY.

Differentiating (4.52) with respect to λ, we define the chromatic aberration for the nonplanar DOE as

$$\frac{df}{d\lambda} = -\left(\frac{f(\lambda) - z_n(r)}{\lambda} + \frac{n}{4}\right).$$

In view of (4.48) in a paraxial approximation we may write

$$\Delta T(r) = -\frac{r_0^2 - r^2}{2cf^2}\left(f(\lambda) - z_n(r) + \frac{n\lambda}{4}\right) \approx -\frac{r_0^2 - r^2}{2f(\lambda)c}\left(1 - \frac{z_n(r)}{f(\lambda)}\right). \tag{4.54}$$

Comparing (4.54) and (4.50), we see that by choosing the shape of DOE surface and its space orientation, it is possible to correct the value of delay between the wave and pulse fronts.

It is obvious that for any kind of diffractive optical element surface the minimal optical path corresponds to the central beam with the coordinate $r = 0$, $L(0) = f$; and the maximal optical path corresponds to the outer beam with the coordinate $r = r_0$ and $L(r_0) = z(r_0) + \sqrt{[f - z(r_0)]^2 + r_0^2}$.

The minimal optical path for the outer beam corresponds to the straight line: $L_{min}(r_0) = \sqrt{f^2 + r_0^2}$. From this it follows that the diffractive optical element surface should be convex towards the focus. Hence, it is impossible to minimize the delay at the point $r = 0$ for any kind of DOE surface by less than $\Delta T(0)_{min} = (\sqrt{f^2 + r_0^2} - f)/c$.

At the same time, from this it follows that for DOE with a non-transparent central zone the minimal delay will be

$$\Delta T_n(0)_{min} = \frac{\sqrt{f^2 + r_0^2} - \sqrt{f^2 + \lambda f}}{c}.$$

Let us define the optimal profile of the diffractive optical element surface [43]. Write down the expression for the optical path of the beam with the arbitrary coordinate r (see figure 4.19) to the point f:

$$L(r) = z(r) + \sqrt{[f - z(r)]^2 + r_0^2}$$

or

$$L(r) = f - \frac{r}{\tan[\varphi(r)]} + \frac{r}{\sin[\varphi(r)]} = f + r\frac{1 - \cos[\varphi(r)]}{\sin[\varphi(r)]} \tag{4.55}$$

where $z(r)$ is the arbitrary surface of rotation, and $\varphi(r)$ is the angle at which the beam arrives at the focus. The propagation time delay is determined from the formula

$$\Delta T(r) = \frac{L(r_0) - L(r)}{c}. \tag{4.56}$$

Upon differentiating (4.56) with respect to r, in view of (4.55), we will find extremes of function from the condition $d[\Delta T(r)]/dr = 0$. After transformations, obtain

$$1 + \frac{r\varphi'(r)}{\sin[\varphi(r)]} = 0. \tag{4.57}$$

Dividing the variables $d\varphi/\sin(\varphi) = -dr/r$ and integrating the obtained expression, in view of $\cos(\varphi) + 1 \geq 0$ and $\cos(\varphi) - 1 \leq 0$ for any φ we obtain

$$\varphi(r) = \arccos\left(\frac{Cr^2 - 1}{Cr^2 + 1}\right) \tag{4.58}$$

where

$$C = \frac{1}{r_0^2} \frac{1 + \cos(\varphi_0)}{1 - \cos(\varphi_0)}$$

is the constant and $\varphi_0 = \mathrm{arctg}(r_0/f)$.

Accordingly, the shape of the optimal DOE surface $z(r)$ will be defined by

$$z(r) = f - \tan[\varphi(r)] = f - \frac{r}{f}. \tag{4.59}$$

It is easily seen that (4.59) imposes constraint on the relative DOE aperture: $\sqrt{D/2} < f$ (D is the diameter). In other words, in the general case, we are noable to attain full correction of the delay between the wavefront and the pulse front. However, it is possible to decrease or correct the delay value. Let us show [41] that expression (4.58) is transformed to $x^2 + y^2 = (z - f)^2/(f/r_0)^2$ and that this corresponds to the canonical equation of cone of the second order with the vertex at the point (x_0, y_0, z_0):

$$\frac{(x - x_0)^2}{a^2} + \frac{(y - y_0)^2}{b^2} - \frac{(z - z_0)^2}{c^2} = 0$$

where $x_0 = y_0 = 0$, $a = b = 1$, $z_0 = f$, $c = f/r_0$.

Therefore, the diffractive optical element surface for different values of the relative aperture is a cone with the vertex at the focal point.

To demonstrate the partial correction of the effect of delay between the pulse front and the wavefront, figure 4.20 shows values of delay for plane and conic DOEs plotted against radial coordinate. Parameters of the diffractive optical element: radiation wavelength 3 mm, diameter 30 mm, and focal length $f = 0.7D$.

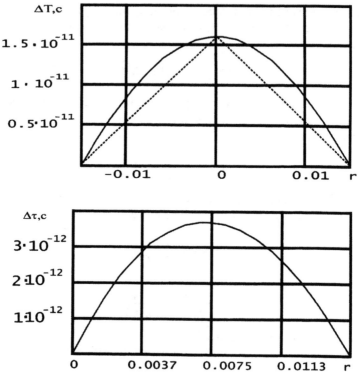

Figure 4.20. Delay between the phase and pulse wavefronts: $\Delta T(r)$ is the time delay for a plane diffractive optical element (solid curve) and a conic element (dashed curve). The bottom plot shows the difference $\Delta \tau(r)$ between the delays $\Delta T(r)$ for the plane and conic DOEs in radius (a half of the plot is shown due to its symmetry).

From analysis of the dependences we see that by fabricating DOE on a conic surface one can reduce the delay between the pulse front and the wavefront by 20–40%.

Thus, by using a nonplanar DOE surface, one can reduce the value of dispersion distortion of femtosecond pulses compared with DOE on a plane surface. In so doing, the optimal shape of DOE surface is conic. It is important factor for the radiovision and tomographic systems based on diffractive optic elements.

Bibliography

[1] Danilov V A, Ionov V V, Prokhorov A M *et al* 1982 'Synthesis of optical elements generating a focal line of arbitrary shape' *Pisma v ZhTF* **8**(13) 810–815
[2] Minin I V and Minin O V 1992 *Diffractive Quasioptics* (Moscow: Research and Production Association 'InformTEI') p 180

[3] Engel A, Steffen J and Herziger G 1974 'Laser machining with modulated zone plates' *Appl. Opt.* **13**(12) 269–273

[4] Goncharsky A V, Danilov V A, Popov V V *et al* 1983 'A solution of the inverse problem of focusing laser radiation onto an arbitrary curve' *DAN USSR* **273**(3) 605–608

[5] Goncharsky A V, Sisakyan I N and Stepanov V V 1984 'On solvability of some inverse problems of focusing laser radiation' *DAN USSR* **279**(1), 68–71

[6] Goncharsky A V 1987 'Mathematical foundations for the problems of synthesizing flat optical elements' *Kompyuternaya Optika* **1** pp 19–31

[7] Goncharsky A V 1987 'Inverse problems of synthesizing optical elements' in *Ill-posed Problems in Science* (Moscow: MGU) pp 275–300

[8] Goncharsky A V, Danolov V A, Popov V V *et al* 1986 'On one problem of synthesizing optical elements' *DAN USSR* **291**(3) 591–595

[9] Goncharsky A V and Stepanov V V 1988 'On one inverse problem of formatting coherent radiation' *DAN USSR* **303**(3) 279–283

[10] Vorontsov M A, Matveev A N and Sivokon' V P 1987 'On calculating radiation focusator in the diffractive approximation' *Kompyuternaya Optika* **1** 74–78

[11] Goncharsky A V, Klibanov M V and Stepanov V V 1988 'Inverse problems in coherent optics. Using the stationary phase techniques in problems of radiation focusing onto a curve' *Zhurnal Vichislitelnoi Matematiki i Matematicheskoi Fiziki* **28**(10) 1540–1550

[12] Babenko S M, Markin A S and Tusnov Yu I 1988 'Numerical modeling of elements of flat optics' (Moscow) p 19. Deposited with *VINITI*, No. 3424, vol 88

[13] Goncharsky A V and Stepanov V V 1986 'Inverse problems in coherent optics. Focusing onto a line' *Zhurnal Vichislitelnoi Matematiki i Matematicheskoi Fiziki* **26**(1) 80–91

[14] Sasikyan I N and Soyfer V A 1987 'Computer optics. Achievements and prospects' *Kompyuternaya Optika* **1** 5–19

[15] Born M and Wolf E 1999 *Principles of Optics* 7th edn (Cambridge University Press)

[16] Goncharsky A V, Morozova G N and Shemkov O Z 1992 'On one problem of focusing laser radiation' *Kvantovaya Electronika* **19**(6) 584–586

[17] Kotlyar V V and Filippkov S V 1993 'Formatters of wavefronts' *Pisma v ZhTF* **19**(18) 5–9

[18] Gerhberg R W and Saxton W D 1972 *Optik* **35**(2) 237–246

[19] Minin I V and Minin O V 1991 'A principle of building elements of diffractive quasioptics for focusing coherent radiation onto an arbitrary spatial configuration' *Proceedings of the First Ukrainian Symposium 'Physics and Technology of mm and sub-mm Radio Waves', Kharkov*, 15–17 October, Part 1, pp 14–15

[20] Soifer V A 1988 'On the computation of a focusator to a coaxial linear segment' in *Optical Recording and Processing of Information* (Kuibyshev: KuAI) pp 45–52 (in Russian)

[21] Sasikyan I N and Soifer V A 1989 'Modans—optical elements for analysing and formatting the transverse mode composition of laser radiation' *Kompyuternaya Optika* **4** 3–9

[22] Kotlyar V V, Nikolsky I V and Soifer V A 'Phase formatters of Hermitian modes in diffraction orders' *Pisma v ZhTF* **19**(20) 20–23

[23] Palchikova I G 1986 'Synthesis of phase structure of kinoform axicons' Novosibirsk: IAE SB AN USSR Preprint, No. 328

[24] Kotlyar V V, Soifer V A and Khonina S N 1991 'Diffraction-based computation of focusators for a longitudinal segment' *Pisma v ZhTF* **17**(24) 63–66

[25] Golub M A, Doskolovich L L, Kazansky N L and Kharitonov S I 1992 'Focusing of radiation to contours composed of straight and curvilinear segments' *Kompyuternaya Optika* **12** 3–8

[26] Doskolovich L L, Kazansky N L and Soifer V A 1993 'Designing two-diffraction-orders focusators' *Avtometriya* **1** 58–63

[27] Golub M A, Doskolovich L L, Kazansky N L *et al* 1992 'Method of matched rectangles for designing focusators to flat areas' *Kompyuternaya Optika* **10/11** 100–110

[28] Goncharsky A V, Danilov V A, Popov V V *et al* 1987 'Device for focusing optical radiation onto a rectangular uniformly illuminated area' Author's Certificate (USSR) No. 1314291 of 26.09.84, published 30.05.87, Bull. No. 20

[29] Minin I V and Minin O V 2002 'New horizons in diffraction quasioptics' *Kompyuternaya Optika* **23** 32–37

[30] Minin I V and Minin O V 2002 'Diffractional lenses and mirror antennas for mm-waves applications' in *6th Russian–Korean Int. Symp. on Science and Technology, Novosibirsk, Russia*, 24–30 June, vol 2, pp 347–350

[31] Minin I V and Minin O V 2001 'Curvilinear diffractional lenses and antennas for mm-wave communications and wireless' in *Proc. Int. Conf. on Electromagnetics and Communications ICECOM-2001, Dubrovnik, Croatia*, 1–3 October, pp 155–158

[32] Zhang W 2000 'The optimal design of printed Fresnel zone plate structure' in *Proc. AP-2000 Millennium Conf. on Antennas and Propagation, Davos, Switzerland*, 9–14 April

[33] Zhang W 1999 'An improved zoning rule of Fresnel zone plate' *Microwave and Opt. Tech. Lett.* pp 69–73

[34] Pedreira A and Vassal'lo J 2000 'Conformed beams using Fresnel's zone flat reflectors' in *Proc. 8th COST 260 Meeting on Smart Antennas: CAD and Technology, Rennes (France)*, October

[35] Pedreira A and Vassal'lo J 1999 Spanish Patent solicitude No. 200001599

[36] Sazonov D, Mishoustin B, Sergeev V, Korenev I and Remizov O 1998 'Radio-holographic antennas for dbs-reception' in *Proc. XXVIII Moscow Int. Conf. on Antenna Theory and Technology, Moscow, Russia*, 22–24 September, pp 403–406

[36] Merkulov K B, Ostankov A V, Pasternak Yu G, Sherstyuk O I and Yudin V I 2002 'Microwave reflector antenna with flat diffractive reflector for telecommunications and satellite TV broadcasting' *Zhurnal Telekommunikatsii* **11** 25–28

[38] Bor Z 1989 *Opt. Lett.* **14** 119

[39] Horvath Z L, Benko Zc, Koracs A P *et al* 1993 *Opt. Eng.* **32**(10) 2491

[40] Piestun R and Miller D A B 2001 *Opt. Lett.* **26**(17) 1373

[41] Minin I V, Minin O V and Lunev A V 2003 *Proc. Russian Sci.-Techn. Conf. on Science, Industry, Defence, NGTU, Novosibirsk*, p 53

[42] Minin I V and Minin O V 1999 *Diffractive Quasioptics and its Applications* (Novosibirsk: SibAGS) (in Russian)

[43] Minin I V and Minin O V 2003 *Proc. Russian Sci.-Techn. Conf. on the First Rdultov Reading* (St Petersburg: BGTU) p 37

Chapter 5

Microwave-range diffractive antennas

Multiple-beam antennas are defined as systems for emission or reception of signals, capable of forming several independent beams with a single aperture. The creation of new scanning antennas in the millimetre and sub-millimetre wavelength ranges requires overcoming technological difficulties and developing novel electronic elements.

The main difficulty in designing multiple-beam antennas with reflectors is how to provide wide-angle scanning or, in other words, how to compensate for the drop in the gain when the irradiator is moved out of the focal position. This problem is solved by using the toroidal mirror [1], by correcting the wavefront of the emitters [2] or by correcting the shape of mirrors [3, 4]. These and other technical solutions make it possible to space the beams by several tens of degrees. Scanning antennas with movable emitters can be designed on the basis of a system of reflectors of multibeam antennas [5].

Lens antennas differ from other types of antennas in simplicity of design, extensive possibilities of formatting the beampatterns of various types, including multi-beam antennas that make it possible to implement a wider scanning sector. These are used much less frequently, in spite of certain advantages. The main reason for this is their relatively high weight and considerable losses of microwave power in the dielectric.

Lens antennas are aperture antennas of optic type. In the general case, a lens antenna consists of an irradiator and a lens. An irradiator must have the phase centre coinciding with the lens' focal point, and must format the beampattern for a required amplitude distribution on the emitting surface and create minimal power loss to energy 'spilling' over lens edges.

As follows from ray-optics principles, such antennas can be used in the entire frequency band from infinite frequency where geometrical optics calculations give exact results, to the longest wavelength where the accuracy of the approximation is the limit imposed by the system's designer.

This broadbandedness of the system includes another degree of freedom that can be used to obtain additional advantages. The loss of quality will then be only that the system will work optimally in a narrower frequency band.

This principle can be transferred to designing reflective and refractive surfaces so as to greatly expand the ranges in working with irradiators displaced relative to the optical axis of the system.

In general, a real lens or mirror must be free of coma-type aberrations to satisfy this last requirement. In addition, manufacturing reflective antennas as zoned mirrors makes possible a reduction of the longitudinal size of the antenna.

Application of diffractive optical elements as antennas allows the engineer to manufacture them on an arbitrary surface: flat, spherical, parabolic, conical etc.

When designing antennas that are meant, for instance, for the reception of satellite TV signals, the following fundamental problems must be solved: how to achieve the minimum possible level of side lobes of the beampattern, increase the coefficient of usage of the aperture area, reduce the overall dimensions, achieve rapid retuning from one correspondent to another and ensure the possibility of flexible design of the external appearance of the antenna.

Small parabolic antennas manufactured in large runs are typically made by die casting of polymer plastics into moulds with subsequent metal coating of antenna's surface. Pressure forming is also used. For instance, the Gründig parabolic antenna [6] was die cast and reinforced with glass fibres. The surface of the plastic is coated with three layers: ground-coat paint, then electrically conducting nickel paint covered with a protective layer of lacquer. The metalized coat was painted over with a matt light-absorbing paint to protect the receiver head from heating when reflecting and focusing light.

Parabolic antennas offered by Channel Master are manufactured from polyester (analogue of lavsan) reinforced with glass fibres. Antennas can be installed on the ground or on the roof or wall of a house. A powder coating safely protects the antenna and its mounting from corrosion. Precise homing on the satellite within its radio visibility is achieved by a rotating support, that is, the entire antenna aperture is rotated.

The shortcomings of the parabolic antennas are: large size, high cost, high wind load, and the need for sufficiently strong design of mounting. Considerable losses are encountered due to the blockage of the aperture by the irradiator and mounting elements and also due to the complexity of simultaneous operation with several subscribers.

Comparative characteristics of the parabolic and diffraction antennas are summarized in table 5.1.

As for the comparative characteristics of beampatterns of the parabolic and diffractive antennas, the following important aspects must be mentioned:

Table 5.1.

Parameter	Parabolic antenna	Diffractive antenna
Optical schematic diagram	Focal point in front of the antenna	Focal point at the rear of the antenna
Blockage	Yes	No
Shape of surface	Fixed, parabolic	arbitrary
Material	Metal	Dielectric
Electromagnetic compatibility	Low	High
Noise immunity	Low	High (the antenna is a frequency filter)
Precision of surface machining	$\pm\lambda/32$	$(\pm\lambda/5\ldots\pm\lambda/10)$
Frequency band	Wide: from 0 to f	Variable
Need for cowling (radome)	Cowling required	Cowling not required
Multi-beam mode	Constrained	Possible in the $\pm(15°\text{–}30°)$ range
Homing to the user	By rotating the entire antenna	Only the receiver moves
Demands on the rotating support mechanism	Increase as antenna diameter increases	Mild

- When a beam in a parabolic antenna is tilted by moving the irradiator, the zeros of the beampattern 'smear over', the main scattering lobe is broadened, the side lobes grow significantly and the gain diminishes.
- The situation is somewhat different with diffractive antennas. Both the width of the beampattern and the amplification change insignificantly while the level of side lobes increases much slower than in the case of the parabolic antenna.

Flat antennas are developed as an alternative to parabolic antennas. These are essentially stripline antenna arrays. The advantages of flat antennas are: compact design, light weight, easy handling and simple installation on house walls. Such antennas readily comply with the interior design of living spaces, both structurally and esthetically, have high manufacturability (printed circuits technology can be used) etc.

Shortcomings of flat antennas are: lower electric parameters in comparison with reflector antennas (for example, lower gain, higher side lobes etc.); the need for tuning and phasing during service life; relatively high level of side lobes; the problem of designing a ferrite gate for electronic control of beampattern has not been solved; high total cost.

An alternative to smooth reflector and flat antennas can be diffractive antennas of lens and reflector types. Such antennas attract interest owing to their structural and maintenance advantages and also their low cost in mass production. Furthermore, asymmetric design of diffractive antennas permits the designer to get rid of aperture blockage by shifting the irradiator.

Another promising but not yet sufficiently developed approach to designing multibeam antennas and scanning systems is the use of electric and magnetic fields to change the refractive coefficient and other optical properties of certain materials.

For instance, an antenna design was suggested in [7] with electronically controlled beam scanning using a controlling lens consisting of phase-shifting sections. Each of the sections contains a cavity with dielectric walls filled with metal particles suspended in liquid dielectric. The antenna is a horn with a correcting lens installed in its aperture. The correcting lens serves to create nearly in-phase distribution in the horn within the aperture. When controlling voltage is fed to the electrodes, uniform electric field forms between them, and metal particles get oriented along its lines of force. The relative dielectric permittivity of the suspension of metal particles then increases and the beam is deflected.

An antenna with electrically controlled beam direction was suggested in [8]; this device contains an irradiator, a unit for controlling and shaping the beampattern designed as a system of magnetized ferrite rods connected through an electromagnetic field and placed within a radiowave-transparent dielectric. The artificial dielectric produced in this way has a variable refractive index depending on the delay caused by each ferrite rod. By changing the electric field, one controls the delay coefficient of each ferrite rod and thus changes the refractive index of the material of the lens. As a result, the beam direction of the beampattern is tilted by a required angle.

Hajian *et al* [9] considered a non-mechanical deflector for fast scanning by a millimetre-wave range beam. The deflector includes a semiconductor within which spatial charge density is controlled by optical irradiation of the semiconductor. It is suggested to project a circular or cylindrical Fresnel zone plate on to the semiconductor. The scanning by the beampattern in the far zone is suggested by displacing the transparency sheet.

As a scanning device, the electronic beam scanner of electromagnetic microwave radiation suggested in [10] can be used. This device can be applied in various radio systems for millimetre and sub-millimetre wavelength ranges by scanning two coordinates in the plane of the aperture. The device contains a rectangular dielectric waveguide whose end faces are tilted, with a metal conducting grid at one endface, placed perpendicularly to the polarization plane of the linearly polarized wave. The wires of the one-dimensional polarized wire grid are arranged with every second line connected and with a distance between two wires about 0.05 of the working wavelength of the waveguide. Metal grids are coated with a ferro-electric layer. This electronic scanning device operates in the following manner. The beam of electromagnetic microwave radiation is directed

normally to the tilted face of the rectangular waveguide. The angle between
the end face and the base of the waveguide is made to be greater than critical
so that the condition of total internal reflection is always satisfied for reflections from the waveguide base. When voltage is applied to the metal grid, the
dielectric permittivity of the ferroelectric coating changes. Consequently, the
displacement of the beam in the plane of the output face of the dielectric
waveguide is proportional to the voltage applied to the metal grid. The use
of zero-inertia ferroelectric coatings makes it possible to increase the
working frequency range to 10 MHz.

The antenna suggested in [11] is based on the travelling wave antenna and
uses electric scanning of the beam based. The travelling wave antenna contains
the feed waveguide and a driving stage. A dielectric waveguide and re-emitting
ridged periodic array are mounted coaxially with this stage. The driving stage
is of horn shape. The dielectric waveguide is a circular-section ferrite rod at
whose free end face conical endcap is installed, and a control coil with U-
shaped magnetic waveguide is installed parallel to the axis of the dielectric
waveguide. One end of this magnetic waveguide is connected to the driving
stage and the other to the conical endcap, while the driving stage and the
conical cap are fabricated of magnetically soft material.

The electromagnetic wave arriving from the feeder waveguide through
the driving stage generates a surface wave in the dielectric waveguide. The
surface wave propagating along the dielectric waveguide successively drives
elements of the re-emitting ridged periodic grid with a phase shift that is
determined by the period of the re-emitting grid and the phase velocity
of the surface wave. By changing the magnetization field and hence the
phase velocity of the surface wave, the phase front of the wave formed by
the re-emitting grid is controlled. Therefore, the deviation angle of the
antenna beam at a fixed frequency is proportional to the magnitude of the
magnetization field.

Scanning devices frequently used in optoelectronic systems [12] can also
be applied in quasioptical systems for the millimetre and sub-millimetre
wavelength ranges: various scanning mirrors, mirror prisms, pyramids,
optical wedges etc.

5.1 Flat antennas

Diffractive antennas find their widest applications as sufficiently efficient
and simple microwave antennas. For instance, a study of a two-level phase
inversion zone plate with aperture ratio equal to unity, $93\lambda_0$ in diameter
and with $N = 24$ zones at a frequency of 140 GHz showed very low level
of side lobes in the **E** and **H** planes of the beampattern [13, 14].

For instance, 45 side lobes were found in the **H** plane, each of
which was 30 dB lower than the main lobe. Of this, 35 side lobes are in a

sector at $\pm 45°$ from the central direction. The level of the first side lobes was $-17.6\,\text{dB}$.

At the edge of the field of view (20°) the distance from the centre of the zone plate to the point where the image was formed varied by not more than 2% of the focal distance. When the source of radiation was displaced by 10° from the normal to the zone plate, the level of the signal dropped by approximately 3 dB, with attenuation increasing to 6 dB for 20° and 10 dB for 32°. The frequency band of this diffractive 'flat' antenna at the 3 dB level was about 15% of the main frequency.

The diffractive antenna analysed in [15–17] is an amplitude-type Fresnel zone plate designed for focusing radiation to an off-axis focal point. In this case the diffractive antenna is a zone plate with zones of elliptic shape. The diffractive efficiency of this antenna is at most 10%.

5.2   Reflector antenna on flat surface

Let us list the main relations describing the geometry of a reflector flat antenna [18]. Consider a flat surface S perpendicular to the axis z and located at a distance f from a point F referred to as the focal point (figure 5.1). Using the techniques of geometrical optics we divide the surface S into zones in such a way that a total optical path length from F to boundary points A_n of each zone and to the plane S_0 satisfy the relation

$$r_n + l_n = r_0 + l_0 + n\lambda/M, \qquad n = 1, 2, 3, \ldots, N$$

where n is the number of the zone, λ is the wavelength and M is a positive integer. The projection of these zones to a plane forms annular regions. The optical path difference for extreme points in each region is then λ/M.

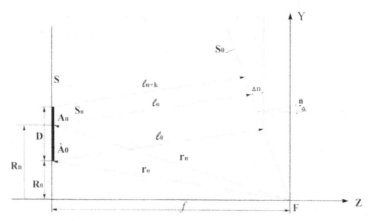

Figure 5.1. A diagram for calculations for a diffractive antenna with off-axis position of the focal point [18].

Taking into account that

$$l_0 = l_n + \Delta_n, \qquad \Delta_n = (R_n - R_0)\sin\gamma$$

where $\sin\gamma = \sin\Theta_0 \sin\varphi$, and φ is the azimuthal angle, we can find Fresnel zone radii R_n from the equation

$$R_n^2 \cos^2\gamma - 2R_n\varepsilon \sin\gamma + f^2 - \varepsilon^2 = 0$$

where f is the focal distance $\varepsilon = (r_0 - R_0\sin\gamma + n\lambda/M)$, $M = 2, 3, \ldots$, and Θ and φ are the angles of the spherical coordinate system.

Consider an ideally conducting screen S_n in a plane S (figure 5.1), irradiated by a monochromatic electromagnetic field from a source located at a point F. The field induces currents in the screen S_n, such that the phases of the secondary fields created by these currents at extreme points of each region on the plane S_0 differ by $2\pi/M$. If the normal n to the plane S_0 determines the required direction of the main maximum of the beampattern of an asymmetric flat antenna, it is necessary to synchronize the secondary fields in the neighbouring zones in plane S_0. The most efficient method for this is the one in which the areas of the screen S_n corresponding to these zones are displaced along the focal axis z by a distance d_n according to the rule

$$d_n = \frac{(n - 1 - M\,\text{entier}\{(n-1)/M\})\lambda}{M(1 + \cos\Theta_0)}$$

where $\text{entier}\{(n-1)/M\}$ is the integral part of $\{(n-1)/M\}$.

Calculations showed [18] that when the diffractive antenna ($f/D = 0.7$, $D = 15\lambda$) is irradiated with a plane wave, the focal point shifts relative to the geometrical focal point towards the bisecting line of the irradiation angle by $(1-1.5)\lambda$ in the direction of the mirror. The size of the focal spot for the diffractive antenna is on average smaller by 10% than that for a parabolic antenna.

It was also shown [18] that to obtain maximum values of field amplitude at the focal point and maximum gain, the optimal ratio f/D of asymmetric flat antennas must be about 0.65–0.70. The width of the beampattern at the -3 dB level is 2.5° at $M = 2$, while for the parabolic antenna this quantity is 2.7°. The conclusion made in [18] was that an optimal choice of parameters of asymmetric diffractive flat antennas makes it possible to achieve gains very close to those for parabolic antennas, at a cost of insignificant increase in the side lobe level.

An improved design of flat diffractive antennas was suggested in [19]. The antenna contains a flat metallic reflector and several concentric metallic circles located in one plane parallel to the reflector at a distance $\lambda/4$. In other words, this antenna is an analogue of a Rayleigh–Wood zone plate. The improvement consists in the absence of any complicated three-dimensional surface of the antenna with profile depth on the order of radiation wavelength.

Table 5.2.

Parameter	Calculated	Measured
Gain (dB)	41	35
Width of the beampattern at the half power level (degrees)	1.5	1.6
Level of side lobes (dB)	−19	−19
Cross polarization level (dB)	–	−25

An experimental study of a diffractive flat reflector antenna fabricated using the circuit board technique was conducted by Huder and Menzel [20]. The antenna consisted of metal rings placed on a dielectric layer $\lambda/4\varepsilon^{1/2}$ thick. Behind the dielectric layer a flat metal screen was added. The antenna parameters were: diameter 125 mm, focal length 80 mm and nominal radiation frequency 94 GHz; a material having $\varepsilon = 2.2$ was chosen as the dielectric. The calculated and experimental parameters of the antenna obtained when the aperture was irradiated at the −10 dB level are listed in table 5.2 [20]. The antenna reflection losses were −15 dB. The total loss in the radiation source was 1 dB, and the losses in side lobes were 4 dB. Losses in the dielectric and annular zones result in reducing the efficiency by 25%.

Two-level design of the antenna (diffractive efficiency ~40%) and a dielectric with sufficiently high dielectric permittivity result in reduced antenna gain.

Shortcomings of this sort can be removed by designing diffractive antennas with four levels of phase quantization, which yields nominal diffractive efficiency of 81%. As the number of phase quantization levels increases further, the actual efficiency of the antenna increases slowly while technological difficulties grow quite steeply. The resulting zone-plate-based antenna can be fabricated by using the technology of multilayer printed circuit boards (figure 5.2(a)).

Foil-coated glass fibre Textolite can be replaced with metal-coated Lavsan film; the thickness of the reflective layer must exceed the skin layer thickness at the given frequency. Glass fibre Textolite, foam polystyrene, and styrofoam can be used for a filler. It is necessary to try to minimize losses to reflection at the dielectric–air interface, that is, to minimize the dielectric permittivity of the filler. An antenna design is possible in which a material with variable dielectric permittivity is used as a filler. In this case, each preceding layer can be chosen from the condition of minimized reflection of radiation from non-metalized parts of the antenna (figure 5.2(b)).

Let us evaluate the tolerances to which the flat surface of the diffractive element must be fabricated and their effect on the antenna's parameters. We assume that each metal-coated zone (a ring) may be located off the plane of the diffractive element, with a certain random shift along the optical axis. In

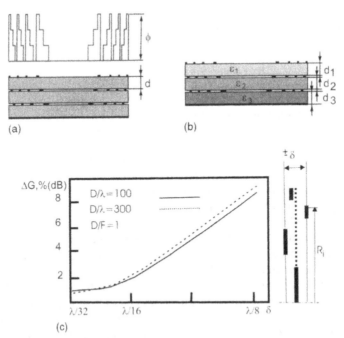

Figure 5.2. (a) Multilayer flat diffractive antenna. (b) Multilayer diffractive antenna with minimized reflection coefficient. (c) Effect of accuracy of fabrication of the antenna surface on antenna's gain.

other words, the plane of each annular zone does not coincide with the antenna's plane but is 'displaced' relative to this plane by a random amount. We will calculate the Fresnel–Kirchhoff diffraction integral. The displacement of each Fresnel zone along the optical axis can be simulated by the Monte-Carlo method. Calculations determined the gain of the antenna.

Figure 5.2(c) shows the results of simulation for antennas with diameters of 100 and 300 wavelengths and aperture ratio $D/F = 1$. The results show that an acceptable deviation from the plane of the antenna is $1/16$ of wavelength.

We will now consider various choices of filler. The dielectric permittivity ε and loss angle $\tan \delta$ of glass fibre plastic are determined to a large extent by the type of the polymer matrix and the composition of the reinforcing filler and depend on the frequency of the electromagnetic field [21].

In the general case ε decreases with increasing frequency of the electromagnetic field. The values of $\tan \delta$ may go through extremal points and vary substantially with frequency. The lowest values of $\tan \delta$ in the range 10^{10}–10^{11} Hz through the glass-reinforced plastics are found in epoxy

Table 5.3.

Type	ε	$\tan \delta$
SPE-24 and glass fibre cloth TS-8.3-KTO		0.01
SPE-25 and glass fibre cloth TS-8.3-KTO		0.008
F-5 and glass fibre cloth T-41-76	4.5–4.45	0.01
F-5 and glass fibre cloth TS-8.3-KTO	3.4	0.01

glass-reinforced plastics based on special binding compounds (tables 5.3 and 5.4), glass-reinforced plastics based on quartz ($\varepsilon \approx 3.4 \pm 0.2$), and glass-reinforced plastics based on alkali-free glass ($\varepsilon \approx 4.5 \pm 0.5$). The frequency dependence of $\tan \delta$ is practically constant in the range from 10^{10} to 1.5×10^{11} Hz. The lowest values of $\tan \delta$ are found in quartz fibre-reinforced plastic in the range from 10^{10} to 1.5×10^{11} Hz (table 5.3).

Fibre-reinforced plastics thus possess relatively high dielectric permittivity and loss tangent angle. To reduce the effect of high values of optical constants of dielectrics on the characteristics of antennas, glass-reinforced plastic must be of minimum thickness. Furthermore, ideally it must only carry a function of supporting those layers of the multilevel zone plate that reflect radiation. That condition is especially important for the millimetre wavelength range.

Various styrofoams can be used as matrices. Physico-mechanical, dielectric and other properties of styrofoams vary in a broad range, depending on the nature of the initial components and the ratio of the polymer to the gas phase (the density of styrofoam).

One common feature of styrofoams is that, as their density increases, their physico-mechanical parameters grow while their dielectric properties deteriorate [22]. In styrofoams with density not greater than 0.1 g/cm^3, their dielectric permittivity remains nearly constant in a wide range of frequencies. One of the advantages of styrofoams is that dielectric permittivity is a linear function of density.

Foam polystyrene styrofoams are characterized as having low thermal stability and elevated inflammability, while polyvinyl chloride styrofoams are inflammable. Polyolefin foams are characterized as having high dielectric parameters and low thermal stability, phenoplastic foams work at high

Table 5.4.

Type	$\tan \delta$
Binding agent SPE-25	$(7\text{–}9) \times 10^{-3}$
Binding agent SPE-24	$(11\text{–}9) \times 10^{-3}$
Binding agent F-5	$(8\text{–}8.5) \times 10^{-3}$

Table 5.5.

Antenna type	Zone boundary radii, r_m (mm)
N1, N2	39.2, 55.7, 68.6, 79.6, 89.4, 98.5, 106.9, 114.8, 122.4, 129.6, 136.6, 143.3, 149.9
N3	27.6, 39.2, 48.1, 55.7, 62.5, 68.6, 74.3, 79.6, 84.6, 89.4, 94.0, 98.5, 102.7, 106.9, 110.9, 118.6, 122.4, 126.0, 129.6, 136.6, 140, 143.3, 146.6, 149.9

temperatures but their brittleness is relatively high, while epoxy foams have high humidity resistance.

In 1988–1989 the authors of this book prepared and studied several antenna models fabricated for working in the millimetre wavelength range. The N1 model was a diffractive reflector antenna with two levels of phase quantization and was fabricated by turning with a digitally controlled lathe. The parameters of the antennas were: nominal wavelength $\lambda = 8$ mm, diameter 300 mm, focal length $F = 190$ mm. In the N1 model antenna the first zone was set to be concave. The N2 model antenna is similar to N1 but the first zone was chosen to be convex. The N3 antenna corresponded to the diffractive reflector antenna with four levels of phase quantization. The accuracy of fabricating the surface of the antennas was about 50 μm. The boundaries of antenna zone radii are shown in table 5.5.

To protect the metal (aluminium) antenna from corrosion and to give it a matt colour, its surface was coated with a protective layer (of about 10 μm in thickness) by anodic spark electrolysis [23]. The micro-arc coat obtained in an aqueous solution of $Na_6P_6O_{12}$ was smooth, matt, homogeneous, without spots or moiré patterns. On aluminium–magnesium zinc alloys it produces a white, slightly greyish colour. The protective coat (Al_2O_3) was deposited at current density 5 A/dm^2 and voltage 350 V. The asperity of the external surface of the antenna coated with aluminium oxide did not exceed 5–7 μm, which is less than the asperity of this surface before coating (figure 5.3(a)).

Measurements of the coefficient of reflection from a metal plate with this coat and without it showed that within the accuracy of measurements the reflection coefficient remained unchanged.

The N4 antenna is similar to N3 but was manufactured with printed circuit board technology, as a multilayer 'sandwich'.

Figure 5.3(a–c) shows the appearance of flat diffractive antennas: (a) two-level, (b) four-level, (c) four-level antenna with 'freely suspended' zones. It should be emphasized that protective coating automatically reduces the asperity of the surface. The densitograms of surface asperity are shown in figure 5.3(a) (in the inset) prior to coating (the upper curve) and after coating (the lower curve). The asperity was reduced from $R_Z = 16$ to $R_Z = 5.9$.

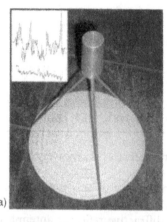

Figure 5.3. (a) A two-level flat diffractive reflector antenna with protective coating [25]. (b) A four-level flat diffractive antenna with protective coating [25]. (c) A flat diffractive antenna with 'freely suspended' zones [25].

An investigation on a flat multilayer diffractive reflector antenna (with four-level phase quantization), fabricated by printed circuit technology of 0.1 mm thick STPA-5-1 glass–textolite sheets (with parameters diameter 300 mm, focal distance 190 mm, nominal wavelength 8 mm) showed that the characteristics of this antenna are comparable with those of parabolic reflector antennas. As a filler for this antenna, foam polystyrene was used.

Measurements showed that the gain of the N1 antenna was 35.9 dB, that of the N2 antenna was 35.8 dB, that of N3 was 37.6 dB, and that of N4 was 37.4 dB. It was established that the gain of diffractive microwave reflective antennas depends on the number of phase quantization levels. The level of side lobes decreases as the number of phase quantization levels increases.

With uniformly illuminated aperture, the level of the first side lobes of scattering did not exceed 17.6 dB for all types of model antennas. The

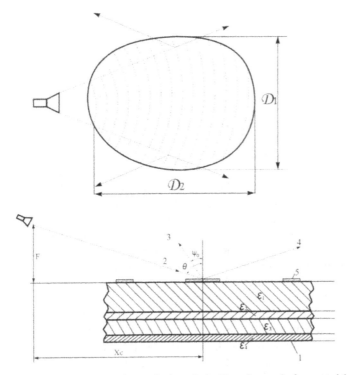

Figure 5.4. Diffractive antenna: 1, metal plate; 2, incident beam; 3, formatted beam; 4, reflected beam; 5, oscillator [24].

beampattern width did not exceed $(1.02–1.04)\theta$ for all types of antennas (θ is the beampattern width as defined by the diffraction limit).

In addition to the simplicity of design of flat diffractive reflector antennas, note that such antennas possess properties of frequency filters. In some cases it becomes possible to improve the conditions of their electromagnetic compatibility.

A novel type of flat diffractive reflector antenna was discussed in [24]. Such antennas were designed as systems of direct satellite television. The antenna is a metal plate coated with four dielectric layers of different dielectric permittivity and different thickness. The oscillators are deposited on to the uppermost surface of the last dielectric layer (figure 5.4).

Experimental tests of the antenna with a reflective diffractive array, formed by 766 oscillators with nominal wavelength of 30 mm, showed that the coefficient of usage of the area of the aperture was 48, 56 and 50% at 9.5, 10.0 and 10.5 GHz, respectively. It was concluded that focusing is maintained within 10% of the frequency band (9.5–10.5 GHz). As the frequency of the incident wave changes, the beampattern shifts by about one degree of arc per one percent of frequency change.

5.3   Lens-type diffractive antennas

Lens antennas differ from other types of antennas in simplicity of design and wide scope of options for forming the beampatterns of different types including multi-beam diagrams.

Focusing properties of dielectric antennas designed for local multipoint circular system (LMCS) communication systems were compared in [61]; the antennas were based on a hyperbolical lens and zone plates (of amplitude type, with two- or four-phase quantization levels).

The hyperbolical lens had the following parameters: diameter 152 mm, radiation frequency 30 GHz, maximum lens thickness 36.6 mm, focal distance 93 mm, $F/D = 0.6$; the dielectric constant of the lens material was $\varepsilon = 2.4$. A specialized optimized irradiator was used for measurements.

The parameters of the zone plates were: diameter 152 mm, $F/D = 0.5$, radiation frequency 29.8 GHz, dielectric constant of the zone plate material (or phase plates) $\varepsilon = 2.5$. The maximum thickness of phase-type zone plates was 15 mm and 20 mm for two- and four-phase quantization levels, and the amplitude-type zone plate thickness was 8 mm. In this case the optimized irradiator was not used for measurements.

Figure 5.5(a) plots the results of [61] to compare the beampatterns of a hyperbolical lens and two-level phase-type zone plate. Note that the beampattern of the hyperbolical lens is shown for the H-polarization; in the case of the E-polarization, the level of side lobes of scattering was higher by approximately 3 dB.

An analysis of the results given in [61] showed that a transition from two phase quantization levels to four increases the gain of a zone plate by approximately 2 dB and reduces the level of side lobes, making it comparable with the side lobe level of a hyperbolic lens. A transition from amplitude-type zone plate to a two-level phase plate increases gain by 4.5 dB.

In view of the differences between the parameters of zone plates and of the hyperbolic lens, the comparative characteristics given above should be considered as approximate. Furthermore, Fresnel zone boundaries were found using the classical formula, which is not optimal for the given value of F/D. We have to assume, therefore, that zone plates with the number of phase quantization levels not less than four are quite competitive against lens antennas.

When a diffractive lens antenna is designed together with an aerodynamic radome of a high-speed object, the requirements applied to it are contradictory. On one hand, the antennas must withstand aerodynamic loads, be resistant to thermal shock and create low drag. For instance, at a speed of Mach number $M = 4.5$, temperature at the deceleration point reaches 1200 °C and pressure rises to about 25 atm. In view of these working conditions a radome–antenna complex fabricated of ceramics or glass–ceramics has a length-to-maximum-diameter ratio of 1.5 to 3 (figure 5.5(b)).

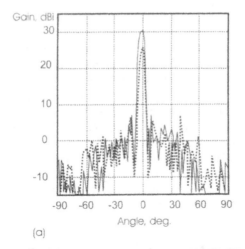

(a)

(b)

Figure 5.5. (a) Comparison of beampatterns of a hyperbolical lens (solid curve) and two-level phase inversion zone plate (dashed curve). Data from Dr A Petosa [61], with his kind permission. (b) Ogival-shape microwave diffractive antenna [25].

One of the obvious advantages of using a radome antenna as compared to the 'radome + antenna' system is that the aperture of the radome antenna can be higher than in the 'radome + antenna' system. This implies that, for the same overall size of the final product, it is possible to achieve narrower beampattern, higher gain and considerable simplification of the design as a whole. On the other hand, for the antenna system to function, the lens antenna must provide considerable width of the scanning sector (at least ±20°). In view of severe pressure to reduce antenna size, it must be a high-aperture design so that the length-to-diameter ratio cannot be higher than 1. If this ratio is high, not only is the scanning sector reduced but the problem of reduced transmission by the lens antenna arises as a result of high incident angles of the electromagnetic wave at the surface of the radome antenna.

A compromise solution for the contradictions discussed above may be a lens antenna having a small length-to-diameter ratio and equipped with an aerodynamic needle to improve streamlining.

An aerodynamic needle combined with an antenna system also allows, in addition to reducing aerodynamic drag for the same dimensions of the radome, less stringent requirements to thermal resistance of the radome material. The minimum wave resistance of the body at speeds corresponding to $M = 2$ seems to fall into the range $C_x \approx 0.2$–0.3. Bodies with aerodynamic needles create roughly twice larger aerodynamic drag than the ogival shaped head section. An aerodynamic needle reduces the height of the main lobe of the beampattern by 4–6%.

5.4   Diffractive radome antenna on a parabolic surface

The equal-signal zone of the field of view of diffractive antennas can be widened by placing them on curvilinear surfaces. For instance, a lens-type diffractive antenna fabricated on a parabolic surface with a generatrix of the type $y^2 = 4px$ and parameters $D/\lambda \approx 16$, $D/F \approx 0.9$ (figure 5.6) provides gain of about 27.5 dB in the 'transmission' mode, with the first lobes on the beampattern in the E and H planes not exceeding $-(18$–$20)\,$dB [25]. If the radiation source shifts by about 25° the gain reduces by less than 3 dB while the level of side lobes remains practically unchanged (table 5.6).

The beampattern width at the $-3\,$dB level was 4.5°. Figure 5.7 shows how the shape of the beampattern changed in scanning over angle.

Figure 5.6. Prototype of a lens-type diffractive microwave radome antenna fabricated on a parabolic surface [25].

Table 5.6.

Gain (dB)	φ (degrees)
27.5	0
25.6	5
24.9	10
25.2	15
26.9	25

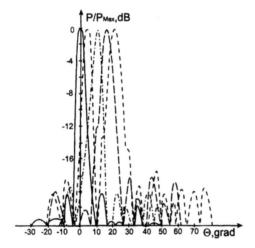

Figure 5.7. Beampattern of a parabolic antenna in the H plane [25].

Figures 5.8(a, b) plot experimental characteristics of irradiation of the antenna's aperture at zero angle and for the deviation of the radiation source by 20° in the E and H planes.

Experimental studies showed that the frequency band of such antennas is sufficient for practical purposes. For instance, the following results were

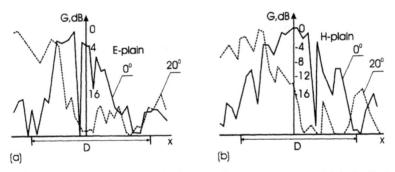

Figure 5.8. The amplitude field distribution over the antenna aperture: (a) in the E plane, (b) in the H plane. Solid curve, on-axis source; dashed curve, off-axis source.

Table 5.7

Δf (%)	0	+2.8	−2.8
G (dB)	27.5	26.2	27.1

obtained for an antenna with $D/\lambda \approx 16$ designed for working at 35 GHz (table 5.7):

The gain G for the antenna with $D/\lambda \approx 16$ was about 27 dB, and for an antenna of the same profile with $D/\lambda \approx 37.1$ the gain grew to 34 dB.

Figure 5.9 shows the experimentally obtained dependence of the level of the nearest side lobe scattering maxima on the beampattern of the antenna with $D/\lambda \approx 16$ on scanning angle, with the aperture irradiated uniformly by E and H waves.

In 'non-flat' diffractive antennas for the millimetre wavelength range, the thickness of the substrate on which the phase-coding profile is fabricated must be taken into account [26, 27]. Tables 5.8 and 5.9 present the results of experimental studies of diffractive lens-type antennas fabricated on a spherical surface. Experimental studies of diffractive antennas were conducted by measuring their beampatterns.

In measuring the beampattern, the far zone condition was satisfied, corresponding to the expression

$$L = 5D_1 D_2 / \lambda$$

where D_1, D_2 are the diameters of effective apertures of the receiving and transmitting antennas. The antenna gain was determined by comparing with a reference source. The antenna parameters were: relative diameter

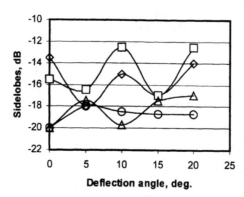

Figure 5.9. The level of first side maxima on the beampattern of the antenna with the aperture irradiated uniformly, as a function of scanning angle: □, E plane, first left maximum; ◇, E plane, first right maximum; ○, H plane, first left maximum; △, H plane, first right maximum.

Table 5.8

Width of beampattern at −3 dB level. Plane H	Antenna type	First minimum (dB)		First maximum (dB)	
		Right	Left	Right	Left
2°00′	A1	−24.3	−22.6	−21.1	−19.9
1°48′	A2	−35.0	−35.0	−26.3	−30.5
1°30′	A3	−30.7	−32.9	−18.6	−21.5
2°06′	A4	−18.3	−23.5	−16.4	−21.0

$D/\lambda = 36$ ($\lambda = 8.1$ mm); aperture ratio D/F, measured from the apex of the antenna was 1.48 for types A1 and A2 of antennas, and 0.98 for types A3 and A4 antennas. The A2 and A3 antennas were designed taking the phase substrate into account, while A1 and A4 were designed in the approximation of infinitely thin diffractive element. The number of phase quantization levels was 20. The shape of the external surface of the antennas was a sphere of radius $R = 37\lambda$ for A1 and A3 and of radius $R = 18.5\lambda$ for A2 and A4 antennas. The antennas were fabricated of Plexiglas.

Taking into account the thickness of the 'refractive' substrate significantly affects the parameters of the antenna's beampattern. Figure 5.10 plots beampattern width at the 3 dB level as a function of the antenna's scanning angle measured for a purely 'diffractive' element (taking no account of the refractive layer and therefore designated as 'diffractive') and for an antenna designed with the refractive layer taken into account ('diffractive–refractive').

The results obtained demonstrate that the antenna gain is independent of the degree of concavity of the antenna's surface and is found to be 34.2 ± 0.2 dB. It was established that taking into account the phase substrate narrows the beampattern of the antenna, reduces the level of side-lobes of scattering and increases the scanning angle of the beampattern to $\pm 20°$.

Note that if the phase substrate is included in the design of such systems, then the reduction coefficient is decreased by a factor of more than 2

Table 5.9

Width of beampattern at −3 dB level. Plane E	Antenna type	First minimum (dB)		First maximum (dB)	
		Right	Left	Right	Left
1°48′	A1	−23.5	−20.5	−16.8	−16.7
1°42′	A2	−20.5	−23.2	−17.0	−19.4
1°24′	A3	−31.2	−35.0	−13.4	−19.8
2°00′	A4	−18.5	−20.0	−18.3	−19.3

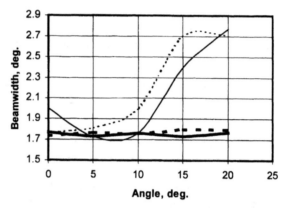

Figure 5.10. Antenna beam width as a function of scanning angle: thin dashed curve, plane H, diffractive; thin solid curve, plane H, diffractive–refractive; thick dashed curve, plane E, refractive–diffractive; thick solid curve, plane E, diffractive.

compared with the approximation of an infinitely thin lens. The corresponding results are given in figure 5.11.

Similar studies on parabolic and conical antennas demonstrated that the gain is independent of the antenna surface shape.

Furthermore, this research showed that the parameters of the beam-pattern of diffractive antennas do not deteriorate at shorter wavelengths; this is confirmed by measurements of the appropriate parameters of antennas possessing the same D/λ ratio, in the 3 mm wavelength range.

Foaming polystyrene of PSV type with density of about $0.5\,\mathrm{g/cm^3}$ (with granule size about 1.0 mm) is a promising material for fabricating antennas;

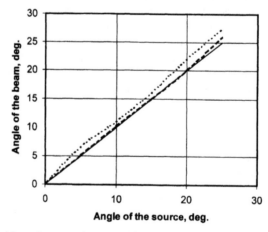

Figure 5.11. Position of antenna beam as a function of irradiation source deviation angle. Short-dash curve, antenna in the infinitely thin element approximation; long-dash curve, with refractive layer taken into account; solid curve, ideal antenna.

it corresponds to a refractive index of $N = 1.30$. Using this material will reduce reflection of electromagnetic radiation at the air–dielectric interface, make the beampattern more symmetric owing to the approximate equality of radiation transmission coefficients in the antenna in the E and H planes, and reduce the height of side lobes.

By applying materials with different refractive indices, it is possible to design an absolutely 'smooth' diffractive antenna and to reduce the effects of both scattering at zone ridges and of 'blocking' by phase zones.

5.5 Diffractive reflector antennas on 'non-flat' surfaces

Consider a surface Σ with a coordinate system (u, v) on it. We define the surface Σ as an array with certain indices of refraction and reflection that are periodic functions of a coordinate, for instance, u (figure 5.12(a,b)). We define S^i as a unit vector that characterizes the incident beam at a point P on the surface Σ, with a as an array period corresponding to the coordinate u. The diffracted beam of order m can be defined by a unit vector S_m [28]:

$$S_m = S^i - m\lambda(\text{grad}_\Sigma u) + gN. \tag{5.1}$$

Here N is the normal to the surface Σ at the point P, g is a scalar quantity and $\text{grad}_\Sigma u$ is interpreted as the gradient of the corresponding quantity in the two-dimensional area Σ.

Consider a reflector diffractive antenna on a spherical surface [29]. Let AP and $A'P'$ be two incident beams parallel to the optical axis of the antenna, P and P' being points on the surface Σ for any two sequentially

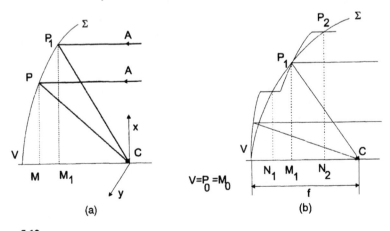

(a)   (b)

Figure 5.12.

located zones of the diffractive antenna (figure 5.12(a)). The zones must be oriented in such a way that the reflected beams arrive at the focal point C. The distances PC and $P'C'$ are equal: $|PC| = |P'C'|$. Therefore, the optical path difference between these beams is $|MM'|$. We assume $|MM'| = \lambda$.

The points P_0, P_1, P_2, lie at the intersection of the spherical surface Σ with the consecutive zones of the diffractive antenna. The points M_1, M_2, $M_3 \ldots$ are projections on to the axis Z of the points P_0, P_1, $P_2 \ldots$ (figure 5.12(b)). Therefore, $P_0 = M_0 = V$ and hence, $VM_1 = \lambda$, $M_1M_2 = \lambda$ etc. As a result of zoning, the generatrix of the surface of the diffractive antenna is described by a piecewise-smooth function. Consequently, the points N_i are projections on to the axis z of the points of intersection of blocking zones with the spherical surface Σ between the neighbouring zones (figure 5.12(b)). Therefore each point N is located between neighbouring points M_{i-1} and M_i.

We introduce the principal rectangular coordinate system $C(x, y, z)$ with the corresponding unit vectors i, j, k, and the axis z pointing along the optical axis of the system. We will also use spherical coordinates (r, Θ, φ), with unit vectors i_r, i_Q, i_φ.

Therefore we assume in expression (5.1) that $a = \lambda$ (period of the array) and $\mathrm{grad}_\Sigma = -i_Q \sin \Theta$, $N = -i_\varphi r$. For $m = 1$ we can rewrite (5.1) as

$$S_1 = S^i + i_Q \sin \Theta - i_r g. \tag{5.2}$$

The scalar quantity g is found from the condition $S_1^2 = 1$. Two solutions are possible for this condition: $S_1 i_r > 0$ and $S_1 i_r < 0$. We obviously have to choose the second case.

An analysis of aberrations of such an antenna shows [28] that the scanning by the beampattern is possible in such devices in a sufficient angular sector; spherical and coma-type aberrations are well corrected in such antennas. Zone reflectors suffer from chromatic aberrations.

Therefore, designing a reflector diffractive antenna on a spherical surface must meet the following two conditions: each zone must focus a parallel beam on to a common focal point, and phase synchronization of all reflected waves must be provided by the method of constructing the zones.

A diffractive reflector antenna on spherical surface was studied in [30] in two wave ranges: $\lambda_1 = 12.46\,\mathrm{mm}$ and $\lambda_2 = 22.2\,\mathrm{mm}$. For an antenna 457.2 mm in diameter, with spherical surface radius $R = 304.8\,\mathrm{mm}$, $R = F$ and with eight zones, gain $G = 36.0\,\mathrm{dB}$ was achieved at 12.46 mm wavelength with beampattern half-width $\Delta\Theta = 1.8°$. The total scanning angle was $35°$ with the voltage standing wave ratio (VSWR) reduced by 2 dB.

Ronchi *et al* [31] considered a zoned cylindrical mirror satisfying the following conditions. The antenna is free of spherical aberrations at axial points while spherical aberration and coma are constant at given aspect angle Ω and aperture angle α for off-axis points.

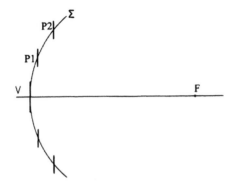

Figure 5.13. Cross-section of a cylindrical mirror.

Three interesting results were obtained for this antenna. First, regardless of the values of Ω and α, the cross-section of the diffraction grating is a parabolic curve. Second, spherical aberration for off-axis points may be negligibly small within the aperture angle O–α for aspect angles within $\pm 20°$. Third, the residual coma may be readily corrected for.

Let us consider a cylindrical mirror whose cross-section is given in figure 5.13.

A diffractive antenna consists of a set of flat plates located on the surface Σ. In the two-dimensional case the following condition holds for the diffraction grating:

$$\sin \alpha_n = \sin \alpha^i + A. \tag{5.3}$$

Here α^i is the angle ($\leq \pi/2$) between the incident ray and the normal to the surface Σ at a point P_1 and A is a quantity depending on the characteristics of the diffraction grating and the coordinates of the point P_1 but independent of α^i (figure 5.14).

Note that if p is the local grating period then $A = n\lambda/p$. The quantity A must be found from the condition that spherical aberration is zero for all values of the aperture α and for a point-like source located at infinity on the optical axis.

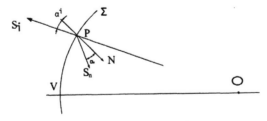

Figure 5.14. The definition of a diffractive antenna surface.

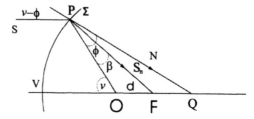

Figure 5.15. The definition of the surface Σ.

In the polar frame of reference, the equation of the surface Σ can be written in the form (figure 5.15)

$$\rho = \frac{p}{1 + \varepsilon \cos \Theta}$$

where p and ε are two positive constants.

The condition that must be satisfied requires that the reflected beam S_n arrive at the focal point F of the reflector antenna in the case when the incident beam S^i is parallel to the axis OV (figure 5.15). In this case, if $OPQ = \varphi$ and $OPF = \beta$, we can write

$$\alpha^{0i} = \Theta - \varphi, \qquad \alpha_n^0 = \varphi - \beta \qquad (5.4)$$

where $\alpha^{0i} = \alpha^i$ and $\alpha_n^0 = \alpha_n$ for $\Omega = 0$. The quantity β can be found from the triangle OPF.

Denoting the distances $|OF| = d$ and $|PF| = \Delta$, we obtain

$$\Delta = \rho + d + 2\rho d \cos \Theta, \quad \sin \beta = (d \sin \Theta)/\Delta, \quad \cos \beta = (\rho + d \cos \Theta)/\Delta. \qquad (5.5)$$

Expressing φ in terms of Θ we arrive at

$$\sin \varphi = \frac{\sin \Theta}{(1 + \varepsilon^2 + 2\varepsilon \cos \Theta)^{1/2}}, \qquad \cos \varphi = \frac{1 + \varepsilon \cos \Theta}{(1 + \varepsilon^2 + 2\varepsilon \cos \Theta)^{1/2}} \qquad (5.6)$$

and finally using (5.4)–(5.6) we find

$$\sin \alpha_i^0 = \frac{\sin \Theta}{(1 + \varepsilon^2 + 2\varepsilon \cos \Theta)^{1/2}}, \qquad \sin \alpha_n^0 = \frac{(\rho \varepsilon - d) \sin \Theta}{\Delta (1 + \varepsilon^2 + 2\varepsilon \cos \Theta)^{1/2}}. \qquad (5.7)$$

Substituting now (5.7) into (5.3) we find the constant A:

$$A = \left[\frac{\rho \varepsilon - d}{\Delta} - 1 \right] \frac{\sin \Theta}{(1 + \varepsilon^2 + 2\varepsilon \cos \Theta)^{1/2}}.$$

Let us derive the necessary conditions for correcting the spherical and coma aberrations. Consider the case when the beam S^i–O is not parallel to

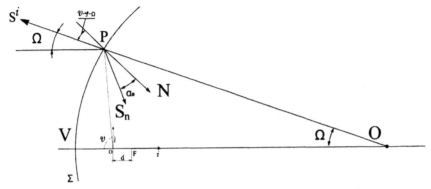

Figure 5.16. The conditions for correcting the spherical and coma aberrations.

the axis OV:

$$\alpha^i = \Theta - \varphi - \Omega$$

where Ω is the field angle (figure 5.16).

We introduce a rectangular frame of reference with the origin at the point O and the axis X parallel to OV. Let i and j be unit vectors on the axes X and Y, respectively. For S_n we have

$$S_n = \cos(\Theta - \varphi + \alpha_n)i - \sin(\Theta - \varphi + \alpha_n)j.$$

Equation (5.3) can be written in the form

$$\sin \alpha_n = \sin(\Theta - \varphi - \Omega) + [(\rho\varepsilon - d)/\Delta - 1]\sin\Theta/(1 + \varepsilon^2 + 2\varepsilon\cos\Theta).$$

These authors derived parametric equations (parameter $\bar{\Omega}$), that define the focal line for the given value of ε and d:

$$\bar{x} = -\rho_0 + p\frac{\cos^3\bar{\Omega}}{2\cos\bar{\Omega} - 1 + \dfrac{\varepsilon\rho_0 - d}{\rho_0 + d}}$$

$$\bar{y} = p\frac{\sin\bar{\Omega}\cos^2\bar{\Omega}}{2\cos\bar{\Omega} - 1 + \dfrac{\varepsilon\rho_0 - d}{\rho_0 + d}}.$$

A reflector diffractive antenna on a parabolic surface with coma-type aberrations corrected is designed similarly to a diffractive antenna on a spherical surface [32]. In polar coordinates, the paraboloid equation has the form (figure 5.17):

$$r = \frac{2f}{1 + \cos\psi}.$$

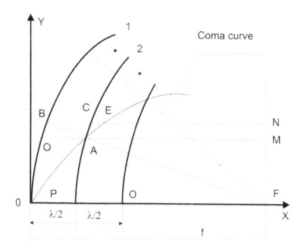

Figure 5.17. A reflector diffractive antenna.

If the focal distance is reduced by a multiple of $\lambda/2$ then the equation of confocal surfaces of the antenna can be written in the form

$$r = \frac{2(f - n\lambda/2)}{1 + \cos \psi}, \qquad n = 0, 1, 2, \ldots.$$

Zoning is carried out to satisfy the coma correction condition $r = f$. The condition of phase synchronization at the focal point for points on the optical axis of the system is written as

$$FO + OF = FP + PF + \lambda = FQ + QF + 2\lambda.$$

The coma curve has a radius equal to the focal distance f, so zoning of the reflector must follow this curve. The first zone boundary is found at the point of intersection of parabola 2 with the coma curve at a point A (figure 5.17). Then the following condition must be satisfied:

$$FA + AM + \lambda = FA + AB + BC + CN, \qquad \text{i.e. } AB + BC = \lambda.$$

However, in contrast to a smooth reflector, a zoned reflector generates areas of geometrical shadow AC, which cause additional scattering of electromagnetic waves by a certain angle $\delta\psi$. These shadow regions can be used for coupling various zones of a reflector antenna into a single design.

At each point A the condition $r = f$ is satisfied so that for parabola 2 of figure 5.17 we have

$$\frac{2(f - \lambda/2)}{1 + \cos \psi} = f \qquad \text{or} \cos \psi_1 = 1 - \frac{\lambda}{f}.$$

The coordinates of point A can be found from

$$y_1 = f \sin \psi = f\{1 - (1 - \lambda/f)^2\}^{1/2} = (2\lambda f - \lambda^2)^{1/2}$$
$$x_1 = f - f \cos \psi_1 = \lambda.$$

The equation of the family of confocal surfaces in Cartesian coordinates can be written as

$$y^2 = 4(f - n\lambda/2)(x - n\lambda/2), \qquad x_n = n\lambda, \qquad n = 1, 2, 3, \ldots .$$

Zone boundaries can be found from the formulas

$$y_n^2 = 4(f - n\lambda/2)n\lambda/2 = 2n\lambda(f - n\lambda/2), \qquad y_{n-1} = y_n$$

$$x'_{n-1} = (n-1)\lambda/2 + n\lambda/2 \frac{f - n\lambda/2}{f - (n-1)\lambda/2}.$$

The height of the step is

$$x_n - x'_{n-1} = h = (n+1)\lambda/2 - n\lambda/2 \frac{f - n\lambda/2}{f - (n-1)\lambda/2}$$

$$= \lambda/2[1 - n\lambda/2(f + \lambda/2)]^{-1}.$$

This last expression implies that as the antenna aperture increases, the step height grows, beginning from $\lambda/2$ on the axis of the system.

We conclude that the rules of zoning a parabolic reflector antenna, with coma-type aberration corrected, include searching for the points where parallel lines $x_n = n\lambda$ intersect the coma curve with a radius equal to the focal distance of the antenna. What is calculated is the distance from the optical axis of the system to the intersection points of the parabolas with a coma curve,

$$Y_n = (2nf\lambda - n^2\lambda^2)^{1/2}.$$

Ramsay and Jackson experimentally studied a zoned parabolic antenna [32] with the following parameters: $D/f = 1, f = 35\lambda, \lambda = 8.7 \, \text{mm}$, the number of zones $N = 4$. The reflector antenna was made of aluminium. This antenna provided 39 dB gain, with the theoretical gain being 40.8 dB. The loss in antenna gain in the scanning sectors $\pm 15°$, $\pm 20.5°$, $\pm 25°$ was 0.5 dB, 1.5 dB, 3 dB, respectively. The width of the beampattern was 2.1°. The best focusing surface was found to be planar.

Dasgupta and Lo [33] theoretically compared the focusing characteristics of a parabolic antenna and coma-corrected zoned antenna whose parameters were $f/D = 0.556, f = 10\lambda$ and the zone number $N = 11$; they determined that if the zoned antenna is irradiated with a plane wave, the intensity at the focal point is less by 4% compared with the parabolic antenna, while the side lobes of a smooth antenna are higher by 60% than for the zoned antenna.

For angles up to ±20°, changes in the gain of the zoned reflector are negligible. For instance, the loss in gain of the diffractive antenna at the angle of 15° was only 0.27 dB, while for the smooth reflector it was 1.36 dB. The difference grows even further as the scanning angle increases.

As the plane wave incidence angle changes, the maximum of field intensity in the neighbourhood of the focal point changes it positions in space, steeper for a smooth reflector than for a zoned antenna.

The ratio of heights of two secondary maxima remains constant for a zoned reflector and equals unity for radiation incidence angles of 20°. For a smooth antenna this quantity varies from 1 to 4.25 as incidence angles change from 0° to 15°. The level of the first side lobes for the smooth reflector is −15.8 dB, and for the zoned antenna is −14.6 dB.

The half-width of the beampattern changes in scanning from 0° to 15° by 30% for the smooth reflector and only by 5% for the zoned reflector.

5.6 Diffractive antenna systems in the millimetre wavelength range, based on the effect of conversion of surface waves to bulk waves [56]

One of the approaches to constructing electrodynamic circuits of various microwave irradiators is the effect of conversion of surface waves to bulk waves on periodic scatters [38–41].

An analysis of scattering of surface waves by diffraction gratings showed that such electrodynamic systems make it possible to format highly directed radiation in space. Various open transmission lines are used as sources of surface waves, for instance, dielectric waveguides of various cross sections. Such antenna arrays are based on holographic principles and contain two main functional elements: an open transmission line represented by dielectric waveguide, and a diffraction grating placed parallel to the lateral surface of the waveguide, for instance, a comb. By varying the characteristics of comb elements and their coupling to the field of the slow-wave open transmission line makes it possible to achieve the required characteristics of beampatterns, to implement wide-angle electromechanical beam scanning, and to produce efficiency as high as 90 to 95%. The polarization of the radiation of the beam is dictated by the type of the wave excited in the open transmission line and the type of scatterer.

Figures 5.18(a,b) show the electrodynamic and mechanical circuit of a 76.5 GHz antenna. The principle of operation of this antenna is to scatter the incident surface wave P_i on a periodic scatterer 1 and send it into the surface wave transmission line 2. The author chose the reflection-type diffraction grating to work on the E-type main wave. The choice of the wave type was dictated by the need for specific horizontal polarization of the antenna when the beam is scanned in the horizontal plane.

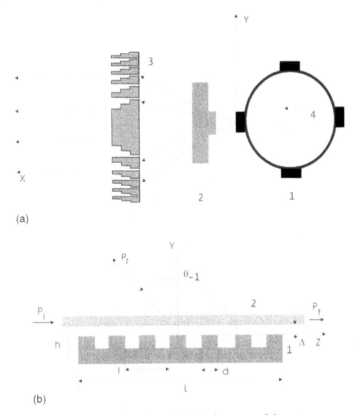

Figure 5.18. (a,b) Electrodynamic and mechanical diagrams of the antenna.

Ridged dielectric waveguides made of Teflon ($\varepsilon = 2.05$) and quartz glass ($\varepsilon = 3.9$) are typically considered for the surface wave transmission lines. The geometrical size of these waveguides provides the c-to-phase-velocity ratio (slowdown) $U = 1.2$ for the Teflon waveguide and $U = 1.5$ for the quartz waveguide. The choice of wave slowdown $U = \gamma/k$ is dictated by the need to minimize losses in the transmission line in the single-mode regime.

Diffraction grating is located at a distance Δ from the surface wave transmission line within its field. If the surface wave of the ridged dielectric waveguide with propagation constant γ in the direction OZ, and the wave with the propagation constant k, scattered by the diffraction grating with period ℓ, i.e.

$$k\ell \cos \theta_n - \gamma\ell = 2\pi n$$

are synchronized in phase, then the spatial orientation of the scattering field is dictated by the angle θ_n between the wave vector k and the axis OZ. Note that the vector k is oriented strictly in the plane XOZ.

The radiation condition given above holds only for negative values of n and for the single-beam mode of operation of the antenna; the value of n is chosen to be $n = -1$. The condition of no harmonics with $n = -2$ is $\ell < 2\lambda/(U+1)$. Therefore, the total field of individual scatters is added up constructively if propagating at an angle θ_{-1}:

$$\theta_{-1} = \arccos\left(U - \frac{\lambda}{\ell}\right).$$

It is quite clear that to scan with a beam, we can change the quantities U, λ and ℓ. Using the parameter ℓ makes it possible to obtain better electro-dynamic characteristics of the scanning antenna. This is achieved by keeping the parameter U and the quantities connected with it plus losses in the transmission line constant, and a fixed chosen level of the side lobes of the beampattern of the antenna in the sector of variation of the angle θ.

A diagram of the mechanically scanning antenna (figure 5.18(b)) assumes the location on a cylindrical surface of such number of arrays that provide either quasi-continuous or stepwise scanning. In the version of the antenna with stepwise scanning increment equal to $1°$, it is necessary to place 30 diffraction gratings, differing in the period ℓ, on the cylindrical surface 4 (figure 5.18).

Calculations of the required periods ℓ for Teflon and quartz wave-guides show that in the single-harmonic mode of radiation, the quartz waveguide provides a scanning sector which is symmetric relative to the normal to the plane YOZ of the antenna.

In the angle of elevation plane, the required beam width is provided by a cylindrical lens 3 (figure 5.18(b)). The lens is made of polystyrene ($\varepsilon = 2.5$), the cheapest material; the absorption can be minimized and the range of the angle of scanning can be increased by fabricating the microwave lens as a special zone plate.

The aspect of matching the transmission line and of the periodic scatterer for it deserves special attention. The reason for this is that for E-type waves the conditions of scattering and the required scatterer parameters differ drastically from those for H-waves. The main problem here is to design a ridged dielectric waveguide with minimal ridge width, maximum moderating coefficient and the required field distribution of the main E-wave, and also to create a periodic scatterer of minimum width. This last factor ensures the minimum possible diameter of drum 4 when 30 diffraction gratings are mounted on it.

To find the power characteristics of diffraction gratings when E-waves are scattered, reflective gratings were studied, having width much greater than that of the waveguide ridge. Attenuation of the transmitted power P_t was studied as a function of the parameter Δ (figure 5.18), that is, the quantity $(-10 \lg P_t/P_i) = f(\Delta)$.

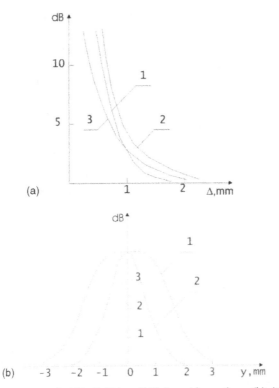

(a)

(b)

Figure 5.19. (a) The curves $(-10\lg P_t/P_i) = f(\Delta)$ for wide gratings. (b) Attenuation of transmitted power as a function of displacement of the diffraction grating.

Figure 5.19(a) plots $(-10\lg P_t/P_i) = f(\Delta)$ for wide gratings with groove depth 0.47 mm (curve 1), 0.7 mm (curve 2) and 0.85 mm (curve 3). The grating length was 120 mm, the groove width 1 mm, the grating period 3 mm. The c/v ratio in the waveguide was $U = 1.2$. These curves show that at the working frequency 76.5 MHz and $U = 1.2$, diffraction grating 2 possesses the groove depth which is the closest to the resonance and therefore the most efficient. This grating provides more than 15 dB of power output P_r if the length is 240 mm and $\Delta = 0.75$ mm, that is, efficiency higher than 97%.

Experimental investigations for narrow gratings of different shapes with grating width comparable with waveguide ridge width (2 mm for Teflon waveguide and 1.5 mm for quartz waveguide) selected the most efficient selective grating. The groove depth of these gratings was $h = 0.75$ mm, the groove width $d = 0.5$ mm. The grating width was 1 mm, the length was 100 mm.

Because of a large wave slowdown in quartz waveguide, it is possible to create a scanning irradiator using it; however, the design of the antenna will require very careful improvement. Obviously, the cost of an antenna with a quartz waveguide will be higher.

The size of the working zones of each grating and the effect of neighbouring gratings on the beampattern of the antenna will be important parameters for the chosen method of controlling the antenna beam. Figure 5.19(b) shows experimental curves for the attenuation of the transmitted power P_t as a function of displacement of the diffraction grating relative to the ridged waveguide along the OY axis. Curve 1 represents the narrow grating in the Teflon waveguide and curve 2 represents the quartz waveguide. The quantity Δ corresponds to the initial attenuation of the signal by 4 dB. With a grating 300 mm long this attenuation will reach 12 dB which corresponds to 93.7% efficiency. The linear displacement of the grating along the OY axis that we consider here does not reflect the actual displacement as it occurs on the generatrix of the cylinder. In a realistic situation, therefore, the effect of neighbouring gratings is smaller.

Nevertheless, this treatment is admissible as it is more stringent. Figure 5.19(b) (curve 1) implies that if $y = 3$ the effect of the above grating on the beampattern created by the neighbouring grating will be of the order of -30 dB for $y = 3$ mm. This corresponds to the maximum on the beampattern of the neighbouring grating. If the gratings are not located consequently according to the ascending (descending) period, the side lobe of the neighbour grating can be shifted by several degrees away from the main lobe of the beampattern.

5.6.1 Specifics in calculating the beampattern of a scanning antenna in the horizontal plane

Until now some important questions remain unanswered, in aspects concerning specifics of diffraction of non-uniform plane waves on lattices of arbitrary finite length. The reason for this is that it is still not possible to construct, for an electrodynamic system, a mathematically rigorous theory of the specific diffraction problem.

An algorithm for computing such antenna characteristics as amplitude distribution over the aperture and efficiency of the irradiator with arbitrary diffraction grating length was suggested in [40, 41]. It is possible to take into account two types of coupling of the dielectric waveguide to the grating in these calculations: uniform and wedge-shaped coupling. Figure 5.18 shows the uniform coupling provided by the gap Δ between the waveguide and the grating. To provide the wedge coupling, a gap δ is introduced into the front section between the waveguide and the grating in addition to the gap Δ.

The initial data for calculations are obtained on the basis of one implementation of the experiment with specific samples of dielectric waveguide and grating in the uniform coupling mode. The coupled waves approximation is used in the computations. Therefore the grating length $\geq 10\lambda$ is necessary for a good correspondence between the calculated and the experimentally obtained curves.

The beampatterns were computed by standard means—using the integral Fresnel–Kirchhoff representation for the fields of the emitting systems in the far zone.

First we consider the case of uniform coupling, with $\delta = 0$ (Figure 5.18). The slow surface wave P_i propagates along the dielectric waveguide. Note that the power transmitted across the effective waveguide cross section equals $P(z)$. Owing to the coupling to the diffraction grating, and the radiation losses, the power transmitted by the waveguide decreases in the direction of propagation of the surface wave. Losses per unit length $(-dP(z)/dz)$ are proportional to the transmitted power and squared field strength of the surface wave on the elements of the grating. The field strength in the dielectric waveguide decreases away from its surface as $\exp(-\alpha\Delta)$, where α is the transversal wave number. The quantity $P(z)$ corresponds to the coupled waves equation

$$dP(z) + KP(z)\exp(-2\alpha\Delta) = 0$$

where K is the coupling coefficient. The solution of this equation yields an expression for the efficiency of the radiating system:

$$\text{efficiency} = \frac{P_r}{P_i} = 1 - \exp[-KL\exp(-2\alpha\Delta)].$$

For a specific dielectric waveguide and a specific diffraction grating, the coupling coefficient is found using the formula

$$K = \frac{\ln 2}{L}\exp(2\alpha\Delta_{0.5}).$$

The quantity $\Delta_{0.5}$ is experimentally measurable. It corresponds to such distance between the dielectric waveguide and grating at which the efficiency equals 0.5. The quantity α is found in terms of the dielectric waveguide slowdown coefficient, $\alpha = 2\pi/\lambda\sqrt{U^2 - 1}$, where λ is the wavelength in free space, $U = \lambda/\lambda_b$ is the slowdown coefficient and λ_b is the wavelength in the dielectric waveguide.

In the case of the non-uniform wedge coupling at the diffraction grating to the dielectric waveguide, the coupled waves equation takes the form

$$\frac{dP(z)}{dz} + KP(z)\exp\left[-2\alpha\left(\Delta + \delta - \frac{\delta}{L}z\right)\right] = 0.$$

This equation takes into account the linear decrease in the 'waveguide-to-grating' as the value of z increases.

Solving this equation, we arrive at the amplitude distribution within the antenna aperture and the systems efficiency:

$$
\frac{\mathrm{d}P(z)}{\mathrm{d}z} \Big/ P_i
$$

$$
= \frac{K \exp\left\{-\dfrac{KL}{2\alpha\delta}\exp\left[2\alpha\left(\dfrac{\delta}{L}Z - \Delta - \delta\right)\right]\right\} \exp\left[2\alpha\left(\dfrac{\delta}{L}Z - \Delta - \delta\right)\right]}{\exp\{-KL/2\alpha\delta \exp[-2\alpha(\Delta + \delta)]\}}
$$

$$
\text{efficiency} = 1 - \exp\left\{-\frac{KL}{2\alpha\delta}\exp(-2\alpha\Delta)[1 - \exp(2\alpha\delta)]\right\}.
$$

The wedge coupling in this electrodynamic system allows various ways of tuning the amplitude distribution on the radiating aperture. For the field strength decreasing to the same value along aperture's edges, we have

$$
-\frac{\mathrm{d}P(z)}{\mathrm{d}z}\bigg|_{z=0} = -\frac{\mathrm{d}P(z)}{\mathrm{d}z}\bigg|_{z=L}.
$$

This condition is met if the distance between the waveguide and the grating is

$$
\Delta = \frac{1}{2\alpha}\ln\left[\frac{KL}{4\alpha^2\delta^2}[1 - \exp(-2\alpha\delta)]\right].
$$

As follows from the above expression, the shape of the amplitude distribution depends on many parameters, of which the decisive one is the dielectric waveguide slowdown coefficient and, hence, both the gradient of decrease of the field of the dielectric waveguide and the position of the grating in this field. In physics terms, the effect of a wedge coupling is that the reduction in 'waveguide-grating' gap compensates for the fall-off of power in the waveguide resulting from radiation losses. Consequently, the amplitude distribution of the field on the emitting aperture becomes more uniform. One specific feature of the electrodynamic system under discussion is that the amplitude distribution can be made nearly Gaussian. The Gaussian distribution ensures the lowest level of side lobes for the same values of aperture, wavelength and width of beampattern as compared with other types of smooth amplitude distribution [43].

The formulas given above imply that the field amplitude distribution in the aperture for the coordinate Z_m has the form [43]

$$
A(Z_m) = \exp\left\{-\frac{KL}{2\alpha\delta}\exp\left[2\alpha\left(\frac{\delta}{L}Z_m - \Delta - \delta\right)\right]\right\}
$$

$$
\times \exp\left[2\alpha\left(\frac{\delta}{L}Z_m - \Delta - \delta\right)\right].
$$

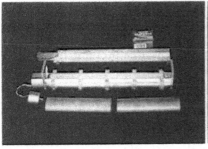

Figure 5.20. Two projections of the antenna.

The expression thus obtained describes the change in field amplitude over the aperture, with account taken of radiation losses determined by the type of coupling of the dielectric waveguide to the grating. The phase distribution is assumed to be linear.

The described method of calculating the beampattern ignores the effect of the diffraction grating on the phase velocity of the wave in the dielectric waveguide, nor does it take into account the change in the effective transversal cross section of the waveguide in the presence of an array. The result is that the coupling coefficient depends on the waveguide-grating distance. From this point of view long gratings with large Δ and δ are preferable.

The design of a scanning antenna includes (figure 5.20):

- open transmission line,
- scanner,
- motor,
- phase correction circuit.

An open transmission line is a metal base on which is mounted a ridged dielectric waveguide with a horn section. The ridged dielectric waveguide of integrated design is made of Teflon ($\varepsilon = 2.05$). The ridge is 2 mm wide and 3 mm high (inclusive of substrate thickness). The geometrical dimensions of the waveguide provide the wave slowdown $U = 1.2$.

The scanner is a cylinder 32.5 mm in diameter and 326 mm long, with 30 reflective diffraction gratings placed on its surface. The gratings are placed at a step of 3.4 mm. The gratings have grooves 0.75 mm deep and 0.5 mm wide. The grating length is 292 mm, the width is 1 mm. Including the gratings, the cylinder diameter is 34 mm. As the cylinder rotates, the period of the gratings that consecutively interact with the field of the dielectric waveguide varies from 3.52 to 2.44 mm. With such grating periods, the beam scanning sector of the beampattern in the horizontal plane is 30°, with a step of 1°.

The cylinder with 30 gratings is placed into a metal housing made of aluminium tube 40 mm in external and 36 mm in internal diameter. On both sides of the cylinder along its axis two bearings are placed, 5 mm in internal diameter and 16 mm in external diameter.

The motor is connected to the scanner through a high-speed connection sleeve. The sleeve ensures coupling and compensation for off-axis deviations by using rubber inserts. The motor power was 1 W with feed voltage 5 V. The motor is placed under the horn coupling to the dielectric wave-guide, as can be seen in figure 5.20. This sleeve also contains an optical sensor for continuous identification and display of the spatial orientation of the antenna beam within a 30° sector.

The phase corrector is to format the 5.6° wide beampattern in the elevation plane. It is a cylindrical lens made of polystyrene ($\varepsilon = 2.5$). The output aperture of the lens is 320 mm × 46 mm. The length of the lens is chosen so as to take into account the wave bundle that is incident at the edges of the scanning sector. The total thickness of the lens is 17 mm. The focal length of the lens measured off its apex is 45 mm. Phase distortions at wavelength $\lambda = 3.92$ mm did not exceed 30° over the entire aperture diameter of 46 mm. The angle at which the flat part of the lens was irradiated was 78.8°. The effective irradiation angle $\leq 60°$ provided an effective aperture of 39 mm.

The antenna dimensions were 383 × 86 × 48 mm.

5.6.2 Electrodynamic characteristics of an experimental scanning antenna

Losses in the ridged geometrical waveguide made of Teflon were 2 dB/m at the working wavelength. Radiation losses in the horn coupling did not exceed 0.1 dB.

Additional losses in the antenna are those in the lens (losses in the lens material and losses to reflection at refraction interfaces) and losses through scattering. Measurements of losses in carefully prepared lenses with hyperbolic profile of the refractive surface gave 1 dB in the range 76–77 GHz.

The beampattern width in the azimuthal plane was 1° (figure 5.21(a)) (for radiation propagating along the normal). The side lobe level is −16 dB and can be reduced significantly by accurately tuning the antenna. The beampattern width in the vertical plane was 5.6° (figure 5.21(b)). The side lobes level was −12 dB.

The antenna gain was 32.8 dB when radiated along the normal. If $\Theta_{-1} \neq 90°$ the antenna gain decreases faster than would be described by the expression $\cos(90° - \Theta_{-1})$, owing to quadratic phase distortions within the lens aperture in two planes.

An analysis of the results given above shows, among other things, that antennas of this type are sufficiently sensitive to the distance between the waveguide and the grating, which imposes considerable technological constraints on the antenna design.

On the other hand, antennas based on conversion of surface waves to bulk waves possess certain advantages. Among them we find both the

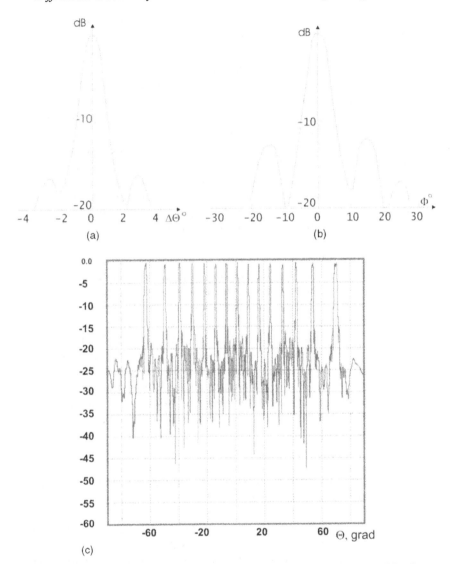

Figure 5.21. Beam width in the azimuthal (a) and vertical (b) planes. (c) Multibeam antenna based on aperiodic diffraction grating [76].

planar design and the possibility of creating various types of beampatterns by using various degrees of freedom. One of such parameters is the distance between grating elements.

The effect discussed above can be used to create multibeam antennas for the millimetre wavelength range [76]. For instance, a six-beam antenna can be implemented for the radiation frequency 76.5 GHz, with the slowdown coefficient 0.83, the step between grating elements 12 mm, and the total number of elements 50. If the distance between diffraction grating elements is reduced by a factor of 2, the number of beams will also be halved to 3; if the distance between grating elements is reduced to 5 mm, the number of beams is 2, and only one beam is left if the inter-element separation is 3 mm.

Another method of controlling the shape of the beampattern is to use not a periodic diffraction grating but an aperiodic one [76]. In this case too a multibeam beampattern can be generated.

Creation of simple scanning devices in which a single scatterer whose motion follows a prescribed law may be a promising way of applying the effect of transformation of surface waves to bulk waves.

To conclude, we will point to certain drawbacks of this type of antennas and to possible approaches to removing these drawbacks. Using a microwave lens to shape the beampattern of an antenna is efficient only if the scanning angle is small. If scanning angles are considerable—up to ± 0–$30°$—the lens become virtually useless and the antenna as a whole cannot support the required scanning angle. The simplest way out of this situation is to install a lens designed to work at a certain 'medium' angle in the prescribed range of scanning angles. This can somewhat smooth the operation of the lens within the entire angle range.

Another drawback of this design is the oscillatory nature of the beampattern of the antennas along its 'wider' side (the narrower wall). To counteract this, a special coupling layer must be added or a coating placed immediately in front of the diffraction grating. A diffraction structure, created on the lens surface in the shape of grooves of prescribed size depending on the dielectric chosen and polarization type, can serve as such a coupling layer.

5.7   Effect of the form of diffractive element's surface on the structure of Fresnel zones

To simplify the analysis, we will consider diffractive elements that transform a plane incident wavefront into a convergent spherical one.

We define the projections of Fresnel zone boundaries on to a plane perpendicular to the optical axis of the diffractive element. For a zone plate fabricated on a spherical surface, zone boundaries can be found for the case of plane wavefront incident on a convex surface using the expression

$$b_n = \sqrt{R^2 - y^2}$$

where

$$y = -\frac{n\lambda}{2} + \sqrt{R^2 + n\lambda(R - F)}.$$

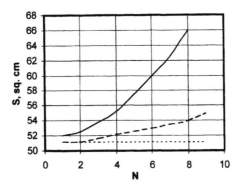

Figure 5.22. Zone area as a function of zone number for a spherical surface: Solid curve, $R = 100$; dashed curve, $R = 200$; dotted curve, $R = 900$.

These formulas show, for instance, that if $R = F$, the expression for the Fresnel zone projection coincides with that of Fresnel zone radii of a flat zone plate.

The area of each nth zone can be easily found from the formula

$$S_n = \sigma_n - \sigma_{n-1}$$

where

$$\sigma_i = 2\pi R\left(R - \sqrt{R^2 - b_i^2} \right).$$

We assume that all diffractive elements discussed below have the same identical parameters:

$$\lambda = 8\,\text{mm}, \qquad D = 200\,\text{mm}, \qquad F = 200\,\text{mm}.$$

Let us consider how the areas of the nth Fresnel zone depend on the zone number for different values of the radius of the spherical surface. The relevant curves are shown in figure 5.22.

The data shown imply that if the curvature of the spherical surface of the zone plate is large (the radius R of the spherical surface is small), the dependence sought is essentially nonlinear. At the same time, the Fresnel zone area for a given type of diffractive element for $R = F$ is independent of the zone number and is a constant. If the radius of the spherical surface is large, the dependence in question is nearly linear, that is, it tends asymptotically to a similar characteristic for flat zone plates.

A diffractive element manufactured on a parabolic surface behaves somewhat differently.

In the case of a zone plate manufactured on a parabolic surface, the expression required has the form

$$b_n = \sqrt{n\lambda f_p}\sqrt{\frac{F + n\lambda/4}{f_p + n\lambda/4}}$$

where f_p is the parameter of the paraboloid.

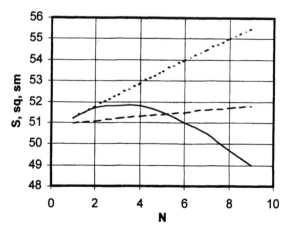

Figure 5.23. Zone area as a function of its number for a parabolic surface: solid curve, $f_p = 50$; dashed curve, $f_p = 150$; dotted curve, $f_p = 400$.

The area of a zone can be found from the expression

$$S_n = \frac{8}{3}\pi f_p \left[\left(1 + \frac{b_n^2}{4f_p^2} \right)^{3/2} - \left(1 + \frac{b_{n-1}^2}{4f_p^2} \right)^{3/2} \right].$$

Figure 5.23 plots characteristic dependence of the area of the *n*th Fresnel zone as a function of zone number for a given type of diffractive element. As can be seen from the data presented, the characteristic is nonlinear if the focal length of the paraboloid is small (the shape of the surface is nearly ogival): Fresnel zone areas first increase with increasing zone number and then begin to slowly decrease, practically following a linear law. If the focal length of the paraboloid is nearly equal to the focal length of the diffractive element, the Fresnel zone areas vary insignificantly as the number of a Fresnel zone increases. If the focal length of the paraboloid is large, this dependence tends to a similar one for a flat zone plate.

The expressions for Fresnel zone radii projections and zone areas for a zone plate fabricated on a conical surface can be written in the form

$$b_n = \sqrt{R_n^2 + \left(\frac{n\lambda}{2\tan(\alpha)} \right)^2} - \frac{n\lambda}{2\tan(\alpha)}$$

where

$$R_n = \sqrt{n\lambda F + \left(\frac{n\lambda}{2} \right)^2}$$

are Fresnel zone radii of a flat zone plate. Correspondingly, the expression for the area of the *n*th zone becomes

$$S_n^c = S_n^{fl} + \pi(2n - 1)\left(\frac{\lambda}{2\tan(\alpha)} \right)^2.$$

Here the superscript 'fl' signifies a flat zone plate and the superscript 'c' stands for a conical plate. We see from the last expression that if the condition

$$\lambda F \gg n\lambda^2/4 \qquad S_n^{fl} = \pi[\lambda F + (2n-1)\lambda^2/4]$$

is met, than the Fresnel zone of a flat zone plate is independent of zone number and remains constant.

The last two expressions clearly manifest the differences between the diffractive optical elements for the microwave, millimetre and optical ranges. In the optical wavelength range the area of each nth zone is approximately constant. Indeed, if $\lambda \approx 0.001$ mm and $n \approx 1000$, the condition $F \gg n\lambda/4$ must be met to ensure constancy of Fresnel zone areas, which is equivalent to $F \gg 0.25$ mm for the indicated characteristic parameters. This condition is practically always fulfilled in optics. In the millimetre and microwave wavelength ranges ($\lambda \approx 3$–30 mm) we have $n \approx 20$–50, which is equivalent to $F \gg 15$–350 mm. This condition cannot always be satisfied in these wavelength ranges and, therefore, the constancy of a Fresnel zone area cannot be maintained.

The fabrication of MWDO elements on an arbitrary surface somewhat changes the condition of Fresnel zone area constancy. For instance, for a conical zone plate this condition can be written in the form

$$F \gg \frac{n\lambda}{4\sin^2(\alpha)}.$$

Figure 5.24 plots characteristic curves for the Fresnel zone area as a function of zone number for various angles of the cone generatrix of the diffractive element. Note that curve N1 corresponds to a flat zone plate. The data shown in the figure clearly demonstrate that as the surface of a diffractive element deviates further from a plane, the behaviour of Fresnel zone areas

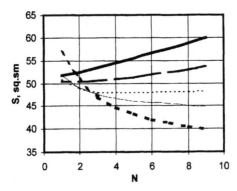

Figure 5.24. Zone area as a function of its number for a conical surface. The cone angle of the conical surface (from top down) is 90, 81.81, 72, 64.28, 51.43 degrees of arc.

of a conical zone plate changes. It is also obvious that zone area will be weakly dependent on its number for a certain apex angle of the generatrix surface of the diffractive element. It must also be mentioned that for this type of surface of the diffractive element, beginning with a certain angle, the dependence under discussion runs contrariwise to a similar dependence observed with other types of surfaces discussed: the Fresnel zone area diminishes as the zone number increases.

By comparing these results for various shapes of diffractive element surfaces, we can see that with the main parameters as indicated, the maximum ranges of Fresnel zone areas for the elements discussed above are:

- for parabolic surface 1.14;
- for spherical surface 1.44;
- for conical surface 2.5.

It follows therefore that the parabolic surface of a diffractive element is least 'sensitive' to changes in the Fresnel zone area on the element's surface when the main parameters of the surface of the diffractive element vary (radius of the sphere, focal length of the paraboloid, cone angle). The most 'sensitive' is, of course, a conical surface.

The main conclusion from the discussion above is that the structure of Fresnel zones of a diffractive element can be controlled by choosing both the shape of the surface of the element and its parameters.

It is not difficult to obtain similar relations for diffractive elements created on an arbitrary surface and transforming a divergent spherical wavefront into a convergent spherical one. However, owing to the unwieldy size of the corresponding relations, we do not display them here.

5.8　Aerodynamic aspects of radome antennas of supersonic aircraft

One of the attractive features of diffractive focusing elements is that they can be created on an arbitrary surface. It is then possible to combine the properties of the element that forms the beampattern and those of the aerodynamic radome, such as an aircraft, in a single design element.

The problem of designing modern radome antennas based on MWDO elements is a complex one. On one hand, it is necessary to create a diagram-forming element with prescribed parameters (beampattern, level of side lobes, scanning angles etc.) and, on the other hand, it must provide the required aerodynamic characteristics. Two approaches are possible to solve this problem. The first approach is based on the diagram-forming properties of the diffractive radome antenna (DRA)—the optimal shape of the antenna is obtained from them, and then the aerodynamic drag is minimized by using these parameters. In the second approach the required aerodynamic characteristics are chosen as the basis, so that the surface shape is determined and then the diagram-forming element is created (synthesized) on the already prescribed shape. The choice of a specific approach follows from the 'priority' of aerodynamic or diagram-forming properties of DRA.

5.9 Radome antenna fabricated on a surface optimized for scanning

It can be shown [44] that to ensure scanning angles up to ±20–$30°$ when shifting the phase centre of the radiation source, the shape of the DRA must be very close to spherical. Besides, the length parameter of the DRA (the ratio of the maximum height of the arc to the base diameter) must not deviate significantly from 1.

On the other hand, the wave resistance of such bodies is rather high in supersonic flight modes corresponding to the Mach number $M = 2$–4.

A way out of this contradictory situation may lie in using a spherical-surface DRA with an aerodynamic needle installed on it. The following problems need solving:

- how to find the effect of aerodynamic needle on the diagram-forming properties of the DRA proper;
- it is necessary to choose an optimal configuration 'DRA + aerodynamic needle' from the standpoint of minimization of the wave resistance Cx of the body;
- it is necessary to evaluate thermophysical parameters of DRA's functioning in order to obtain the requirements to the DRA material.

These problems were investigated by numerical experiments [45] using a software package [46]. The method of computation and the structure of the complex were discussed in [47].

We will point only to the main conclusions essential for designing DRA, without going into the finer details about the behaviour of aerodynamic characteristics of the configurations involved.

We introduce a parameter $l = 2L/(1 - d)$, where l is the relative length of the needle and d is its diameter in units of the DRA diameter. Depending on the value of l, three main modes of streamlining around the body can be singled out [48]:

- 'short needle': the distance of separation of the shockwave (SW) from the streamlined body is greater than the needle's length. In this case the pattern of flow is mostly determined by the shape of the DRA;
- 'middle-length needle': the length of the needle is sufficient to change the shape of the leading density front, transforming it from a nearly rectangular (in the neighbourhood of the symmetry axis) to oblique;

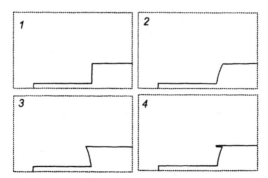

Figure 5.25. Aerodynamic configurations of bodies.

- 'long needle': the needle length is greater than the distance of separation of the SW from the body. Here the main density front step forms in the stagnation zone in front of the body.

At supersonic flight speeds ($M > 1$) the minimum value of the wave resistance coefficient Cx is found when the third of the listed streamlining regimes is realized. In the general case, the total 'DRA + needle' length is limited by the overall dimensions constraints, so that in choosing the optimal configuration it is necessary to minimize at least three parameters: l, d and the degree of blunting of the cylindrical body (the shape of the DRA). Numerical studies showed that in the case of a spherical shape of the DRA with $R \approx D$, the optimal value of l for velocity corresponding to $M = 2$ is $l \approx 3.00$ ($l = 1.56$, $Cx = 0.79$; $l = 3.00$, $Cx = 0.31$; $l = 3.67$, $Cx = 0.38$). When velocity increases to $M = 4$, the minimum value of Cx increases to 0.4 (for optimal l).

To minimize the aerodynamic drag of a body while retaining sufficient diagram-forming and scanning properties of the DRA, configurations were investigated shown in figure 5.25. From the point of view of aerodynamic characteristics, the optimal version is configuration 4. The reason is that in this configuration the gas flow in the region of contact of the cylindrical body and the DRA is always directed towards the axis of the body. By controlling this flow (e.g. by adding an aerodynamic 'visor' shown in figure 5.25 (4), it is possible to control the shape of the stagnation zone in the region of the reverse circulation flow. The optimum value of Cx ($Cx \approx 0.23$) for $M = 2$ is observed for the following parameters of the 'visor': length $0.13D$ and thickness 10% of its length.

As an example, figure 5.26 plots pressure contour lines in streamlining of a body shown in figure 5.25 (configuration 4). The position of the shockwave front was found from the gradient of the corresponding contour lines. The contour lines are numbered in the order of increasing pressure. The change of pressure between the neighbouring contour lines is constant and equal

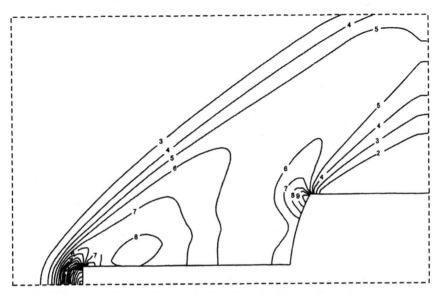

Figure 5.26. Pressure contour lines in the flow for one of the configurations.

to one twentieth of the difference between the maximum and minimum values (within the field chosen for computation).

Another feature in the aerodynamics of such configurations must be mentioned in view of the requirements to the DRA material. We know [49] that if $0.3 < l < 3$, non-stationary periodic modes of streamlining in the 'blunted body–aerodynamic needle' configuration are observed. Figure 5.27 plots the

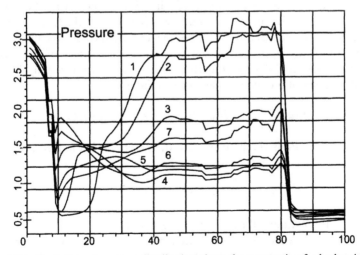

Figure 5.27. Evolution of pressure distribution along the generatrix of a body with time. Numbers by the curves indicate time sequence.

Table 5.10.

T (K)	510	460	470	565	600	550	490	480	510	565
Step	1200	1500	1800	2100	2400	2700	3000	3300	3600	3900

distribution of pressure along the generatrix of the surface as a function of time, conclusively showing this effect. The oscillatory mode of streamlining leads, among other things, to thermal loads on the DRA surface also having a periodic behaviour. Investigations showed that at flight speeds corresponding to $M = 2$ the range of temperature fluctuations in the region of coupling of the aerodynamic needle to the DRA reaches $\Delta T \approx 200\,K$ in the established optimal configuration.

For instance, table 5.10 gives temperature at a fixed point on the DRA surface as a function of time. Temperature is given in kelvins, and time in arbitrary units. We see that temperature is a periodic function of time.

The absolute value of temperature (averaged) along the surface of DRA was $T \approx 600\,K$ for $M = 2$, and $T \approx 1400\,K$ for $M = 4$. Such specifics of the aerodynamics of configurations discussed pose additional requirements to the selection of the DRA material.

5.10 Evaluation of the effect of aerodynamic needle on focusing properties of DRA

The presence of an aerodynamic needle in the aperture of the lens radome antenna results in a deformation of the directional properties of the antenna because the aperture contains shaded (blocked) regions and because energy is scattered by these regions and by the aerodynamic needle.

The following approach serves to evaluate the effect of the electro-dynamic needle on DRA's beampattern. If the aperture is axially blocked, the beampattern of a circular aperture has the form [50]:

$$F(u) = \int rf(r)J(ur)\,dr - \eta^2 \int rf(r\eta)J(ur\eta)\,dr.$$

Here $\eta = Ra/R$, Ra is the blocked disk diameter, $2R$ is the DRA diameter, and $u = 2\pi R \sin(\vartheta/\lambda)$. Since $\eta \ll 1$, we have

$$F(u) = \int rf(r)J(ur)\,dr - 0.5\eta^2 \Lambda(u\eta) \qquad (5.8)$$

where $\Lambda(z)$ is a cylindrical function of the first kind.

Estimates based on (5.8) show that blocking of the aperture increases side lobes and decreases the width of the beampattern. If the aperture blocking is less than 5%, the side lobes grow by less than 0.5 dB.

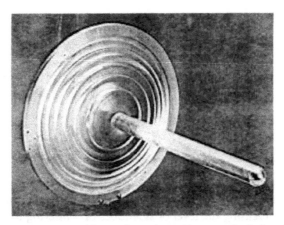

Figure 5.28. Radome antenna with aerodynamic needle on a spherical surface [44].

To evaluate the effect of aerodynamic needle on the diagram-forming properties of DRA experimentally, MWDO elements were used, created on a conical [50] (half-cone angle 35°), spherical ($R \approx D$ and $R \approx D/2$) [44] and parabolic [51] surfaces. Figure 5.28 is a photograph of a DRA on a spherical surface with a radius $R \approx D$ and optimal aerodynamic needle. Numerical simulation of the diagram-forming properties of DRA was carried out using a suitable algorithm of calculating the diffraction integral [26].

When conducting experimental studies, the far-zone condition was satisfied, and nonuniformity of the phase distribution did not exceed $\pi/8$. The DRA gain was found by comparing with a reference source. The minimum level of detected signal was -35 dB. The total error of measurements is shown in table 5.11. The study was run in the millimetre wavelength range.

DRA of two types were studied in the experiments. For the first type of DRA, the phase structure was computed in the approximation of an 'infinitely thin' element. For the second type of DRA, the thickness of dielectric material was taken into account in calculations of the phase structure; it included a thickness of a 'technological' layer on which the phase-coding profile was actually fabricated [44].

Table 5.11.

Level of signal (dB)	Error of signal level measurement (dB)	Error of angular position (minutes of arc)
-10	± 0.54	11.8
-20	± 1.44	19.5
-30	± 3.73	45.4

Table 5.12.

Gain (dB) (without needle)	Gain (dB) (with needle)	Half of scanning angle (degrees)
27.5	26.9	0
25.6	25.3	5
24.9	25.0	10
25.2	24.8	15
26.9	26.1	25

The results of experimental studies confirmed that the effect of the aerodynamic needle on the diagram-forming properties of DRA manifests itself only in increasing the level of side lobes of scattering by 0.6–0.9 dB in the entire range of the investigated scanning angles (see table 5.12).

Experimental studies and a comparison with the result of [26] established that the DRA gain is independent of the degree of concavity of its surface. With a given degree of concavity of the surface, the DRA gain is independent of its shape (cone, parabola, sphere). Furthermore, taking into account the 'technological' phase layer in finding the phase-coding DRA profile leads to a narrowed beampattern of the DRA, to a reduced level of scattering side lobes, to increased scanning angles and a smaller reduction coefficient of the antenna by a factor of more than 2 in comparison with 'infinitely thin' DRA. This happens because taking the refractive layer into account significantly affects the position of phase zone boundaries on the DRA surface in comparison with the case of 'infinitely thin' diffractive element (see chapter 3). Figure 5.29 shows radii of phase inversion zones on the conical DRA surface as functions of zone number. We see that taking the refractive layer into account significantly affects the phase structure of the diffractive element. A consequence of this is the fact that if the DRA diameter is fixed, the calculated gain is somewhat higher if the infinitely thin element approximation is used.

Moreover, if a spherical wavefront is incident on a planar diffractive lens-type antenna or if the antenna is on a non-flat surface, it is necessary to take into account and correct the height of the phase increment of the phase-coding layer. For instance, it is readily shown using as an example the incidence of a spherical wave formed by an axial pointlike radiation source on to a flat diffractive element, that for two-level phase quantization the height of the phase step is

$$h_N = \frac{\lambda}{2(\sqrt{n^2 - \sin^2 \alpha_N} - \cos \alpha_N)}, \qquad \alpha_N = \arctan(R_N/A).$$

If the wavefront is plain, this expression transforms into a familiar formula

$$h = \frac{\lambda}{2(n-1)}.$$

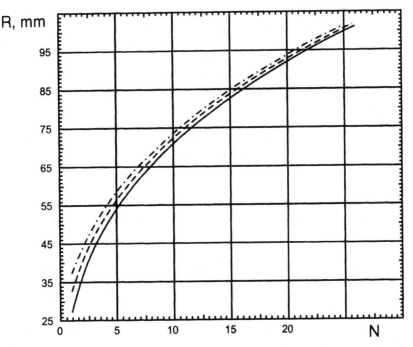

Figure 5.29. Zone radii as functions of the refractive layer thickness for conical diffractive element with parameters $2\alpha = 70°$, $nl/\lambda = x$: ——, $x = 0.00$; – – – –, $x = 1.73$; – · – · –, $x = 3.47$.

Similar expressions are easily obtained for an arbitrary curvilinear surface of the diffractive antenna.

Measurements of the reflection coefficient of the DRA surface showed that to within measurement accuracy the reflection coefficient does not exceed 15% (the value of the standing wave coefficient $K_{st}U$ varies by 3.5%).

5.11 Creation of DRA on an optimal aerodynamic surface

Investigations showed that the so-called ogival surface shape is optimal for the aerodynamic shape of the DRA. With this type of DRA surface, it is possible to reach minimal values of Cx in a wide range of velocities. However, the beampattern-forming properties of such DRA are hardly acceptable.

To test this statement, the DRA of the ogival shape (figure 5.5(b)) was tested experimentally. Experimental studies of this type of diffractive antenna showed the results outlined below. The comparison of a beampattern parameters for DRA of ogival shape with the characteristics of DRA on spherical, conical and parabolic surfaces (without aerodynamic needle) can be summarized as follows. The DRA gain of the ogival surface is less than the gain of DRA on spherical and parabolic surfaces by an average of 3 dB; the level of the first maximum on the beampattern increases in comparison with the two shapes above by 6–8 dB; the level of the first minimum increases by 10–12 dB. Furthermore, it appears that with ogival-shaped DRA it is not possible to achieve scanning angles of the beampattern of more than ±2–5 degrees of arc.

These features mostly stem from large incidence angles of the electromagnetic wave on the DRA surface and fairly large diffraction angles. It is probably possible to reduce this effect by applying special coupling coatings.

5.12 Application of diffractive radome antennas in automobiles and satellite antennas

Systems of direct satellite TV broadcasting (DSTVB) that send TV programs from a satellite directly to home TV sets [52, 53] are becoming more and more widespread. An important element of the receiving equipment is the antenna whose parameters are to a very large extent decisive for the quality of reception. At the moment DSTVB systems mostly use parabolic reflector antennas and their modifications—parabolic antennas with displaced irradiator. As a rule, irradiators of parabolic reflector antennas are placed at the focal point of reflectors, that is, in the path of the propagating radio waves, thus leading to additional losses; blocking by the irradiator power supply, by irradiator mounting parts and by the frequency converter which is typically combined with irradiator, also produce negative effects. Parabolic reflector antennas have substantial dimensions, low wind stability, undesirable design constraints and insufficient electromagnetic compatibility.

In view of more demanding requirements to the stability of satellites on heliostationary orbits and their electronic compatibility, it becomes necessary to design antenna systems that allow utilization of a single antenna for simultaneous reception of signals from several correspondents or for fast scanning by the antenna beampattern with stationary aperture. Such an antenna device which at the same time functions as an aerodynamic radome can be developed on the basis of diffractive antennas [26, 54]. The main advantages of such antennas would be:

- A system of detectors placed along the focal surface of the stationary antenna can be used for simultaneous reception of signals from several satellites (signal sources).
- The application of lens-type antennas permits two effects to be achieved at the same time: using this antenna as an aerodynamic radome for lowering wind loads, and improving the operating conditions for the

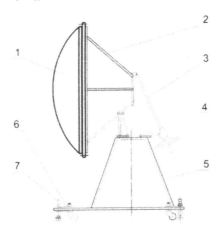

Figure 5.30. Pilot diffractive antenna for reception of satellite TV signal: 1, radome antenna; 2, 3, 5–7, mounting parts; 4, adjustment screw.

reception unit by protecting it from the aggressive factors of the surrounding environment.

• It becomes possible to design the external appearance of the antenna almost arbitrarily.

• The use of reflector parts of MWDO elements as antennas makes it possible to reduce their effective scattering surface (ESS) through frequency selectivity on non-operating frequencies. Such diffractive reflector antennas can be used to solve problems of electromagnetic compatibility, for instance, in radio–technical complexes in which several antenna systems are installed for different frequency ranges, in the immediate vicinity of one another.

Figure 5.30 is a diagram of a pilot design of diffractive antenna for reception of satellite TV signals (foam-polystyrene, $n = 1.3$, diameter 1.2 m). Such antennas make it possible to improve their aesthetic perception and their harmony with the exterior of buildings, structural and architectural complexes.

Another promising approach to DRA application is as follows. Along with traditional approaches to improving automobile technology, intensive work on a new class of onboard radio equipment is being done now in designers' offices: radiolocation systems and sensors in a car. Along with earlier radar sensors for measuring velocity and distances, creation of multifunctional systems of radar 'vision' was started in a number of countries (USA, Germany, Russia, France, Japan, Korea). Such systems, possessing high information capacity, make it possible to generate radar images of the road, of other cars and of the surrounding landscape elements, and also to measure all the necessary parameters of motion of the vehicles. These systems will improve driving safety, will lead to a system of automatic

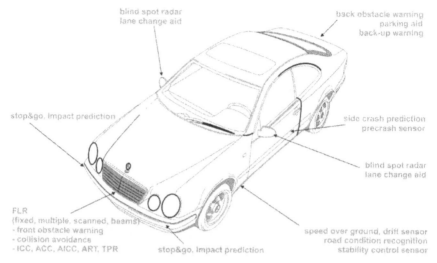

blind spot radar
lane change aid

back obstacle warning
parking aid
back-up warning

stop&go. impact prediction

side crash prediction
precrash sensor

blind spot radar
lane change aid

FLR
(fixed, multiple. scanned, beams)
- front obstacle warning
- collision avoidance
- ICC, ACC, AICC, ART, TPR

stop&go. impact prediction

speed over ground, drift sensor
road condition recognition
stability control sensor

Figure 5.31. Concept of placing warning sensors in a car as envisaged by DaimlerChrysler AG [56].

driverless control of the vehicle, of driving in limited or zero optical visibility conditions, and will certainly create a number of additional possibilities. According to the concept of DaimlerChrysler AG (figure 5.31), the modern automobile will be virtually 'larded' by various safety driving sensors, one of which will be the front view radar.

One of the main components of the system is the antenna. At this moment, various developers formulate different requirements to antenna's parameters.

For instance, DaimlerChrysler AG proposes the following requirements to the antenna of the automotive radar [56]:

- Scanning angles range: $-15°$ to $+15°$
- Scanning step: $0.5°$ to $1°$
- Beam width (at -3 dB level):
  1. Azimuthally: $<1°$
  2. Elevation angle: $3°$ to $10°$
- Level of side lobes: <-25 dB
- Polarization: linear horizontal
- antenna thickness: <10 cm
- Power emitted: <1 W
- Frequency range: 76–77 GHz
- Cost of antenna (assuming production of 100 000/year): $\sim$\$10

Requirements formulated by the Korean company LG to the automotive radar antenna are somewhat different:

1. Method of beam scanning: fixed beam for target irradiation and three beams triggered to signal reception mode
2. Parameters of the antenna and of the beampattern:
   (a) central frequency (GHz): 76.5
   (b) gain (dBi): 32 (for frequency band ±500 MHz)
   (c) level of side lobes (dB): −20
   (d) frequency band (MHz): ±500
   (e) polarization: linear horizontal
   (f) beam width: in azimuth (3.0°) and elevation angle (3.5°) (at −3 dB level)
3. Overall dimensions
   (a) Dimensions (mm): 106 mm × 95 mm × 17 mm
   (b) Weight (g): 210
   (c) Material of aerodynamic radome: polystyrene
   (d) Wind load (km/h): 300
   (e) Temperature range (°C): −45 to +85.

A number of companies are working on a pilot automotive radar. For instance, Phillips Research Laboratories, Redhill, England, announced as early as 1995 the development of a pilot car radar [57] with the following basic parameters:

- Basic frequency: 76.5 GHz
- Modulation: FM
- Type of beam scanning: frequency
- Beam parameters: by azimuth −1.5°, by elevation angle 6°
- Scanning rate: 100 MHz/degree
- Antenna gain: 33 dB
- Number of beams 3 (5), selected
- Scanning range: ±4° (azimuthally).

General requirements are simplicity of fabrication and minimal cost. Special research established [58, 59] that the front vision radar cost should not exceed $92 with annual production of 100 000, and the cost of the antenna proper must not exceed 10% of the radar cost. Note that the aerodynamic radome is a separate part which is envisioned as a component of the antenna. It is possible to satisfy these sufficiently stringent requirements to antenna parameters of minimum cost by combining the functions of the beam-formatting element with those of a vehicle's part. A simple MWDO-element-based multibeam antenna of lens or reflector type can be incorporated into a part of automobile's design, for instance, as part of the hood, without disrupting its design [55, 56]. Figures 5.32 and 5.33 show two versions of designing such DRAs as part of automotive radar project.

The main advantages of this approach are: simplicity of manufacturing; low cost; integration of the functions of the aerodynamic antenna radome, a

Figure 5.32. Radome antenna of an automotive radar fabricated as an element of bumper [56]: 1, shape of bumper prescribed by automobile design; 2, aerodynamic radome—diffractive antenna; 3, phase profile of the diffractive antenna—stiffening ribs; 4, radio-transparent bumper material.

diagram-forming device and a vehicle's part. The results are an improvement in the parameters of the beampattern of the antenna proper (as a result of eliminating the aerodynamic radome that adds losses and distorts the beampattern), reduced overall dimensions of the radar; substantial drop in the radar cost through elimination of some of its parts etc. The formulated requirements to the antenna of the automotive radar are met in the range of scanning angles up to ±25–30 degrees of arc.

Figure 5.33. Diffractive antenna on freely suspended zones integrated with the manufacturer's emblem [56].

5.13   Integration of diffractive element and aperture diaphragm

Certain possibilities of improving the scanning properties of antennas whose focusing element is a diffractive lens (DL) with a structure placed on a spherical or aspherical surface, were shown in [71] on the basis of a combined geometrical optics and wave approaches.

Development of automotive radar antenna involves parallel solution of two problems: making the effective aperture of the antenna independent of the orientation of the beam axis and reducing aberrations to a level at which the beampattern is sufficiently close to diffraction-constrained within the maximum possible range of scanning angles. The first problem can be solved by using an aperture diaphragm of variable diameter, and the second by moving this diaphragm to the object space, that is, into the space between the irradiator and the focusing element. Note that the focusing element should have non-zero spherical aberration because this is the only case in which moving the pupil permits influencing all field aberrations (beginning with the coma).

The direction of the beam axis of the antenna consisting of the irradiator and a diffractive lens can be controlled by displacing the irradiator relative to the lens axis. However, it is preferable to use a chain of irradiators mounted in the front focal plane of the lens antenna (a matrix of irradiators can be used in a more complicated radar that serves to obtain a two-dimensional image of obstacles). In this case it is easy to realize any of the above-described modes of radar functioning; no mechanical devices are needed and inertia ceases to be a factor. The only significant drawback of this solution is the discreteness of possible scanning angles as dictated by the step of the string of irradiators.

When the sequence of irradiators is coupled to the diffractive lens, the effective aperture of the antenna can be made independent of the scanning angle and the pupil can be moved out in the following way. The axes of the beampatterns of irradiators of the string must intersect at the same point on the beam axis at a distance t that the pupil must be moved to. The beam width of each irradiator must depend on its distance from the axis of the diffractive lens and be chosen from the condition of constant effective aperture of the antenna as a whole.

Antennas were designed in two stages. First, design parameters of the antenna were found using geometrical optics techniques [72, 73], and then numerical experiments were run to study the diagram-forming properties of the antenna in the scalar approximation.

The geometrical optics computation of antenna parameters was run in the following manner. From each irradiator a divergent bunch of rays was sent to the diffractive lens, with the aperture ratio that ensured constant prescribed width (in the meridional plane) of the bunch of rays that diffracted on the lens structure into the working order of diffraction. It is the width of

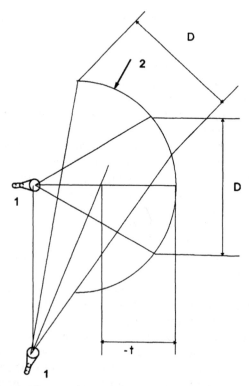

Figure 5.34. Schematic diagram of an antenna with a diffractive focusing element: 1, irradiators; 2, diffractive lens. 2001 IEEE. Reprinted, with permission, from [71].

the diffracted beam (which is collimated in the first approximation) that determines the effective antenna aperture D for the scanning angle at which this beam propagates (figure 5.34).

Aberration-caused divergence of the diffracted beams was reduced by optimizing it to an acceptable level in the scanning angle range that is maximally possible for this arrangement. In designing an antenna that includes a diffractive lens and whose structure is placed on a spherical surface, optimization was run using such parameters as surface curvature c, distance from the apex to the entrance pupil t, and coefficients of the aspherical deformation of the recording eikonal b_{2p+3} of the diffractive lens. It was assumed that the distribution of the spatial frequency structure of this diffractive lens is described by an expression of the type

$$\Omega(\rho) = \frac{1}{\lambda_0}\left[\Phi\rho - 2\sum_{p=0}(p+2)b_{2p+3}\rho^{2p+3}\right] \qquad (5.9)$$

where ρ is the distance to the optical axis, $\Phi = 1/f$ is the optical focal power of the diffractive lens and f its focal length, λ_0 is the recording wavelength

Table 5.13. Design parameters and characteristics of the antenna incorporating a diffractive lens on a spherical surface.

Antenna parameters	Values of parameters for two values of the ratio of the effective aperture to the working wavelength D/λ	
	30	60
F (mm)	148.64	297.13
cf	-1.3373	-1.3203
$B_3 f^3$	1.5714	1.1970
$b_5 f^5$	-5.4624	-5.2262
$b_7 f^7$	10.5055	14.099
$b_9 f^9$	-7.1143	-14.300
t/f	-0.5470	-0.603
$\pm \omega$ (degrees)	36	27.7
2θ (degrees)	4.66	2.33

which in this case is equal to the working wavelength λ of the radar and b_{2p+3} are the coefficients of the aspheric deformation of the recording eikonal [72, 73].

When designing an antenna that includes a diffractive lens whose structure is placed on an aspherical surface, coefficients of aspherical surface deformation, σ_3–σ_7, in the equation of the surface

$$cz - 1 + \sqrt{1 - (c\rho)^2} - \tfrac{1}{8}\sigma_3(c\rho)^4 - \tfrac{1}{16}\sigma_5(c\rho)^6 - \tfrac{5}{128}\sigma_7(c\rho)^8 = 0 \qquad (5.10)$$

were added to the parameters listed above.

Optimization was run by the method described in [74], using the functions Q_1 and Q_4 that evaluate the quality of optimization via the parameters of the ray scattering diagram [75]. Computations were run for two values of the ratio of the antenna's effective aperture to the working wavelength: $D/\lambda = 30$ and $D/\lambda = 60$.

The results of optimization are given in tables 5.13 and 5.14. In these, ω is the maximum admissible scanning angle and 2θ is the diffraction limit of the beam angular width (angle between the first zeroes).

Among other things, we see in tables 5.13 and 5.14 that the range of acceptable values of scanning angles is considerably wider for $D/\lambda = 30$, and is limited in this case only by the fact that if we go beyond the limiting angle, some of the rays emitted by the irradiator will be blocked by the edge of the diffractive lens (shown in figure 5.34)

Furthermore, the data shown in tables 5.13 and 5.14 also imply that a transition from spherical to aspherical surface of the diffractive lens antenna improves the scanning angles by increasing them by about the width of the beampattern 2θ. At the same time, the technological difficulties

Table 5.14. Design parameters and characteristics of the antenna incorporating a diffractive lens on aspherical surface.

Antenna parameters	Values of parameters for two values of the ratio of the effective aperture to the working wavelength D/λ	
	30	60
f (mm)	152.71	307.13
cf	-1.0313	-1.0133
$b_3 f^3$	3.0671	2.0654
$b_5 f^5$	-22.008	-15.339
$b_7 f^7$	51.168	47.566
$b_9 f^9$	-19.404	-76.002
$\sigma_3 f^3$	18.0	43.146
$\sigma_5 f^5$	-19.711	-344.63
$\sigma_7 f^7$	0	950.18
t/f	-0.46	-0.4878
$\pm\omega$ (degrees)	40.9	31
2θ (degrees)	4.66	2.33

in fabricating an aspherical surface are considerably higher and more severe than for a purely spherical shape.

An analysis of the diagram-shaping properties of the diffractive lens, the applicability of geometrical optics techniques in this case, and the acceptability of the results obtained were checked by computing the Fresnel–Kirchhoff integral. The series of numerical experiments conducted for this purpose confirmed, on the whole, the tendency of the main characteristics of diffractive lenses to vary, and also identified certain quantitative differences. We must immediately mention that in computing a diffractive lens by the geometrical-optics method, we do not take into account the constraints of the diffractive efficiency of the diffractive lenses and a finite (small) number of zones within the diffractive lens aperture. Furthermore, both the geometrical optics and the wave simulation were carried out in the approximation of infinitely thin diffractive lens structure. When calculating the diffraction integral in the first approximation, losses were taken into account that were caused by creating the lens of a dielectric material. At each point of integration on the lens surface, the incidence angle was found for a flat zone plate as well, and reflection losses and absorption in the material of the diffractive lens were calculated. It was assumed that the dielectric is characterized by the following optical constants: refractive index 1.5, absorption coefficient 0.001. This procedure is equivalent to prescribing a specific nonuniform irradiation distribution over the diffractive lens aperture.

Since the concept of 'quality of diffractive lens' is not strictly defined, it was necessary to completely control the shape of the diffractive lens when

Table 5.15. Additional design parameters and characteristics of antennas.

Parameter	Spherical diffractive lens		Aspherical diffractive lens	
	$D/\lambda = 30$	$D/\lambda = 60$	$D/\lambda = 30$	$D/\lambda = 60$
Field of view, n_φ				
Geometrical optics	7.7	11.9	8.8	13.3
Wave	7.1	11.0	8.3	12.9
Number of complete kinoform zones in the diffractive lens structure	16	32	13	23
Maximal design diameter of the diffractive lens in units of λ	53	118	49	88
Width of the last zone in units of λ	1.50	1.43	1.60	1.57

running numerical experiments: its width at the half-power level and the first minima, the level and type of placement of the first type maxima, and reduction of gain in the main maximum. As for the main maximum, there is a generally accepted tolerance for this parameter: the drop in gain as a function of electric field strength in the main maximum must not exceed $-3\,\mathrm{dB}$.

As follows from the calculations, the level of side lobes of the beam-pattern in the entire range of scanning angles was less than $-17\,\mathrm{dB}$, and the loss of gain in the main maximum—less than $-3\,\mathrm{dB}$ in the range of scanning angles up to 30°. We remind the reader that when evaluating the limiting scanning angle from the ray scattering diagram for a given version of the antenna, the obtained result was $\pm w = 31°$ (see table 5.14).

From the standpoint of operation of automotive radar as a whole, not only the absolute values of limiting admissible scanning angles $\pm w$ are important, but also the number of resolved elements in this field of view, that is, the fields of view evaluated in units of width of the antenna's beam-pattern: $n_\varphi = \pm w/2\theta$. The corresponding data are given in table 5.15, for the geometrical optics and for wave computations. The results shown demonstrate that an increase in the lens aperture leads to a significant increase in n_φ, although the transition from a spherical diffractive lens surface to an aspherical one hardly affects the value of n_φ.

At the same time, the design parameter of the diffractive lens on an aspherical surface is considerably smaller than on a purely spherical one. In addition, the width of the last kinoform zone in the structure of the diffractive lens is wider for the aspherical lens than for a spherical one.

An analysis of the characteristics of antenna systems discussed above demonstrates that the obvious 'price' for the substantial increase in scanning angle achieved by moving the aperture diaphragm into the space between the irradiator and the diffractive lens is an increase in the design diameter of the lens (see figure 5.34).

We have thus discussed one of the approaches to increasing scanning angles of diffractive lens-type antennas. The characteristics achieved with this approach in the basic parameters (scanning angle and distortions of the diffractive objective lens) met the requirements formulated for antenna systems of automotive radar, among other devices. We must also point out that the approach of two-stage optimization of diffractive lenses suggested in the paper cited above (optimization based on geometrical optics and adjustment of the characteristics based on wave simulation) is sufficiently efficient. Further improvement of limiting antenna parameters is possible by applying wave optimization.

5.14   Selected problems facing the creation of automotive radar

In general an automotive front view radar, when evaluated in terms of the type of problems solved and information extracted, can be subsumed under one of the following types:

1. Radar for measuring the automobile–obstacle separation distance and approach speed. This is in fact a collision prevention system.
2. Multifunction radio vision system allowing generation of images of the road, automobiles and obstacles.
3. A radar that makes it possible not only to generate radio images of various objects but also to identify and classify them.
4. A radar that makes it possible to implement 'autopilot' automobile driving.

One of the important parameters of a system of automatic classification of automobiles using a millimetre wavelength-range radar is the quality of the image generated. Ideally, this quality must be equivalent to that of the optical image. At the same time, cars in the millimetre wavelength range are essentially three-dimensional objects (targets) and, from the standpoint of the radio image generated, are an ensemble of 'sparkling dots' whose parameters change with time.

To evaluate the requirements to the automotive radar of the third type (of those listed above), special research was carried out using a radar with the following characteristics:

- Frequency: 35 GHz
- Modulation type: pulse; pulse length 60 ns
- Emission power: 70 mW
- Antenna type: reflector diffractive antenna with a small revolving reflector located at the focal point

- Beampattern width: 1.2°
- Scanning angle: ±12°
- Gain of the antenna: 37.8 dB.

Functionally, the radar consisted of three units: antenna unit, analogue-to-digital converter, and the unit for processing and visual presentation of data. To protect the antenna from corrosion and the effects of visual-range radiation on antenna characteristics, it was coated with a special layer whose properties are described below.

Typical images of two types of automobiles, (A) a heavy Kamaz lorry and (B) a small Moskvich car, shown in [77] and an analysis of the data given there shows that depending on the mutual orientation of radar's antenna and the target automobile, the radio image of the vehicle changes with time and modifies its structure. Furthermore, the radio images of automobiles that differ in class, size and structure are found to be quite similar.

This result becomes clear if we take into account the following factors. In the millimetre wavelength range an automobile is a strongly reflecting object. Its radio image is obtained as a set of so-called 'sparkling dots' reflected from various parts of the vehicle. The number and distribution of these 'sparkling dots' are dictated by the shape of automobile's surface and the resolving power of the radar.

The fact that radio images of an automobile vary from one moment of time to another and appear to be 'quasi-random' is demonstrated by the results given in [77]: successive radio images of the same automobile do not correlate among themselves, while the corresponding autocorrelation functions have very different structure.

Generation of high-quality dynamic radio images of automobiles and obstacles required for the identification and classification of radio targets is a very complex and separate problem [62]. It would be possible to reduce the effect of 'sparkling dots' by using a broadband frequency signal. However, this approach is hardly possible owing to the narrow frequency band allocated to the automotive radar [60].

5.15 Omnidirectional zone-plate-based antennas

We witness huge progress now in the development of micro- and pico-cell communication lines in city and countryside condition that work in the millimetre wavelength range [65–67]. At the same time systems of wireless broadband radio access are being developed for the millimetre wavelength range (LMDS: Local Multipoint Distribution Systems; MVDS: Multipoint Distribution Systems; MMAC: Multimedia Mobile Access Communication; PCS: Personal Communication Systems; UHTS: Universal Mobile Telecommunication Systems etc).

A large number of very different millimetre-range antennas are required for the operation of such systems. Firstly, they must incorporate antennas for base stations with azimuthally circular or sector-organized beampattern. Secondly, they must include narrow-beam antennas for reception terminals.

A successful design of the omnidirectional millimetre-range antenna must meet the following obvious requirements: it must be simple, low cost and efficient.

Biconical antennas can be used as omnidirectional antennas for base stations.

A biconical antenna consists of two metal cones directed apex-to-apex, and a dielectric lens installed between the cones and the driving circuit. The size of the biconical horn aperture determines the width of the beampattern in the elevation plane.

A zone plate replacing the dielectric lens in a biconical antenna on a cylindrical surface will reduce the weight of the entire antenna and suppress losses to absorption of millimetre radiation in the lens material, and will also reduce the longitudinal size of the antenna. Furthermore, the frequency characteristics of the diffractive lens increase the noise protection of the system. Note that by choosing the zone structure of the diffractive lens it is possible to shape the beampattern of a special form, for instance, it is possible to design antennas with a beam somewhat tilted downwards and thus extend the zone of coverage of LMDS cells.

The internal structure of a biconical antenna is shown in figure 5.35(a). Antennas for different polarizations can be designed using identical diagrams and may differ in the choice of wave driving devices. Antennas with vertical polarization use coaxial waveguide driver for the TM_{00} wave in the radial line, and antennas with horizontal polarization use a slotted waveguide driver for the TE_{11} wave with rotating field structure. Biconical horn is fabricated as two coaxial aluminium cones.

A modified millimetre-range omnidirectional antenna in the horizontal plane consists of a flat diffractive reflector antenna and an irradiator designed as a conical horn.

The gain of the omnidirectional antenna can be evaluated using the following expression:

$$G(\text{dB}) = 10\lg(113.889/\theta).$$

Here θ is the beampattern width in the elevation angle plane.

The horn is installed at the apex of the reflecting cone which, together with a diffractive reflector (the zone plate), creates the beampattern which is omnidirectional in the horizontal plane. By choosing the cone angle different from 90°, it is possible to control the tilting of the beampattern in the vertical plane.

Increasing the aperture of a conical horn narrows its beampattern and increases its gain. On the other hand, increased aperture of a conical horn

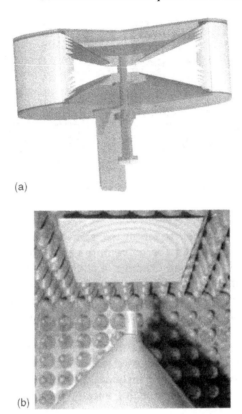

(a)

(b)

Figure 5.35. (a) Biconical antenna with diffractive lens for LMDS systems. (b) Omnidirectional antenna with conical reflector based on amplitude-type reflector zone plate.

reduces the area of the conical reflector, resulting in wider beampattern and reduced gain of the antenna as a whole. A conical horn with aperture diameter 1.4–1.5λ and length 8λ was chosen from optimizing for minimal overall dimensions and maximum gain. The diffractive antenna was chosen with the aperture ratio $D/F = 1.79$ (figure 5.35(b)).

To reduce the dimensions of the omnidirectional antenna, the space between the diffractive antenna (the zone plate) and the conical reflector is filled with a dielectric with low losses at microwave frequencies. It also plays the role of a protecting radome (figure 5.36).

Another modification of the omnidirectional antenna is created by replacing the diffractive reflector by a diffractive transmission-type lens and by shifting the conical horn-type reflector from inside the antenna to the area of the focal point of the lens (figure 5.37). In this design, the entire area of the conical reflector is used to the maximum, and the effect of the wave reflected from the flat diffractive reflector on to the conical

Figure 5.36. A version of an omnidirectional antenna with a zone plate for the 60 GHz range.

horn is reduced. As a consequence, the gain of the antenna as a whole increases and the beampattern becomes more symmetric in the vertical plane.

Figure 5.38 shows, as an example, a characteristic beampattern of an omnidirectional antenna using a zone plate and a conical reflector.

When a diffractive reflector antenna is used with an off-axis focal point, it becomes possible to design an omnidirectional antenna that completely utilizes the entire surface of the conical reflector and does not suffer from mutual blocking by the antenna and the reflector (figure 5.39).

The same principle can be used to design antennas with *sectoral* organization of beampattern, for instance, by replacing a conical reflector with a segmented one, or a cone with truncated pyramid etc. (figure 5.40).

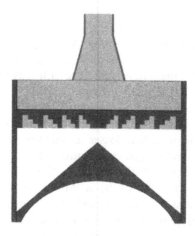

Figure 5.37. Omnidirectional millimetre-range antenna with phase-inversion zone plate.

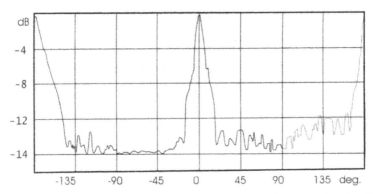

Figure 5.38. Typical beampattern of an antenna with conical reflector for the 60 GHz range.

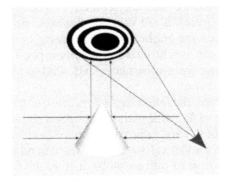

Figure 5.39. Omnidirectional antenna with flat diffractive reflector and off-axis focal point.

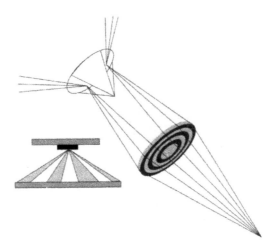

Figure 5.40. Diffractive antenna with segmented beampattern.

5.16   Omnidirectional antennas on cylindrical surface

Let us consider a zone plate fabricated on a cylindrical surface (figure 5.41).

In all papers known to the authors of this book on the study of the properties of cylindrical zone plates for microwave range (e.g. in [68–70]), the method used to compute the boundaries of Fresnel zones was that of geometrical optics, while zone boundaries were chosen by using the classic expression for flat zone plates,

$$R_i = \sqrt{r\lambda i + (\lambda i/2)^2}$$

where r is the radius of the cylindrical surface and i is the number of the Fresnel zone boundary.

At the same time a question arises: to what extent is it justifiable to use the above expression that was derived for a flat surface for computing Fresnel zone plates in the cylindrical geometry? General physical arguments proving that this approach is not valid are discussed below.

When calculating the zones of a cylindrical zone plate, it is essential whether the phases are synchronized at a certain point in space where the cylindrically diverging wave propagates and what are the corrections due to the aperture limitation.

In the general case, the following factors cause differences between zone radii of cylindrical and flat zone plates. Consider a plane wave incident on a screen with a hole. In the first approximation the field behind the screen is a sum of a plane and a spherical waves, and there is no phase shift between these waves. In the case of diffraction on a ring, the following waves add up on the cylindrical surface behind the screen: spherical ones that result from the boundedness of the aperture, and a divergent cylindrical one whose phase is shifted in the asymptotic limit relative to the spherical

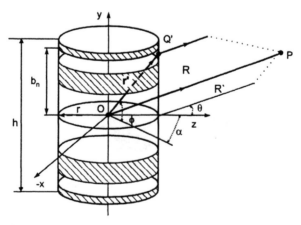

Figure 5.41. Schematic diagram of a cylindrical zone plate.

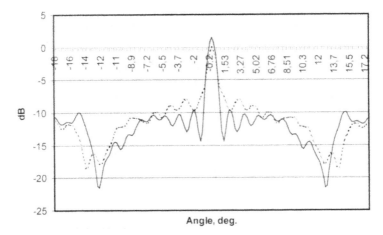

Figure 5.42. Beampattern of a cylindrical zone plate. Solid curve, proposed technique; dashed curve, conventional technique.

waves by $\pi/4$. It is this phase shift that causes the reduction of the Fresnel zone size for the cylindrical zone plate as compared with classical values.

In calculating Fresnel zone boundaries on a cylindrical surface, we applied the following general technique. We calculated the field distribution on the surface of the antenna along its generatrix and obtained, in accordance with the extrema of this function, the Fresnel zone boundaries. As a result of these calculations, we observed discrepancy between Fresnel zone boundaries calculated using the wave diffraction theory and those obtained using geometrical optics.

In one particular case it was possible to fit an empirical analytical formula to calculate Fresnel zone boundaries on cylindrical surface:

$$R_i = \sqrt{r(i\lambda - \lambda/4) + (i\lambda/2 - \lambda/4)^2 - \lambda^2/32 + i\lambda^2/8 + \lambda^2(i-1)\lambda/(212r)}$$
$$+ \lambda^2/220r.$$

The results of calculating the beampattern of a cylindrical zone plate are shown in figure 5.42. The following parameters were chosen for the cylindrical zone plate: antenna height $h = 30\lambda$, $r = 10\lambda$, $\lambda = 8$ mm (figure 5.41).

The frequency band of this antenna is about 12.4%. The calculation of the parameters of the cylindrical antenna showed that if this method is used to find Fresnel zone boundaries or they are found from the empirical analytical curve it is possible to increase the gain and to reduce the height of side lobes in comparison with calculations based on geometrical optics. The main results are summarized in table 5.16.

Note that it is possible to increase the efficiency of an amplitude-type zone plate on a cylindrical surface [69] by placing inside the plate a solid

Table 5.16. Comparative parameters of cylindrical antennas.

Parameters of cylindrical antenna	Relative level of the main maximum	Beampattern width (degrees)	Angle to the first side lobe (degrees)	Height of the first side lobe (dB)
$r = 10\lambda$, $h = 3r^*$	1	0.98	±2.15	−8.88
$r = 10.4\lambda$, $h = 3r$	1.541	0.96	±2.07	−9.1
$r = 11\lambda$, $h = 3r$	1.268	0.98	±2.11	−9.78
$r = 12\lambda$, $h = 3r$	1.635	0.96	±2.12	−9.68
$r = 15\lambda$, $h = 3r$	2.412	0.95	±2.07	−9.47
$r = 15.4\lambda$, $h = 3r$	3.14	0.94	±2	−8.88
$r = 15.4\lambda$, $h = 3r^*$	2.34	0.92	±2	−7.88
$r = 18\lambda$, $h = 3r$	3.135	0.93	±2.07	−9.2
$r = 30\lambda$, $h = 3r$	4.988	0.92	±2.07	−8.58

* Fresnel zone boundaries calculated using geometrical optics.

metal cylinder [70] thus converting the binary amplitude-type zone plate into a two-level one.

5.17 Omnidirectional antennas on arbitrary surface of revolution

By presenting the generatrix of the antenna surface in the general form

$$r(h) = r_0(1 + Bh + Ch^2 + Dh^3)$$

we can obtain, depending on the coefficient with h, various forms of diffractive antenna surfaces. Three typical forms of such antennas are given in figure 5.43.

Figure 5.44 presents the corresponding beampatterns of the antennas. The solid curve traces the beampattern obtained using the technique described above, and the dashed curve traces the beampattern of the corresponding antenna whose Fresnel zone boundaries were determined using the geometrical optics method.

An analysis of the results given above shows that in all cases of simulation of parameters of diffractive antennas fabricated on curvilinear

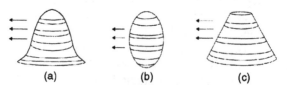

(a)   (b)   (c)

Figure 5.43. Three types of omnidirectional diffractive antenna surfaces: (a) $B = C = 0$, (b) $B = D = 0$, (c) $C = D = 0$.

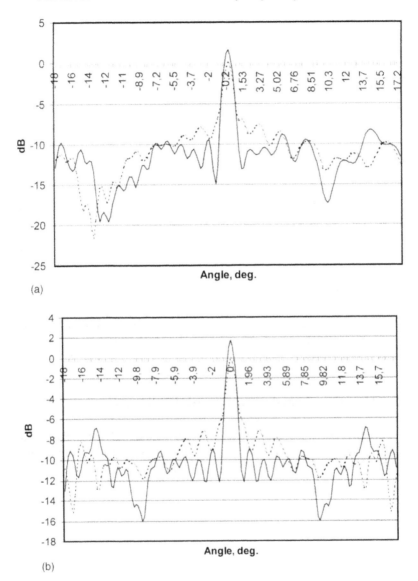

Figure 5.44. (a) Beampattern of the antenna shown in figure 5.43(a). (b) Beampattern of the antenna in figure 5.43(b).

surfaces, the application of the wave approach allows, as compared with geometrical optics approach:

- to increase antenna's gain,
- to lower the height of scattering side lobes,
- to improve the symmetry of beampattern.

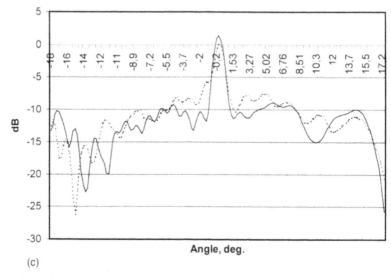

(c)

Figure 5.44. (c) Beampattern of the antenna in figure 5.43(c).

In the general case, the application of the geometrical method of calculating Fresnel zone boundaries on the surface of a three-dimensional diffractive element is not correct.

5.18 Omnidirectional antenna with flat facets

An omnidirectional antenna based on the cylindrical zone plate can be replaced with an antenna consisting of flat surfaces, for example, shaped into a hexagonal prism (figure 5.45). This greatly simplifies the fabrication

Figure 5.45. Omnidirectional antenna with six flat facets.

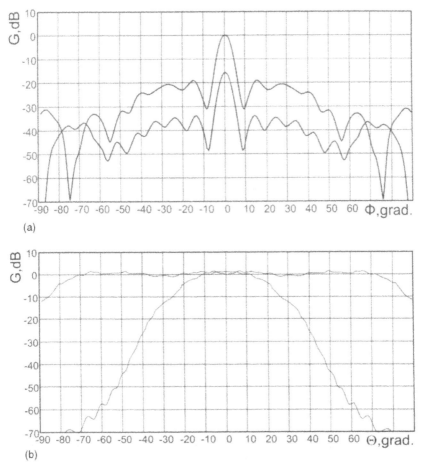

Figure 5.46. Beampatterns of the antenna with flat facets in the vertical (a) and horizontal (b) planes.

of the antenna and allows masking its appearance as an element of the original design, for instance, as light fixtures in a perimeter-fence protection systems [78].

Let us briefly discuss the focusing properties of such an antenna. Figure 5.46(a) shows the beampattern of the antenna in the vertical plane and figure 5.46(b) in the horizontal plane. The upper curve in these figures corresponds to three facets, and the lower curve corresponds to one facet (i.e. to a one-dimensional zone plate).

The figures show that an increase in the number of facets from one to three increases the gain by approximately 16 dB and substantially improves the uniformity of the beampattern in the horizontal plane from ±10° to ±70°.

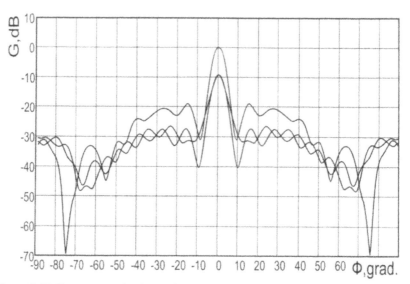

Figure 5.47. Beampattern in the vertical plane for three wavelengths: nominal (upper curve) and nominal ±5%.

Figure 5.47 shows how the shape of the antenna's beampattern in the vertical plane changes when the radiation wavelength deviates from the nominal value by ±5%.

The main effect of deviation of radiation wavelength from the nominal value is a substantial drop in antenna's gain at the nominal focal point (up to 10 dB).

An investigation of the effect of displacement of the radiation source showed that this antenna is sufficiently 'sensitive' to changes of this sort. The main changes in the shape of the beampattern as the radiation source is displaced from its nominal position can be described as broadened beampattern and reduced gain.

5.19   Dielectric reflector zone plate

All types of diffractive reflector antennas based on zone plates contain, to the best of our knowledge, a metal reflector as part of the design. A metal reflector may be flat or profiled. In [63] a completely dielectric reflector zone plate was suggested (figure 2.6 of chapter 2).

In [64] the phenomenon of total reflection of the incident electromagnetic wave from a corrugated interface between dielectrics was considered in the special case of total reflection observed in the reflection from the main interface of these dielectrics.

Let us consider the interface between two dielectrics having dielectric permittivity ε_1 and ε_2, corrugation period d and amplitude A_0 described in the simplest case by the formula

$$A(x) = A_0 \sin(2\pi x/d).$$

Let $\varepsilon_1 = \varepsilon > \varepsilon_2 = 1$ and assume that a wave of unit amplitude and frequency ω is incident on to this interface from the denser medium at an angle ϑ; for the sake of specificity we also assume that the component of the electric field along the corrugation is

$$E^i = e_z^0 \exp(ik\sqrt{\varepsilon}\sin\vartheta x - ik\sqrt{\varepsilon}\cos\vartheta y).$$

Here $k = \omega/c$, and c is the velocity of light.

When a plane wave is scattered by a corrugated interface between two dielectrics, harmonics with amplitudes R_n and T_n are generated in the first and second media:

$$E_m^{(1)} = R_n \exp\left[i(k\sqrt{\varepsilon}\sin\vartheta + 2\pi n/d)x + i\sqrt{k^2\varepsilon - (k\sqrt{\varepsilon}\sin\vartheta + 2\pi n/d)^2}y\right]$$

$$E_n^{(2)} = T_n \exp\left[i(k\sqrt{\varepsilon}\sin\vartheta + 2\pi n/d)x - i\sqrt{k^2 - (k\sqrt{\varepsilon}\sin\vartheta + 2\pi n/d)^2}y\right].$$

If $k\sqrt{\varepsilon}\sin\vartheta = \pi/d$, the minus one harmonic propagates towards the incident wave. If $\sin\vartheta > 1/\sqrt{\varepsilon}$ or $d/\lambda < 0.5$ and $\sin\vartheta > 1/3$ or $d/\lambda < 1.5/\sqrt{\varepsilon}$, only localized waves exist in the optically less dense medium, and only the mirror-reflected and minus first harmonics propagate as bulk waves in the optically denser medium.

This effect was used experimentally [64] in the 8 mm wavelength range. The material of the dielectric reflector was Teflon ($\varepsilon = 2$). The corrugation facets were equilateral triangles with a base of $d = 3.6$ mm and height $2A = 7.2$ mm. The measured reflection coefficient for the frequency 35.7 ± 0.5 GHz was 0.95–0.98.

The effect of replacement of a metal reflector with a resonance dielectric reflector is an increase in the electromagnetic compatibility of the reflector antenna with other devices, increased noise protection (as a result of higher frequency selectivity of the reflector), reduced electromagnetic conspicuousness of the antenna and removal of some redundant metal parts. As a result, this antenna cannot be detected by many magnetic detection devices.

Bibliography

[1] Kreutel R A 1976 'Multiple-beam reflector antenna' Satellite Commununication Advanced Technolology, paper AIAA/CASI, 6th Communication Satellite Systems Conf., pp 441–449

[2] Kumuzawa Hiroyuki *et al* 1979 '*A circularly symmetric dual-reflector multibeam antenn*' in *Int. Symp. Antennas and Prop, Seattle, Washington*, vol 3, pp 748–751

[3] Sazonov D M, Gridin A N and Mishidstin B A 1981 *Microwave Devices* (Moscow: High School Publisher) (in Russian)

[4] Dragone C 1983 'Unique reflector arrangement with very wide field of view for multibeam antennas' *Electron. Lett.* **19**(25/26) 1061–1068

[5] Krichevsky V 1982 'Beam scanning in offset Cassegrain antenna' in *Int. Symp. Dig. Antennas and Propagation, Albuquerque, New York*, 24–28 May, vol 1, pp 257–260

[6] Korchagin Yu A 1990 *Individual Antennas for Reception of Satellite TV* (Voronezh: VTU) p 112

[7] Avdeev S M, Bei N A and Tokarev B E 1983 USSR Patent 117886, MKI HOIQ3/44 'Antenna with electronic scanning'

[8] Avdeev S M, Bei N A and Morozov A N 1987 *Lens Antennas with Electronic-Control Bandwidth* (Moscow: Radio i Svaz) (in Russian)

[9] Hajian M, Kcizer W P M N, Reits B J and Ligthart L P 1997 'Concept of a scanning beam antenna at 35 GHz based on photoconductivity technology' in *20th ESTEC Antenna Workshop on Millimeter Waves: Antenna Technology and Antenna Measurement, ESTEG, Noordivijk, The Netherlands*, 18–20 June, pp 55–66

[10] Lubetskii N B and Masonov V V 1989 USSR Patent 1665438, MKI HOIQ3/44 'Device for electronic scanning'

[11] Zaitsev E F and Fedotov A M 1982 USSR Patent 1072154, MKI HOIQ3/26 'Leaky-wave antenna'

[12] Miroshnikov M M 1977 *Theoretical Foundations of Optoelectronic Devices* (Leningrad: Mashinostroyeniye), p 600

[13] Black D N and Wiltse J C 1987 'Millimeter-wave characteristics of phase-correcting Fresnel zone plates' *IEEE Trans. Microwave Theory and Techniques* **MTT-35**(12) 1122–1129

[14] Garret J E and Wiltse J C 1990 'Performance characteristics of phase-correcting Fresnel zone plates' in *IEEE MTT-S Int. Microwave Symp. Dig., Dallas, TX*, 8–10 May, vol 2, pp 797–800

[15] German Patent 3536348, MKI H 01 Q 15/14. Fresbek'sche Zonenplatte zum Fokussieren von langs einer Achse einfallender Mikrowellenstrahlung

[16] Guo Y J, Sassi I H and Barton S K 1994 'Offset Fresnel lens antenna' *IEE Proc. Microwave Antennas and Propagation* **141**(6) 517–522

[17] Gawronski W and Hellstrom J A 1994 'Antenna servo design for tracking low-earth-orbiting satellites' *J. Guid. Contr. and Dyn.* **17**(6) 1176–1184

[18] Leschuk I I and Tsaliev T A 1995 'Fresnel antennas with offset irradiator' *Radioelectronika* **9** 37–43

[19] German Patent 3801301, MKI H 01 Q 15/23. Fresnel'sche Zonenplatte des Reflektor für eine Mikrowellen-Sende/Empfangsantenne

[20] Huder B and Menzel W 1988 'Flat printed reflector antenna for mm wave applications' *Electronic Lett.* **24**(6) 318–319

[21] Gurtovik I G, Sportsmen V N, Gusev A A *et al* 1989 'Technology of manufacturing and processing of glass-reinforced plastics' in *VNIISPV Proceedings, Moscow, Investigation of Frequency Dependence of the Properties of Glass-Reinforced Plastics in a Wide Range of Microwave Frequencies*, pp 66–69

[22] Gurtovik I G and Sportsmen V N 1987 *Glass-reinforced Plastics for Radiotechnical Applications* (Leningrad: Khimiya) p 160

[23] Chernenko V I, Snezhko L A and Papanova I I 1991 *Technology of Coating by Anodic Spark Electrolysis* (Leningrad: Khimiya) p 128

[24] 1991 'Novel flat reflector diffractive antenna' *Radiotehnika Sverhvysokih Chastot. Ekspress Informatsiya* (Moscow: VINITI) vol 46 pp 24–31

[25] Minin I V and Minin O V 1989 'Focusing elements of diffractive quasioptics and their applications' in *3rd USSR School on Propagation of mm and Sub-mm Waves in the Atmosphere, Khar'kov*, pp 251–252

[26] Minin I V and Minin O V 1991 'Antenna systems in the mm wavelength range based on elements of diffractive quasioptics' in *Radiotechnical Systems for mm and Sub-mm Wavelength Ranges: Research Proceedings* (Khar'kov: Inst. Radiofiziki i Elektroniki AN Ukrainy), pp 120–122

[27] Minin I V and Minin O V 1994 'Elements of diffractive quasioptics, Part I. Fundamental properties' *Avtometriya* 3 110–120

[28] Ronchi I and Toraldo di Francia G 1958 'An application of parageometrical optics to the design of a microwave mirror' *IRE Trans. Antennas and Propagation* **AP-6**(1) 129–133

[29] Kornbleet S 1976 'Microwave Optics' in *The Optics of Microwave Antenna Design* (London, NY: Academic Press)

[30] Provenceher J H 1960 'Experimental study of a diffraction reflector' *IRE Trans. Antennas and Propagation* **AP-8**(3) 331–336

[31] Ronchi L, Russo V and Toraldo di Francia G 1961 'Stepped cylindrical antennas for radio astronomy' *IRE Trans. Antennas and Propagation* **AP-9**(1) 68–74

[32] Ramsay J F and Jackson J A C 1956 'Wide-angle scanning performance of mirror aerials' *Marconi Review* **122**(19) 119–140

[33] Dasgupta S and Lo Y T 1961 'A study of the coma corrected zoned mirror by diffraction theory' *IRE Trans. Antennas and Propagation* **AP-9**(2) 130–139

[34] Andreenko S D, Devyatkov H D and Shestopalov B P 1978 'Antenna gratings for the mm wavelength range' *Dokl. AN SSSR* **240**(6) 1340–1343

[35] Kalinichev V I and Kuranov Yu V 1991 'Diffraction of surface waves on a grid of metal rods and analysis of the dielectric leaky wave antenna' *Radiotehnika i Elektronika* **10** 1902–1909

[36] Zaitsev E F, A.c. No. 1072154. 'Travelling wave antenna' A N Fedotov. publ. 07.02.84, Bulletin No. 5

[37] Seiler M R and Mathena B M 1984 'Millimeter-wave beam steering using diffraction electronics' *IEEE Trans. Antennas and Propagation* **AP-32**(9) 987–990

[38] Andreenko S D, Evdokimov A P and Sidorenko Yu B 1984 'Scanning antenna for the mm range' in *Propagation and Diffraction of Radio Waves in the mm Wavelength Band*, Collected Work, Kiev, pp 208–212

[39] Andreenko S D, Evdokimov A P, Kryzhanovskii V V *et al* 1988 'Scanning antenna for the airplane radiometry system in *Radiophysical Methods and Devices for Probing the Environment in the mm Wavelength Range*, Collected Work, Kiev, pp 154–160

[40] Evdokimov A P, Kryzhanovskii V V and Sidorenko Yu B 1991 'Computation of beampattern of the dielectric waveguide-array emitting system' in *Proc. 1st Ukraine Symposium on Physics and Technology of mm and Sub-mm Radio Waves*, Collected Work, Khar'kov, Ukraine, 15–17 October, p 325

[41] Kryzhanovskii V V, Provalov S A and Sidorenko Yu B 1991 'Energy characteristics of a dielectric waveguide-diffraction grating system' *Radiophysics and Electronics of the mm and Sub-mm Ranges*, Collected Work, Har'kov, pp 104–113

[42] Tamir T (ed) 1975 *Integrated Optics* (Berlin, Heidelberg, New York: Springer)
[43] Barton D K and Ward H R 1969 *Handbook of Radar Measurement* (Englewood Cliffs, NJ: Prentice-Hall)
[44] Minin I V and Minin O V 1992 *Diffraction Quasioptics* (Moscow: Research and Production Association 'InformTEI')
[45] Minin V F, Minin I V and Minin O V 1992 'The technology of numerical experiment' in *Proc. Int. Symp. on Intense Dynamic Loading and its Effects, Chengdu, China,* 9– 12 July, pp 431–433
[46] Dushin V R, Minin I V, Minin O V and Fedotov V G 1987 'Software package for three-dimensional problems of gas flow past objects' in *Numerical Methods of Mechanics of Continuous Media*, Part 2, Krasnoyarsk, pp 66–67
[47] Dushin V R, Minin I V, Minin O V *et al* 1989 'Three-dimensional supersonic flow of real gas around axisymmetrical bodies' Vestnik MGU, ser. 1, No. 4, pp 41–49
[48] Davydov Yu M, Korobitsyn G P and Postnikov V G 1979 'Flow of fluid past blunted bodies with needles and cavities' *Inzhenerno-fizicheskii Zhurnal* **37**(4) 712–716
[49] Antonov A N, Elizarova T G, Pavlov A N *et al* 1989 'Mathematical simulation of oscillatory modes in fluid flow past bodies with a needle' *Mathematical Simulation* **1**(1) 13–23
[50] Minin I V and Minin O V 1990 'Elements of VHF diffractive quasioptics' *Pribory i Tehnika Eksperimenta* **6** 201–202
[51] Minin O V and Minin I V 1988 'Diffractive lenses on parabolic surfaces' *Computer Optics* **3** 8–15
[52] Gvozdenko A A 1992 'Directly broadcasting satellite services' *Zarubezhnaya Radioelectronika* (Moscow: Radio i Svyaz) **4/5** 81–110
[53] *New Scientist* (GB) **127**, No. 1733, 8 September 1990, p 42
[54] Minin I V and Minin O V 1991 'Elements of diffractive quasioptics and mm-range systems using them' in *Radiotechnical Systems for the mm and Sub-mm Wavelength Ranges*: Collected papers (Har'kov: Inst. Radiofiziki i Elektroniki AN Ukrainy) pp 102–109
[55] Minin I V and Minin O V 1994 'Elements of diffraction quasioptics, Part 2. The main applications' *Optoelectronics, Instrumentation and Data Processing* 63–69
[56] Minin I V, Minin O V, Russer P *et al* 1998 'Cost effective production of multi-directional antennas in the mm-wave regime' Final Scientific Report on Contract No. G005830106 of DaimlerChrysler AG, p 71
[57] 'TH am—Automotive Radar' 1995 Phillips Research Lab. 20th Int. Conf. on IR and mm-Waves
[58] Raffaelli L 1995 ARCOM Workshop 'Microwave Vehicular Technology', IEEE MTT-S Conf., Orlando, FL, May
[59] Rose T 1996 MA-COM Workshop 'Microwave MMW Technology for Intelligent Vehicles', MTT-S, San Francisco, 21 June
[60] European Radiocommunications Committee, ERC Decision of 22 October 1992
[61] Petosa A and Ittipiboon A 1999 'Dielectric lenses for LMCS applications' in *Proc. Int. Conf. on Electromagnetics in Advanced Applications, ICEAA–99, Torino, Italy,* pp 273–276
[62] Minin I V and Minin O V 1999 'New methods to identification and classification of the targets in the car radar imaging systems' in *50th Vehicular Technology Conference VTC 1999—Fall, 19–22 September, Amsterdam, The Netherlands,* pp 1924–1928

[63] I V Minin and O V Minin 2000 'The dielectric non-metallic reflecting ZPF Antennas' in *Proc. 25th Int. Conf. on IR and mm-waves, Beijing, China*, 12–15 September

[64] Ilasov S N and Koposova E V 1993 'Full antireflection at corrugated interface between dielectrics. Quasioptical dielectric Echelette resonators' *Pis'ma v ZhTF* **19**(19) 80–86

[65] Lane J and Norbury J 1980 'Propagation factors and communication systems in the band 30–300 GHz' in *IEEE Conf. on Radio Spectrum Conservation Techniques, London*, No. 188, pp 97–102

[66] Bystrov R P, Petrova A V and Sokolov A B 2000 'Millimeter waves in communication systems' *Radioelectronika* **5** 18–25

[67] Lyubchenko V E, Sokolov A V and Fedorova L V 1998 'Millimeter wavelength range communication lines in local information networks' *Radiotekhnika* **12** 68–75

[68] Hristov H 1999 'Variety of cylindrical Fresnel zone plate antennas' in *IEEE Int. Antennas and Propagation Symp., Orlando, FL, Symp. Digest*, July 11–16, vol 2, pp 750–753

[69] Ji Y and M Fujita 1996 'A cylindrical Fresnel zone antenna' *IEEE Trans. Antennas and Propagation* **44** 1301–1303

[70] Ye C F and Tan S Y 2000 'A reflective half-cylindrical Fresnel zone plate antenna with low backward radiation for wireless LAN', *Microwave and Optical Technology Lett.* August

[71] Greisukh G I, Ezhov E G, Minin I V and Minin O V and Stepanov S A 2001 'New possibilities of diffractional antennas for car radar' in *Proc. 2001 CIE International Conference on Radar, Beijing, China*, 15–18 October, pp 649–652

[72] Bobrov S T, Greisuh G I and Turkevich Yu G 1986 *Optics of Diffractive Elements and Systems* (Leningrad: Mashinostroyeniye) p 223

[73] Greisukh G I, Bobrov S T and Stepanov S A 1997 *Optics of Diffractive and Gradient-Index Elements and Systems* (Bellingham, WA: SPIE Press) p 414

[74] Greisuh G I, Stepanov S A and Ezhov E G 1999 'Triple glued radial-gradient objective lenses' *Opticheskii Zhurnal* **66**(10) 92–96

[75] Bobrov S T and Greisuh G I 1985 'Mutual correlation of numerical criteria in evaluation of image quality' *Opt. i spektr.* **58**(5) 1068–1073

[76] Minin I V and Minin O V 1999 *Diffractive Quasioptics and its Applications* (Novosibirsk: SibAgs)

[77] Minin I V and Minin O V 1999 'Unsteadies of the car radar imaging in dynamics' in *50th Vehicular Technology Conference VTC 1999—Fall, Amsterdam, The Netherlands*, 19–22 September, pp 2048–2051

[78] Minin I V and Minin O V 2002 'Diffractional lenses and mirror antennas for mm-waves applications' in *6th Russian–Korean Int. Symp. on Science and Technology, Novosibirsk, Russia*, June 24–30, vol 2, pp 3347–350

Selected bibliography

Baggen L C J and Herben M H A J 1993 'Design procedure for a Fresnel-zone plate antenna' *Int. J. Infrared and Millimeter Waves* **14**(6) 1341–1352

Baggen L C J, Jeronimus C J J and Herben M H A J 1993 'The scan performance of the Fresnel-zone plate antenna: A comparison with the parabolic reflector antenna' *Microwave and Optical Technology Lett.* **6**(13) 769–774

Black D and Ngueyen C 1999 'Metallic zone plates for sectoral horns' in *IEEE Int. Antennas and Propagation Symp., Orlando, FL, Symp. Digest*, 11–16 July, vol 2, pp 730–733

Delmas J-J, Toutain S, Landrac G and Cousin P 1993 'TDF antenna for multisatellite reception using 3D Fresnel principle and multiplayer structure' in *IEEE Antennas and Propagation Symp. Digest, Ann Arbor, MI*, 29 June–2 July

Du H-p 2001 'A new reflector antenna based on the Fresnel principle' *J. China Univ. Posts and Telecommun.* **8**(2) 66–68, 78

Guo Y J and Barton I H 1994 'Offset Fresnel zone plate antennas' *Int. J. Satellite Communications* **12** 381–385

Guo Y J and Barton S K 1992 'A high-efficiency quarter-wave zone plate reflector' *IEEE Microwave and Guided Wave Lett.* **2**(12) 470–471

Guo Y J and Barton S K 1993 'Fresnel zone plate reflector incorporating rings' *IEEE Microwave and Guided Wave Lett.* **3**(11) 417–419

Guo Y J and Barton S K 1993 'Multilayer phase correcting Fresnel zone plate reflector antennas' *Int. J. Satellite Commun.* **11** 75–80

Guo Y J and Barton S K 1995 'Analysis of one-dimensional zonal reflectors' *IEEE Trans. Antennas and Propagation* **43**(4) 385–389

Guo Y J and Barton S K 1995 'Phase correcting zonal reflector incorporating rings' *IEEE Trans. Antennas and Propagation* **43**(4) 350–355

Guo Y J, Barton S K and Wright T M B 1991 *Efficiency and Sidelobe of Multiphase Fresnel Zone Plate Antennas* (London, Ontario: IEEE AP-S) pp 182–185

Herben M and Hristov H 1999 'Some developments in Fresnel zone plate lens antennas' in *IEEE Int. Antennas and Propagation Symp., Orlando, FL*, 11–16 July, *Symp. Digest*, vol 2, pp 726–729

Hristov H D and Herben M H A J 1995 'Quarter-wave Fresnel zone planar lens and antenna' *IEEE Microwave and Guided Wave Lett.* **5**(8) 249–251

Minin I V and Minin O V 1999 'Antennas of mm-range based on the quasioptical diffraction elements for the communication systems' in *50th Vehicular Technology Conference VTC 1999—Fall, Amsterdam, The Netherlands*, 19–22 September, pp 3040–3046

Minin I V and Minin O V 1999 'Diffraction Elements of the mm-waves Planar Integral Optics' USNC/URSI National Radio Science Meeting, Orlando, FL, 11–16 July, p 287

Minin I V and Minin O V 1999 'Low cost multibeam antennas in the mm-wave regime for radar systems of transport means' USNC/URSI National Radio Science Meeting, FL, 11–16 July, p 327

Minin I V and Minin O V 1999 'New class microwave antennas based on the elements of the diffraction quasioptics: advantages and applications' in *IEEE AP-S Int. Symp., Orlando, FL*, 11–16 July, AP-8-002, pp 754–758

Minin I V and Minin O V 1999 'New conception of the low cost multibeam antennas in the mm-wave regime for radar systems of transport means' in *50th Vehicular Technology Conference VTC 1999—Fall, Amsterdam, The Netherlands*, 19–22 September, pp 2053–2056

Minin I V and Minin O V 2001 'Curvilinear diffractional lenses and antennas for mm-wave communications and wireless' in *Proc. Int. Conf. Electromagnetics and Communications ICECOM-2001, Dubrovnik, Croatia*, 1–3 October, pp 155–158

Minin I V, Minin O V, Stepanov S S *et al* 2001 'New possibilities of diffractional antennas for car radar and wireless' in *Proc. Int. Conf. Electromagnetics and Communications ICECOM-2001, Dubrovnik, Croatia*, 1–3 October, pp 360–363

Onedera T and Hoashi T 1999 'A design method of offset Fresnel zone plate reflector as receiving antenna' in *IEEE Int. Antennas and Propagation Symp., Orlando, FL,* July 11–16, *Symp. Digest,* vol 2, pp 742–745

Sazonov D 1999 'Computer aided design of holographic antennas' *APURSI* pp 738–741

Sluijter J, Herben M H A J and Vullers O J G 1995 'Experimental validation of PO/UTD applied to Fresnel-zone plate antennas' *Microwave Optical Technol. Lett.* **9** 1–13

Van Houten J M and Herben M H A J 1994 'Analysis of a phase-correcting Fresnel-zone plate antenna with dielectric/transparent zones' *J. Electromagnetic Waves and Applications* **8**(7) 847–858

Wilcockson P C 1994 'Evaluation of flat reflectors for space and ground station antenna applications' Prepared for the European Space Agency under Purchase Order 142600, pp 1–46

Wiltse J C 1997 'Dual-band Fresnel zone plate antennas' Proc. of SPIE, vol 3062, Orlando, 22 April, pp 181–185

Wiltse J C 1998 'Recent developments in Fresnel zone plate antennas at microwave/millimeter wave' *Proc. SPIE,* vol 3464, San Diego, 23 July, pp 146–154

Wiltse J C 1999 'History and evolution of Fresnel zone plate antennas for microwaves and millimeter wave' in *IEEE Int. Antennas and Propagation Symp., Orlando, FL, Symp. Digest,* 11–16 July, vol 2, pp 722–725

Wiltse J C 1999 'Second-generation zone plate antenna design' *Proc. SPIE,* vol 33765, Denver, 19 July, pp 287–294

Wiltse J C 2000 'Advanced zone plate antenna design' *Proc. SPIE,* vol 4111, San Diego, 1 August, pp 201–209

Wiltse J C 2001 'The stepped conical zone plate antenna' in *Photonic and Quantum Technologies for Aerospace Applications III* (ed E Donkor, A K Pirich and E W Taylor), *Proc. SPIE,* vol 4386, pp 85–94

Y J Guo, I H Sassi and S K Barton 1994 'Multilayer offset Fresnel zone plate reflector' *IEEE Microwave and Guided Wave Lett.* **4**(6) 196–198

Yamauchi J and Nakano H 1999 'Theoretical and experimental investigations of focusing properties of the Fresnel zone lens and plate' in *IEEE Int. Antennas and Propagation Symp., Orlando, FL, Symp. Digest,* 11–16 July, vol 2, pp 746–749

Ye C F and Tan S Y 2000 'A reflective half-cylindrical Fresnel zone plate antenna with low backward radiation for wireless LAN' *Microwave and Optical Technology Lett.*

Zhang W and Jiang G 1999 'Rewiew on the layered Fresnel zone plate lens antenna' in *IEEE Int. Antennas and Propagation Symp., Orlando, FL, Symp. Digest,* 11–16 July, vol 2, pp 734–737

Chapter 6

Applications of diffractive optical elements

The present chapter briefly discusses the main fields from x-rays to the centimetre wavelength band where elements of diffractive optics are applied. There can be no doubt that such a review cannot cover all the aspects and specifics of employing the diffractive focusing elements in the corresponding fields of science and technology; hence, we only outline the main areas where these elements are used and spend more time with certain unconventional and, from our point of view, more interesting applications of these elements.

6.1   X-ray diffractive optics

The need to focus x-ray radiation arose simultaneously with the discovery of the penetrating ability of this electromagnetic radiation band. X-ray focusing elements are required for creation of x-ray microscopes and high-resolution telescopes, x-ray lithography, for recording, transmission and processing of images, for local chemical analysis, x-ray spectroscopy, x-ray interferometry [1–5] etc.

It is very difficult to apply methods of designing and building classical focusing elements created for the visible, infrared and microwave bands to the x-ray wavelength band [3, 4]. The reason for this is that the refractive index of all known materials for soft x-ray radiation ($\lambda = 0.5$–$10\,\mathrm{nm}$) differs very little from unity while the absorption coefficient of most materials is high. The known solutions of the problem of focusing x-ray radiation use various types of diffraction phenomena that are observed in the interaction of radiation with matter. Such are multilayer x-ray mirrors and elements of diffractive optics.

Table 6.1 gives a comparative description of constraints on various types of x-ray elements [1, 6, 7].

Table 6.1.

Type of optics	Glancing incidence optics, 1952 $\varphi_{kp} = \Delta N$	Fresnel optics, 1952 $r_N = Nf\lambda + N^2\lambda^2/4$	Bragg–Fresnel optics, 1986 $N - 1 < [kr]^2/r_1 + [kr]^2/r_2 \leq N$
Chromatic aberration		λ	None
Maximum resolution	Limited by strong aberrations $\delta \approx 1\,\mu m$	$\delta \approx 2\lambda/(1 - N)^{1/2}$ 50–10 nm	Dictated by state-of-the-art technologies and may reach 10 nm
Maximum aperture angle φ_{max}	$\varphi_{max} < \varphi_{kp}$	$\varphi_{max} = 0.5(1 - N)^{1/2}$	Dictated by state-of-the-art technologies
Feasibility of amplitude and phase modulation	Impossible	Impossible	Feasible
Feasibility of manufacturing various x-ray elements	Focusing and scattering elements are feasible	Lenses and gratings are feasible	Lenses, modulators and mirrors are feasible

Multilayer x-ray mirrors are formed of alternating layers of materials that differ greatly in electron density. For this purpose one layer incorporates elements with low atomic number (boron, carbon) and the other heavy metals. In the microwave band, analogues of multilayer mirrors are anti-reflecting coatings and multilayer absorbers that are used to reduce their reflectivity [8]. When multilayer x-ray mirrors are fabricated, extremely rigid requirements are applied to the quality of the underlying substrate; this is an obstacle to creating high-resolving-power devices [9]. The required period of layer repetition in multilayer x-ray mirrors comes to only several nanometres and the corresponding layer thickness is measured in several atomic radii of the material employed.

Another approach to creating focusing elements for x-ray frequencies is the development of x-ray optics based on Fresnel and Rayleigh–Wood zone plates. Applications of Fresnel zone plates are limited by their low diffraction efficiency. To increase this efficiency, it is necessary to switch to phase-type diffractive focusing elements: Rayleigh–Wood zone plates and kinoforms.

An analysis of the limitations of the approximation of diffraction on a 'thin' object in the x-ray band showed that the limiting resolution of flat optic elements, taking into account the specific properties of materials in the x-ray band, is at best 300–500 Å [10].

Elements of planar optics, such as Fresnel zone plates and Fourier optics, possess limited resolution in the nanometre wavelength band. This happens because in the shortwave part of the x-ray band, $\lambda < 30\,\text{nm}$, the thickness of the material required to create the necessary amplitude and phase contrast becomes much greater than the wavelength so that the 'thin' optics approximation becomes invalid.

The topology of Fresnel zones of focusing elements was found by solving the variational problem of optimizing the profile of diffractive x-ray optics elements made of arbitrary material; optimization was carried out to achieve maximum diffraction efficiency [11–13].

An analysis showed that for a real refractive index of the material, that is, for a purely phased element, an optimal solution points to the constant-thickness Rayleigh–Wood plate, and for transparent materials, to the amplitude-type Fresnel zone plate with somewhat modified size of open zones. In the case of a complex refractive index, the zone plate profile can hardly be physically implemented.

For a zone plate with rectangular-profile zones and thickness d, the efficiency σ_N of focusing to the Nth order can be found from the formulas [14]

$$\sigma_N = \begin{cases} 0.25\{\gamma_1^2 + \gamma_2^2 - 2\gamma_1\gamma_2\cos[2\pi d(\delta_2 - \delta_1)/\lambda]\}, & N = 0 \\ \{\gamma_1^2 + \gamma_2^2 - 2\gamma_1\gamma_2\cos[2\pi d(\delta_2 - \delta_1)/\lambda]\}/(N\pi)^2, & N \text{ odd} \\ 0, & \text{otherwise.} \end{cases}$$

Here $\gamma_i = \exp(-2\pi d\beta_i/\lambda)$, β_i and δ_i are the imaginary and real parts of the refractive index of two neighbouring zones, $n_i = 1 - \delta_i - j\beta_i$, $i = 1, 2$.

A novel approach to creating focusing elements was suggested in [15, 18]: combining Bragg diffraction on a crystal with diffraction on an artificially created structure. The Bragg diffraction on a perfect crystal produces a spectrum of diffracted rays in a wide range of angles (about $100''$) dictated by the artificial superlattice. An investigation of the properties of a one-dimensional Bragg–Fresnel zone plate on a silicon single crystal [16] with central zone half-width $r_1 = 10\,\mu\text{m}$, last zone width $0.5\,\mu\text{m}$, focusing element aperture $200\,\mu\text{m}$, focal distance $39\,\text{mm}$, basic radiation wavelength $0.154\,\text{nm}$ and profile height $2.5\,\mu\text{m}$ (which is larger than the extinction depth ($1.53\,\mu\text{m}$) to which radiation penetrates into the crystal), achieved resolution of at least $7.5\,\mu\text{m}$ was, limited in this case by the resolving power of photographic emulsion.

The efficiency of focusing achieved by diffractive elements in the x-ray band of wavelengths was improved by synthesizing multilayer x-ray mirrors with elements of diffractive optics [1, 19]. Such are devices known as Bragg–Fresnel elements. The structure of an ideal Bragg–Fresnel focusing element resembles the structure of a three-dimensional hologram [20]:

- The resolution and shape of the diffraction maximum is imposed by the three-dimensional Laue function. The resolution of an object's point on

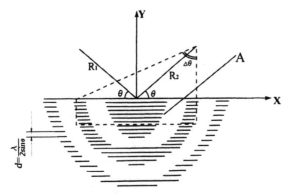

Figure 6.1. The shape of bulk Fresnel zones with constant period of multilayer medium [3]. A: the region of Bragg diffraction; $\Delta\Theta$: the width of the angular range in which the Bragg condition in satisfied.

the image is determined by both the transversal and longitudinal dimensions of the hologram.

• Each surface of the multilayer medium reflects the wave incident on the focal point regardless of the radiation wavelength. The ideal Bragg–Fresnel lens has no chromatic aberrations.

An important property of Bragg–Fresnel elements is that the image is formed within the Fresnel region, that is, diffraction phenomena are described by taking into account quadratic terms in the small-angle approximation of the Fresnel–Kirchhoff integral.

Figure 6.1 illustrates the form of bulk Fresnel zones with constant period of multilayer medium (A is the region of Bragg diffraction and $\Delta\Theta$ is the width of the angular range in which the Bragg condition in satisfied).

The main difference between bulk Bragg–Fresnel elements and elements of planar optics is that the former use coherent Bragg scattering by specific layers of matter. This leads to increased diffractive efficiency, to reduced chromatic aberration, and to resolution constrained at the current technologic level by the diffraction limit. When Bragg–Fresnel bulk elements are fabricated, all the latest achievements in creation of smooth surfaces, multilayer interference layers and sub-micron size elements are employed.

To fabricate elements of diffractive x-ray optics, methods of microelectronics technology are applied: scanning electron lithography, multilayer technology and dry etching [3, 4]. The scanning electron lithography is the most efficient tool for creating Fresnel zone plates and complex kinoform optics [21, 22].

Impressive progress has been achieved in manufacturing condenser and wide-aperture zone plates using holographic techniques [5].

The technology of fabrication of diffractive x-ray elements is tied to the x-ray lithography—a method of transferring an image with high resolution using soft x-ray radiation. The interest in x-ray lithography stemmed from the principal possibility of increasing the resolution of tiny details of structure in comparison with the traditional optical lithography [23]. The practical limit on the resolution of optical lithography is 0.3–0.5 μm. The resolution is drastically improved by reducing the wavelength by several orders of magnitude. However, the current demands on the technology of sub-micron lithography gradually exhaust the potential of this technique because the size of elements of structure approaches the limit of resolution of contact x-ray lithography (about 0.1 μm). A natural way of overcoming these difficulties is exactly as in optical lithography: in converting from contact to projection techniques of image transfer. This transition can be implemented by using either holographic methods or special devices for focusing x-ray radiation. Therefore, serious improvements in projection and scanning x-ray lithography also require progress in high-resolution focusing x-ray elements.

A new and promising technology for fabricating microelements was actively pursued in the past ten years: LIGA-technology (LIGA is an abbreviation of the German terms Lithographie, Galvanoformung, Abformung). In its classical interpretation, the LIGA technology includes three processes: 'deep' x-ray lithography using synchrotron radiation, electroplating and microforming.

Using the first process, the drawing of a chosen microelement is transferred from an x-ray template to a thick layer of photoresist, which is typically a polymer material sensitive to x-ray radiation. In contrast to conventional x-ray lithography for production of computer chips when layer thickness rarely exceeds 1 μm, in this case we deal with x-ray exposure of materials to considerable depth: tens, hundreds and even thousands of microns. After exposure and processing of the photoresist what is formed is essentially a three-dimensional polymer structure of the desired product. The transversal sizes of the elements may be only a few microns but their depth, as has already been mentioned, may reach hundreds of microns or even several millimetres.

Since we discuss large-depth exposure, it is natural that x-ray radiation wavelengths required for LIGA technology are shorter than for x-ray lithography and microelectronics. Typical wavelengths for conventional x-ray lithography are about 10 Å and for a LIGA process they are in the region of 2 Å; still 'harder' radiation will be required to expose centimetres-deep layers. Harder x-ray radiation will require a greater thickness of absorption coating on x-ray templates (typical values for LIGA templates are 15–30 μm Au, as against 1 μm for conventional x-ray processes). X-ray exposure for micron-size structures with high aspect ratio will also require a high degree of collimation and a high intensity of x-ray beams; this undoubtedly implies that deep x-ray lithography can only be done with synchrotron radiation.

At the second stage of the LIGA technology, microgalvanoplastics produces a complementary metallic replica of the polymer microstructure of the first stage. The resulting metal matrix is used at the third stage as a mould for the replication of microproducts by microforming into plastic, ceramic, metallic and other materials.

We cannot say that the LIGA technology is the only one that allows manufacturing of this type of microstructure. A number of other technologies exist, are being developed and are currently applied: large-thickness silicon technology, various techniques of micromechanical processing, processing by laser beams etc. All these techniques are to a large extent competitive but are frequently found to be complementary.

The dependence of the focal distance of Fresnel zone plates on wavelength λ was used to create a spectrometer to generate spatially separated wavelength-selected images of the source [3, 24]. Condenser zone plates with large apertures and with the number of zones up to 1000–10 000 are used to monochromatize 'white' synchrotron radiation [24]. Two zone-plate-based monochromators (with zone numbers 630 and 1196 and 2.3 mm aperture) achieved the degree of monochromatization $\Delta\lambda/\lambda \approx 100$ and 250, respectively.

A Bragg–Fresnel lens was used to focus the output of a high-power pulsed x-ray source based on the Z pinch [25]. The lens consisted of 121 W/Si layers with a period $2d = 70$ Å, of size 0.5 mm×0.5 mm, focal length 7 cm, nominal wavelength $\lambda_0 = 1$. The size of the image of the 'hot' point did not exceed 20–30 μm.

Two linear Bragg–Fresnel lenses for two-dimensional focusing of x-ray radiation ($\lambda = 1.54$ Å) were studied in [26]; the possibility of using them in systems of microprobes of synchrotron radiation sources was discussed.

Owing to very low radiation damage caused by soft x-ray radiation, it is possible to study objects for which the application of electron microscopy is very much restricted, for instance biological objects [27, 28].

Scientists of the London College and Brookhaven National Laboratory (USA) designed for the laboratory of the research council (in Cheshire County) an x-ray microscope that generates snapshots of life cells of human tissue with resolving power of about 50 nm; it may be improved in the future to 10 nm.

The microscope uses soft x-ray radiation which is less destructive for life cells than electron microscope beams by a factor of 10 to 100. A special Fresnel zone plate was developed for focusing x-rays; it consisted of a number of concentric rings with gaps between them, made of a carbon polymer opaque for x-ray radiation. The resolving power of the microscope is determined by the width of the outer ring of the plate.

The soft x-ray radiation for the microscope is generated in the 12 m diameter synchrotron in which electron beams move along spirals in magnetic field. X-rays focused by the Fresnel zone plate on to a spot are incident on the tissue sample which is mechanically scanned while irradiated. Collagen is used as a sample; it is present in the connective tissues of bones, tendons and skin. It is expected that this investigation will help solve the problem of ageing and treat such diseases as cancer or osteoporosis.

Having passed through the sample, x-rays enter a gas-filled chamber with wiremesh detector. X-ray-ionized gas atoms and the released electrons reach the detector and create electric pulses fed to a computer. By processing these signals, an image of the irradiated sample is obtained. First snapshots were rather fuzzy. More distinct images with resolution as indicated above and contrast improved by a factor of 1000 were obtained at the Brookhaven National Laboratory synchrotron by using a gold zone plate fabricated by International Business Machines (USA).

A zone plate was also used in the Göttingen microscope which worked from 1979 to 1982 on the Orsay synchrotron source in France [3]. The resolution of this instrument was 0.5 nm (500 Å). The design of a scanning x-ray microscope in which the NSLS synchrotron x-ray radiation source was used with Fresnel zone plates as focusing systems was described in [27, 29]. The wavelength of the radiation used was 3.2 nm. The efficiency of the focusing element was found by comparing the flux entering the first order focal region with the flux incident on the zone plate; the losses in the substrate (about 30%) were taken into account. The measured efficiency of the five zone plates was in the interval from 0.75 to 2.9%. Images were obtained of biological specimens and artificial microstructures at a resolution of 75–100 nm. The time of data collection for 256×256 elements of the image was about 5 min.

6.2 Diffractive elements for visible and infrared bands

The progress in diffractive optical elements was spurred on first of all by the all-pervading applications of lasers in various fields of science and technology. These elements are developed for laser-using technological units in industry, medicine, in optical instrument industry, fibre telecommunications and optical processors; all the same, there is still much to be learnt about the potentials of diffractive optics.

Diffractive optical elements are the most widely used for focusing monochromatic radiation as analogues of classic lenses. A subject of interest is the application of axial diffractive lenses for designing projection objective lenses [30]. Studies of diffractive lenses as components of optical systems showed that they greatly differ from their refractive analogues. The aberration expansion of a diffractive lens, that is, the presentation of its monochromatic aberrations as a sum of third, fifth and higher orders of smallness, is a series that converges much faster than in the case of a refractive surface. Furthermore, it is possible to control spherical aberration of a diffractive lens by changing the law describing the sequence of annular lines in its structure. An axial diffractive lens is characterized by coinciding coefficients of certain field aberrations. For instance, in the third order of smallness the coefficients of astigmatism and field curvature are equal, in the fifth order nine types of aberrations are described by six coefficients etc.

In two-lens diffractive objectives, it is possible to cancel out all mono-chromatic aberrations of third order whatever the objective's magnification is. In refractive optics, a similar problem is solved by using three refractive surfaces, two of them aspherical. In asymmetric three-lens diffractive objectives, all aberrations of third and fifth order are simply absent. For instance, a diffractive lens for 10.6 μm and diffractive efficiency 96% was fabricated and investigated in [31].

In integrated optics, focusing elements are among the basic components that are created by using integrated circuit technology. Focusing systems form one of the main functional units of integrated optical spectroanalysers, correlators and some other systems for optical processing of information [32, 33]. The main requirements for the elements of integrated optics are the manufacturability of the structure of the focusing element and the possibility of mass production without compromising the main characteristics. Planar diffractive elements do satisfy these requirements.

One of the serious problems encountered in creating telescope systems that format radiation with divergence of $\varphi < 1$ μrad is how to compensate for distortions of the wavefront that are caused by various factors, such as tolerances in fabrication of elements, misalignments, vibrations etc. An experimental check was carried out [35] to test the possibility of compensating for the distortions in a telescope with a composite main mirror coated with a diffractive optical element. Experiments were run at wavelength $\lambda = 1.064$ μm. The diffractive element was a hologram that corresponded to the pattern of interference of a plane wave propagating along the axis of the telescope, and a spherical wave with curvature centre lying on the optical axis of the mirror. An investigation of functioning of such an optical system demonstrated the possibility of synchronization of the radiation from individual elements of the composite aperture of the main mirror and of cancelling out the distortions.

Thermal laser processing of materials and labelling of items involve high-power CO_2 lasers that heat the surface of the metal; heating is followed by rapid quenching and tempering, which improves wear resistance of manufactured parts.

The application of diffractive elements for laser labelling of metals seems to have been first proposed by Ageshin *et al* [35]. These authors also discussed one of the possible principles of designing an element that would focus radiation on to a ring. To focus laser radiation to a specific point or curve, focusators were suggested in [36, 37]. In [36] a focusator to a linear segment was proposed for thermal laser hardening of metal surfaces. An element was used that focused to a linear segment 400 μm wide and 20 mm long, with this segment lying at a distance of 300 mm from the focusing element. Experiments on thermal laser hardening conducted using such focusators demonstrated a high degree of uniformity over the depth of the tempered layer as compared with hardening using a defocused laser spot. Copper galvanoplastic copies equipped with cooling of focusing elements can work for a considerable time at an emission power of 2 kW. Focusators can be used for laser annealing, surface doping of metals and welding of polymers [37]. In the visual range, focusators to a straight segment can be used to illuminate slits of spectral instruments and linear emission receivers, to format a prescribed direction diagram for laser radiator, to pump capillaries in dye lasers etc.

In principle, diffractive optical elements can be used in diode-laser arrays, in light telecommunication systems, fibre-optics communications, optical memory, endoscopes, facsimile machines, printers [42] and compact disks. Elements of diffractive optics are encountered in such systems as infrared cameras and sensors of ultraviolet radiation. Inclusion of diffractive lenses improves the working characteristics of these devices and at the same time reduces the number of optical elements in them, also reducing their overall dimensions and cost.

Furthermore, owing to their high quality [29], diffractive lenses can be unbelievably small—merely several microns in diameter—and thus can be used as focusing elements in radiation sensors, in scanning devices of lasers printers, or as converters in telecommunication networks.

Diffractive lenses can be constructively integrated with other devices of microelectronics [43] thus leading to sensors capable of collecting and processing of data, for instance, in vision systems for robots and satellite navigation systems.

Scientists of the 3M corporation developed elements that can replace the eye lens of patients suffering from cataract. Conventional lenses are widely used in such surgery but their focusing abilities are limited and do not allow a patient to view nearby and remote objects with equal clarity.

G Footey, senior researcher of 3M Corporation, and his colleagues developed a method of combining elements of diffractive optics and conventional lenses. When light enters the eyeball, such elements focus roughly one half of the light flux on to the retina surface and the other half deeper into the retina. The eye and the brain can concentrate at one focal point while ignoring the other, thus providing either short- or long-distance vision. Diffractive lenses were implanted to several thousand people in fifty countries.

Companies like Rockwell International Corp, AT & T, Polaroid Corp, General Electric Co, 3M Co, Texas Instruments Inc, Honeywell Inc, Perkin-Elmer Corp. and some others [44–48] work on the diffractive optics technology of formatting matrices of microscopic optical elements in integrated optics circuits in order to manipulate light beams instead of wires and transmit enormous volumes of data. This technology also permits fabrication of optics for telescopes, contact lenses [42], matrices of coupled laser micro-resonators and multiplexers of laser beam.

Optical diffractive elements are combined with conventional lenses in order to reduce chromatic aberrations [49–51] which arise because of the dispersive properties of glass and other refractive materials: the dispersion of diffractive elements has the sign opposite to that of most glasses.

Diffractive optical elements are used and developed as correctors of wavefronts for systems of certification for aspherical optics [52], in diffraction interferometers [53], in medicine (ophthalmic surgery) [54], for diagnostics of dispersive systems [55], in spectroscopy [56], in various scanning devices [57], to read data off optical disks [58–62], in interferometers for controlling the shape of parts, to measure deviations from straight line and planarity [63], in devices for visual indication [64], for assembling and alignment of complex optical systems [64] and in laser-based Doppler lidars [65].

Diffractive focusing elements may find attractive applications as concentrators of solar energy [66], for instance, when used in space vehicle's power units with a Stirling engine. When developing such concentrators, it is necessary to take into account that the Sun is not a pointlike source of light and that, when viewed from the Earth, its angle diameter is $32'$, so that its image has finite dimensions in the focal plane. The following requirements must be met in developing a concentrator:

- high optical precision,
- low weight,
- structural reliability in space,
- small volume when folded,
- reliable unfolding mechanism,
- strength of material in the conditions of space light,
- low cost.

It was established [66] that a Fresnel concentrator with a cavity receiver is an optimal version since it provides a number of advantages:

- possibility to design a light metal structure,
- planar shape facilitates various types of compact folding of the concentrator,
- possibility to use a simple and reliable mechanism of unfolding (springs, hinges, power cylinders etc),
- high optical precision of elements obtained through high-precision templates,
- all-metal structure is more stable with respect to certain factors of space environment (micrometeorites, ultraviolet radiation and solar flashes),
- convenience of running various tests such as folding and unfolding.

A rigid concentrator consisting of 'petals' may have a maximum diameter of about 10–12 m, with the power of the unit reaching up to 3 kW.

A Fresnel concentrator (or Fresnel mirror) has characteristics very similar to those of a paraboloid. It consists of a number of concentric annular surfaces all in one plane. The surfaces of the rings are set at such angles that a beam of parallel rays is reflected to one and the same point in the focal plane. The surface of each ring can be a part of a paraboloid. In a simplified version a paraboloidal segment can be replaced with a spherical or conical one.

Note that a paraboloid and a Fresnel concentrator have approximately the same ability of concentrating energy on to a given area. However, the maximum flux density is higher in a paraboloid than in the Fresnel concentrator. This difference stems from two factors: (1) the conical surface of Fresnel rings is only an approximation to a paraboloid, (2) the total energy reflected by the Fresnel mirror to the focal plane is less than the total energy reflected by the paraboloid with the same focal distance, as a consequence of losses due to mirror design.

One possible method of fabricating solar concentrators for space vehicles is galvanoplastics. This method makes it possible to process a massive rigid template with high precision, to prepare a very thin reflecting surface free of stress, to apply an all-metal concentrator design, and to use the same template for manufacturing several concentrators.

The method consists essentially of galvanic deposition of thin metal foil (film) on to an exact copy of the Fresnel surface. Then the foil on the template is attached to a support structure. After this, the whole device is removed from the template and covered on its working side with a highly reflective layer.

The problem of protecting observers' eyes from laser radiation arose immediately after lasers were invented, in view of the development of laser-using military-type equipment and instruments. Night vision systems, TV equipment and other photo receivers sensitive to a specific wavelength range face the same danger from laser radiation.

It was reported [67] that protective eyeglasses were developed in the USA that use diffractive optics which reflect laser radiation. This essentially improves transmission in the allowed spectrum range as compared with colour filters that inevitably curtail the visibility of targets in the day time and especially in the twilight period. Special attention was paid to developing protective equipment that prevents eye damage to observers using optical instruments (range finders, target designators and communication systems) by laser radiation of different wavelengths at the same time. Protection equipment will be made as a mask or protective visor using holographic (multilayer) and diffractive optics which provide reliable protection against laser radiation at several wavelengths.

A new type of indicator is proposed to be installed on the windscreen as part of the sighting and navigation infrared 'Lantern' system [68, 69]; the system uses diffractive optical elements which provide considerably better optical characteristics and larger field of view as compared with conventional indicators. The indicator displays information required to detect and identify targets and to support the flight in the land profile tracking mode, using proper symbols, and also video data from the infrared system of frontal viewing. The main requirements of the 'Lantern' indicators are:

- horizontal and vertical fields of view 25 × 17° (instantaneous) and 25 × 20° (overall);
- contrast ratio 1.2:1 with background brightness 30 426 kD/m^2;
- transmission coefficient 70%;
- secondary reflections, 2% of screen brightness;
- solar noise, 3% of source brightness;
- horizontal revolution of 400 pairs of lines, vertical resolution of 480 active rows (modulation level 10%);
- the accuracy of superposition of symbols on the screen 1.0–7.5 mrad, accuracy of optical superposition 1.0–0.6 mrad;
- shift of image on the combining element of the indicator 0.25 mrad.

These characteristics are considerably better than similar characteristics of the windscreen indicators of conventional systems. For instance, standard indicators of the A-10 and F-16 planes have a field view of only 12°, and attempts to extend it by increasing the collimating objective lens may lead to unacceptable rise in the mass and overall dimensions of the indicator.

By analogy to classical holography the windscreen indicator for the 'Lantern' system is sometimes referred to as 'holographic'; this is not quite true, however, because in contrast to holography it is not the imaginary image of the object that is reconstructed but the given optical properties (reflectivity, focal power etc.) of the diffractive element. This indicator uses the following properties of the diffractive element:

- only the light rays incident on the element at a prescribed angle are allowed through, while all the rest are reflected;
- the system reflects light of certain wavelengths and is transparent to the rest of the spectrum;
- large aberrations are created if the directions of the incident and reflected light are very different.

These properties of diffractive elements allow considerable improvement of the optical characteristics of windscreen indicators, namely:

- to increase transmission of incident light flux;
- to improve reflectivity in the formation of images from a projection CRT on the combining element;
- to create collimated output light flux and prescribed focal power.

Diffractive optical elements will allow the creation of a high-quality combined image of the behind-the-cockpit space and the image generated on the screen by a monochromatic projection CRT; the wavelength emitted by the phosphor layer of the CRT screen must be equal to the working wavelength of the diffractive element. The diffractive element is placed between two glass plates, and there is no need in a collimating element. Compared with conventional systems, secondary reflections and solar background noise in a diffractive windscreen indicator should not appreciably affect the quality of the image: the indicator circuit does not contain reflecting glasses and hence antireflective layers can be used without problems in the corresponding optical elements of the indicator.

To control the phase of coherent light beams, flexible control mirrors can be used [70]; their surface can be given specific shape by controlling voltages applied to the mirror. A flexible computer-controlled mirror makes it possible to change the intensity in the focusing zone according to a dedicated program. One promising possibility [70] is a combination optical formatter consisting of a kinoform and a flexible mirror. These elements very successfully complement each other. Bending the flexible mirror generates a relatively smooth optical surface with the required depth of profile. The task of the diffractive element is to provide the required spatial resolution with a relatively small profile depth. In this case the number of zones becomes low and the efficiency of the diffractive element increases (the spatial discretization of the space function becomes less important).

The diffractive elements described above, based on Fresnel zone plates, have isotropic profiles and are capable of implementing only amplitude-phase transformations of the light field. Zone plates with anisotropic profiles [71] permit transformation and separation of light by polarization state [72]. Anisotropic profile zone plates can be fabricated by recording on a polarization-sensitive material the total pattern of the vector summation of spherical waves with different polarizations.

In contrast to an isotropic zone plate, which is at the same time a negative and a positive lens depending on the state of polarization of the transmitted light, an anisotropic-profile zone plate is characterized by well-pronounced asymmetry. If it transmits light with a specific circular polarization, it works as a positive lens; but once the polarization is flipped, it turns into a negative lens. For non-polarized or linearly polarized transmitted light, it functions simultaneously as a positive and a negative lens.

Such anisotropic focusing elements can be applied to circular polariscopes in problems of optical processing of information [71].

6.3   Diffractive elements in optical computers

Optical computers [73–75] promise very high speed of computation. The surge in computer speed (as compared with conventional computers of today) will be mostly achieved by greatly parallelized mode of operation.

For instance, about 10^6 light beams in a common light beam can interact simultaneously with 10^6 elements of a logic matrix. However, at this moment the technology of optical computers is still at a level that conventional computers reached 30 to 40 years ago.

At the same time, optical components offer a number of advantages, including high speed of operation owing to the higher speed of travel of photons, simpler implementation of Fourier transforms that are necessary, for instance, in radioelectronic warfare and in radioastronomy, and to the promise of solving the problem of inter-electrode and inter-circuit contacts that constitute an obstacle to increasing the level of integration of individual chips etc.

Final state devices are developed for promising optical computers, capable of commutation of the final states. Such devices include bistable elements that switch the states, one state opening the transmission of light beams at certain wavelengths and the other state blocking this light.

In addition to parallelism of data processing, optics has another resource for increasing the speed of optical computers: the polychromatism of radiation. Each of the millions of light beams functioning in parallel may contain about a thousand spectrally distinguishable monochromatic components. Each of them is in principle capable of interacting simultaneously and differently with the same logical cell. On a single cell, it is possible to use 16 wavelengths simultaneously and realize a complete set of 16 logical functions. The results of interactions of other radiations with the same cell (repeating the same functions) can be treated as multiplexing, with simple separation of the results using dispersive elements. In this way, optical computers of future generations may have their speed increased by two to three orders of magnitude by using the polychromatic nature of the light beam.

The implementation of optical digital computers mostly depends on the creation of optical logic elements (optical analogues of electronic gates) that carry out various logical operations (AND, OR etc.) that would go beyond the speed of microelectronic devices and their degree of integration, also reducing cost and power consumption. At the moment, a number of optical switches have been created, among which the most promising are optical interference filters, optical etalons (OLE, Optical Logic Etalon), bistable SEED that are devices with their own electro-optic effect (Self Electro-optic Effect Device) and the so-called QWEST (Quantum Well Envelope State Transition) devices.

The switching time of optical switches reaches tens of nanoseconds and might be reduced to 0.1 ns.

Diffractive (dispersive) elements can be used for spectrally selective addressing of signals, can be applied in polychromatic optical processors, can serve as a basis for polychromatic logic elements etc.

Planar diffractive elements have a number of specific features, the main ones being the one-dimensional nature of the element and the need to take into account the properties of the substrate on which the element is placed. From the mathematics standpoint, the first of these factors is easy to take into account: for this, integration over angle in calculating the diffraction integral is replaced with a sum of two values of the integral for $\varphi = 0$ and $\varphi = \pi$ (in the symmetric case). In other words, it is necessary to compute not the double integral but the sum of two single integrals.

Thus, for the planar analogue of the zone plate it will be sufficient to calculate the sum of two single integrals of the following type:

$$U(p_2) = \sum_{i=1}^{2} \int \frac{\exp[ik(r_i + S_i)]}{r_i S_i} \left(\frac{z_2}{S_i} - \frac{z_1}{r_i} \right) R \, dR$$

where $p_2 = (x_2, 0, z_2)$ are the coordinates of the observation point; $p_1 = (x_1, 0, z_1)$ are the coordinates of the radiation source;

$$A = -z_1, \qquad B = z_2$$

$$r_1^2 = (z_1 - z)^2 + (x_1 - R)^2, \qquad S_1^2 = (z_2 - z_1)^2 + (x_2 - R)^2$$

$$r_2^2 = (z_1 - z)^2 + (x_1 + R)^2, \qquad S_2^2 = (z_2 - z)^2 + (x_2 + R)^2.$$

Taking into account the effect of substrate requires a drastic modification of equations; consequently, we ignore the substrate in this section of the chapter. When describing the numerical results we assume that the effect of substrate on the properties of the corresponding element is very small.

The frequency and focusing properties of diffractive elements fabricated on a curvilinear surface are determined, among other things, by the degree of convexity (concavity) of the surface. We can expect, therefore, that placing of planar elements of integral optics on non-flat curves will provide the possibility of controlling both the frequency and the focusing properties. As an example, consider the main properties of a 'conical' diffractive element (figure 6.2) [76]. The initial data for such 'planar conical' diffractive element were: $D/\lambda \approx 35$, $D/A \approx 0.4$, $D/B \approx 0.8$; the number of phase

Figure 6.2. Diffractive planar element on a conical surface [76].

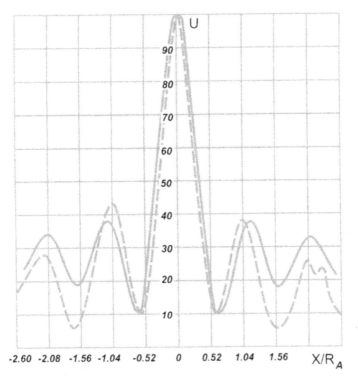

Figure 6.3. Field intensity distribution transversally to the optical axis of a planar diffractive element [76]: —— theory; – – – experiment.

quantization levels was two; element of phase inversion type was chosen. Its properties were investigated numerically and experimentally in the 4 mm wavelength band.

Figure 6.3 shows the distribution of field intensity in the focal region of a diffractive element transversally to its optical axis; it was obtained experimentally and by numerical computation of the diffraction integral. The transition from three-dimensional element to two-dimensional structures increases the level of side lobes. Further investigation of the properties of such elements showed that the frequency and focusing properties are maintained in a wide spectral interval, just as they are for three-dimensional diffractive elements.

Therefore, the investigation of the planar two-dimensional diffractive element fabricated on a conical surface and a comparison of the results obtained with the characteristics of similar three-dimensional diffractive elements allow us to draw a conclusion that such planar elements possess adequate frequency characteristics, sustain focusing ability in a wide spectral range on wavelengths that differ from the nominal wavelength, and are diffraction-constrained systems.

Diffractive planar elements fabricated on a non-flat surface make it possible to considerably enrich the 'pool of devices' of integrated optics and to design elements with novel properties and potentials. This can be illustrated most clearly using as an example optical elements for optical polychromatic computers. For instance, the 'conical' diffractive element discussed above can be used as a nonlinear device for polychromatic radiation. Frequency characteristics for such elements are determined by the extent of concavity (convexity) of the surface of the element and by the direction of incidence on to it. Therefore, when working on a wavelength $\lambda \neq \lambda_0$, the position of the focusing area in space (the amount of its displacement) should depend on the direction of incidence of the radiation. Hence, it is possible to distinguish between a signal incident on the 'tip' of the element from that falling on its 'base' by placing radiation receivers at the corresponding points in space.

It is just as easy to organize logic elements in a similar fashion. Let us consider as an example a diffractive element that focuses radiation emitted from one point on to two points. Such elements are two-dimensional analogues of a three-dimensional element that provides focusing of a point source to a ring. We will use this element 'the other way around'—we let it focus radiation from two pointlike sources to a single point. Then, if we change the wavelength of radiation emitted by one of these two pointlike sources, the area of focusing will change its position in space: it will move transversally to the optical axis of the element. This situation was computer-simulated and shown in figure 6.4 where solid and dashed curves show the field intensity distribution across the optical axis for two different wavelengths. This is effectively a polychromatic logic element: if radiation frequencies at its two ends are identical, NO is the output; if input frequencies are not identical, no output signal is generated (or appears depending on the position of radiation receiver). In other words, logical elements 'END' and 'OR' are realized.

The possibility of applying diffractive elements as elements of optical computer operating with radiations of different spectral composition are discussed in more detail in [77]. For instance,

- **Nonlinear element.** One of the most important elementary cells present in any computer is some sort of 'black box' with nonlinear properties. A diffractive element fabricated on an arbitrary surface, for instance, surface of revolution of second order or its segment, can serve as such a cell. The main feature of such devices is a certain non-singlevaluedness in the manifestation of frequency characteristics with respect to the direction of incidence of the electromagnetic wave on to the apex of its surface when the device functions in a non-nominal mode.
- **Analogue adder with fixed weights.** A specially synthesized diffractive element can function as an adder in the optical process. For example, it was shown theoretically and experimentally that a diffractive focusing

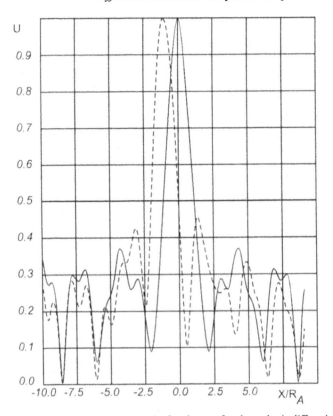

Figure 6.4. Field intensity distribution in the focal area of a planar logic diffractive element for two states of the control input [76].

device in a prescribed spectral band is feasible. The principle of synthesizing such an element is that the profile of the element surface and the corresponding phase function are found by solving the inverse ill-posed problem starting with prescribed spatial and frequency characteristics. This element enables mixing of radiations of different spectral composition in a prescribed volume of space.

- **Parallel processing.** Parallel processing of signals may lead to substantial progress in the productivity of computers and their structures. The essential point here is that optical radiations of different compositions do not interact and therefore some operations can be carried out simultaneously on several wavelengths. Furthermore, it is possible to synthesize a diffractive element that allows 'splitting' of fluxes of radiation incident on this element with the same or different wavelengths and separate them in space: to direct each beam focused by a diffractive element into its designated volume or separate a single beam into several beams with prescribed weights (operations of division and subtraction).

- **'OR', 'AND' elements.** An 'OR' element outputs a signal if any one of its inputs receives a signal (state I). The element discussed above that focuses radiation from two off-axis sources to a common focal point allows creating an 'OR' logic element. As a consequence of the frequency characteristics of the diffractive elements, a change of the wavelength of one of the input radiation waves will result in a shift of the focusing area in space, so that the signals will not add up. Similarly, if the diffractive element is designed to work with radiation at identical wavelengths, the adding up condition will be violated if the wavelength of one of the waves changes, so that the 'OR' element is realized.
- **Controlled switch.** The principle of designing logical 'AND', 'OR' elements makes it possible to implement a controlled switch. For this purpose, one of the inputs of the diffractive element is a controlling one, while the element itself has several outputs located in the region of focusing of the radiation and corresponding to the value of wavelength at the controlling input.

A control gate can be implemented in a similar manner [77]. Optical gates (figure 6.5) can be used in various fields of science and technology: in integrated optics, in input (output) devices for optical signals, in devices

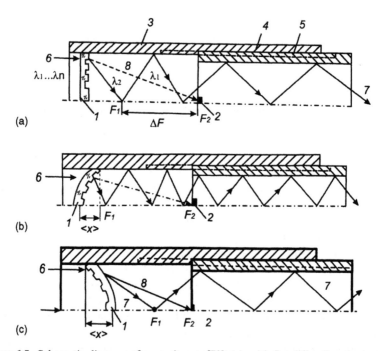

Figure 6.5. Schematic diagram of an optic gate [79]: (a), with flat diffractive element; (b), with convex diffractive element; (c), with concave diffractive element. 1, diffractive element; 2, absorbent of radiation; 3–5, parts of the housing; 6, incident radiation; 7, transmitted

that protect a driver from blinding by headlights of oncoming vehicles [78] etc. To ensure selective properties of optic gates and the possibility of substantial adjustment of their spectral range it was suggested [79] to fabricate them as diffractive elements on curvilinear surface. Such elements possess frequency characteristics (longitudinal chromatic aberrations) which depend, among other factors, on the orientation of the surface of the element relative to the direction of incidence of the radiation. By choosing the profile of the element and its orientation it is possible to control the spectral properties of the gate.

The application of diffractive optics in optical computers using multifrequency discrete signals or continuous spectral radiation within a certain prescribed range may lead to devices that have no analogues among today's computers, that is, optical devices with several levels of stable states—note that a transistor has only two levels of the output signal. The principle of functioning of such devices may be the use of the corresponding modes or harmonics of the radiation whose selection is implemented by the selection of the number of phase quantization levels of the diffractive element. Development of such optical systems may lead to drastic changes in the logical approach to designing computational systems as such because devices will be created that allow execution of several consecutive switchings.

Photolithography methods are currently used to fabricate elements of diffractive optics. This process typically goes through the following consecutive operations [80–82]. Using the prepared photographic template, a photoresistive mask is created on the surface of an optically transparent material, and then the substrate is subjected to ion-chemical etching through windows on the photoresist. As a result, after the photoresist is removed, a three-dimensional profile is left on the substrate. To create a multistep profile, the process described above must be repeated as many times as there are steps in the element profile. Standard photolithographic equipment designed for photographic templates of integrated semiconductor circuits is used for preparing such photographic templates. A new photolithographic method for creating highly efficient diffractive optical elements was suggested in [82], which does not require several sets of photographic templates and their superposition. Using the technology of binarized halftone images makes it possible to work with a single rasterized photographic template and the projection optical system of a photolithographic setup used as a filter of high spatial frequencies to prepare diffractive optical elements with continuous phase profile and high diffractive efficiency. This binarization of halftone images further simplifies the procedure of fabricating diffractive elements as a result of applying a conventional template with two-grade transmission instead of a halftone photographic template. This method was applied to preparing diffractive elements with better than 80% diffractive efficiency.

The method of photorasterizing has advantages in comparison with the multilevel one when preparing diffractive optical elements with zone sizes smaller than 10–20 μm and overall elements size below 10–15 mm.

However, the best results in preparing photographic templates were achieved by using special laser plotters [83–85]. Such plotters enable one to fabricate photographic templates up to 300 mm in diameter with minimum zone width down to 1 μm.

6.4 Elements of diffractive optics in systems of millimetre band radiovision

Methods of direct quasioptical radiovision and radioholography can be used in systems of microwave band radiovision to form high-quality radio images of objects [86]. The formation of radio images in the millimetre and submillimetre wavelength bands has certain specific features, namely: the sizes of the image-forming systems and of the objects of observation are comparable with wavelength, so that the diffraction nature of the image must be taken into account in calculating the structure of the image.

Both in the millimetre and optical wave bands, dielectric lenses and mirror microwave antennas are used to format radio images of objects [87]. The application of these focusing elements does not solve the problem completely since objective lenses with very large apertures—on the order of several meters—are needed to obtain high-quality radio images (even in the shortwave part of the millimetre band). Fabrication of such radio lenses involves considerable technological problems because the more practical ones are lenses with a refractive coefficient of 1.3–1.6 and focal length equal to the aperture [88].

The thickness of the lens is several tens of percent of aperture size. Therefore the mass of such a radio objective is considerable. Energy losses connected with absorption of the transmitted radiation in the lens material are high because even the best dielectric materials such as fused quartz, Teflon, polystyrene and polyethylene have considerable losses in the millimetre wave band; hence, the transmission of a realistic radio objective with lens aperture $D/\lambda \approx 250$ is only about 10%. Using mirrors to generate radio images is constrained by the fact that the object and the image are on the same side of the focusing system.

Promising analogues of lenses in the microwave band are radio objectives based on diffractive elements, namely on zone plates [76, 89, 90].

When building a real system for generating radio images of objects with the resolution depth greater than given by a conventional image, one must scan the object in three coordinates. For instance, using a mechanical scanning into the depth of the object makes it difficult, and sometimes impossible, to obtain the entire radio image of the object in real time.

A realistic system of visualization of three-dimensional objects in the millimetre wave band must provide scanning of a volume of space of at least $(10^5 - 10^8)\lambda^3$, and it is required that objects whose characteristic sizes come to several wavelengths must be reliably identified in this volume. Therefore, the system of visualization must provide resolving power in the object space of about 5–6 mm. In classical systems of image generation, that is, in systems that use lenses and mirrors as image formatting elements, high transversal resolution (relative to the optical axis of the radio objective lens) is achieved at high values of numerical aperture. However, as the resolution on the object increases, the resolution depth of the lens (and therefore, the longitudinal resolution) decreases, and if we take into account that

$$\delta_{\text{longit}} \approx 2\lambda(F/D)^2, \qquad \delta_{\text{transv}} \approx 1.22\lambda(F/D)$$

it is not difficult to arrive at the following estimate:

$$d_{\text{longit}} \approx 1.3_{\text{transv}}/\lambda.$$

A contradiction thus arises: trying to increase resolution on the object in the transverse direction, the resolution depth of the image-formatting systems decreases following square law, that is, the problem of generating a radio image of a three-dimensional object whose extension in the longitudinal direction is several tens of wavelengths, becomes practically unresolvable in this approach because of the high spatial resolution.

6.4.1 Generation of radio images of objects in the millimetre band

We shall consider now how to obtain radio images of objects with resolution depth greater than that provided by the optical system.

The problem may be solved in this case by applying the so-called layer-by-layer scanning of the object. The essence of this technique is that at each given moment of time a flat two-dimensional radio image is constructed of one layer of the object or of an individual point within the resolution depth of the transmitting lens with high transversal resolution. The total radio image of a three-dimensional object is then reconstructed by summing up individual layer images with assigned weight coefficients. A layer-by-layer scanning of a three-dimensional object can be implemented, for instance, using the method pointed to in [91].

For instance, three-dimensional information is reconstructed in several spatial zones arranged stepwise into three-dimensional space. The zones are then displayed sequentially one after another for a short interval on the controlling video device. Therefore, the observer is offered not only the general view of an object observed in a single plane of image—as we have in cinema or television—but also information on spatial depth which is used as additional information by quantizing over depth. It is expedient to choose the time sequence of the displayed two-dimensional flat images in such a way that the observer (owing to the inertia of eye vision) perceives the sequence as one total image.

In a continuous process of creating a large number of images with various positions of layers, an object is created and poorly defined details are suppressed by filtering through a filter with predominantly high-frequency characteristics. By adding up these filtered signals, the total image of a three-dimensional object is created and the corresponding non-filtered image is added to the filtered one. Filtration here can be implemented with a filter with a linearly growing frequency curve—because it is assumed that in continuous focusing and summation (integration), the contrast at high heterodyne frequency is greatly diminished. For instance, if a distribution of intensity of light dots over black background is processed and then this image is integrated, a pointlike image with a wide halo appears.

Mathematically this method can be described as follows.

An out-of-focus part defocused to $\alpha = 0.5K \sin^2 u$ produces an image with the transfer function

$$D(f, \alpha) = \frac{\sin(2\alpha|f| - \alpha f^2)}{2\alpha|f|}.$$

Here U denotes the aperture, $k = 2\pi/\lambda$ and f is the heterodyne frequency.

If filtering is $F(f) = 2|f|/\pi$, we have

$$D_F = \frac{\sin(2\alpha|f| - \alpha f^2)}{\pi\alpha}.$$

With information on α we have

$$\langle D_F \rangle = \int_{-\alpha_1}^{\alpha_1} D_F(f, \alpha)\, d\alpha = \frac{2S_i[\alpha_1(2|f| - f^2)]}{\pi}.$$

If α_1 is larger than the depth of the object, the transfer function refers not only to the middle plane but also to all planes. This method enables us to add up about 32 out-of-focus layer-by-layer images of the object.

A change in the focal distance of the optical system that uses diffractive elements is implemented by changing the frequency of radiation emitted by the irradiator. To implement spatial selection of the signal reflection by the object and the signal sent by the irradiator, and for automatic 'tracing' of the region of focusing along one of the coordinates, for instance, along the optical axis, it is advisable to use either off-axis diffractive elements or diffractive elements with off-axis position of the focusing region. The former option is preferable for the radio vision because diffractive elements then retain their focusing ability in a wider frequency band than elements with off-axis position of the focal point.

Figure 6.6 shows the principal diagram of the designed and tested device for generating radio images of three-dimensional objects scanned over the depth of the objects by focused radiation, with the signal reflected from

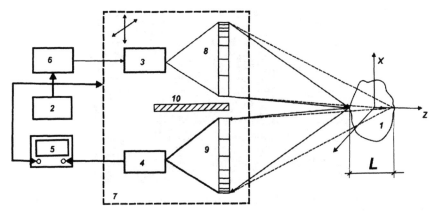

Figure 6.6. Principal schematic diagram of a setup for generating radio images of essentially three-dimensional objects [205]: 1, object; 2, sawtooth generator; 3, 4, source and receiver of microwave radiation; 5, video control device; 6, controlled power supply; 7, three-dimensional scanning device; 8, 9, sending and receiving off-axis zone plates; 10, screen.

inside the area of focusing being received automatically. The device works by using frequency characteristics of off-axis diffractive elements (in this particular case, off-axis zone plates). It is then possible to scan electronically the scene of interest over its depth and automatically trace the position of the focusing region; this improves the quality of the resulting radio image (mostly by improving the signal/noise ratio and through selective screening of defocused layers of the images).

The device works in the following manner. Radiation source 3 irradiates the transmitting radio lens 8 that focuses radiation to a point on object 1 whose position within the scene is dictated by the current wavelength. Radiation reflected by object 1 from the focusing region of the radio lens is received by radio objective lens 9 and is fed into radiation receiver 4. Row and frame scanning are done by moving the object.

Figure 6.7 gives the radio image, obtained by using the frequency characteristics of the objects, of two Cyrillic letters 'С' and 'Я' located along the optical axis of the system at a distance of 10 times the longitudinal resolution of the diffractive element.

When diffuse objects are observed with coherent background illumination, the quality of the image deteriorates because of the formation of interference bands and speckle-structures. It was shown in [92] that speckling degrades resolution by a factor of 5 to 6. Much attention is currently paid in optics, radio optics and acoustics to removing such noise [93].

In working with microwave and laser radiation, the spatially non-coherent illumination can be created by continuous motion of nonuniform scatterer either at the output of the irradiator [92] or at the input of the objective lens that formats the image [94, 95]. One example of such a scatterer for the

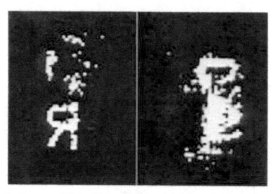

Figure 6.7. Radio image of flat objects obtained by frequency ranging using MOWD elements [203].

visible wavelength band was the aqueous solution of milk, in which diffuse illumination resulted from the Brownian motion in the suspension [96]. It is also possible to use a revolving phase modulator created as a parallel slightly tilted glass plate. According to [96], the rotation speed of the modulator must satisfy the inequality

$$v \gg \frac{F\lambda}{r\,\Delta t\,a}$$

where v is angular velocity of the modulator, F and a stand for the focal distance and aperture of the system that forms the image, λ is the radiation wavelength, Δt is the temporal resolution of the observation system or exposure time, and r is the distance from the centre to the point at which the beam passes through the modulator.

It was shown [94] that by using a revolving random amplitude mask placed at the input of the objective lens and choosing the exposure time it is possible to completely remove the speckling. The choice of the mask is a special problem because this mask does not always remove speckle completely. The distribution of average intensity created by a pointlike source in the focal plane of an optical system was theoretically and experimentally studied in [95]. It was shown that the mean intensity distribution essentially depends on the transparency of the mask and the number of holes in it:

$$\langle I(\rho)\rangle = \left(\frac{\pi a_0^2}{\lambda F}\right)^2 \left[\frac{2J_1(2\pi a_0\rho/\lambda F)}{2\pi a_0\rho/\lambda F}\right]^2$$
$$\times \left\{1 + (N-1)\left[\left(\frac{2J_1(2\pi R_0\rho/\lambda F)}{2\pi R_0\rho/\lambda F}\right)^2\right.\right.$$
$$\left.\left. - \left(\frac{r_0}{R}\right)^2 \frac{2J_1(2\pi R_0\rho/\lambda F)}{2\pi r_0\rho/\lambda F}\right]\right\} \tag{6.1}$$

where $r_0 = 1/(2\pi n_0)^{1/2} + 2a_0$ is the mean distance between holes, n_0 is the average number of holes per unit area of the mask, ρ is the vector radius at the focal plane, R_0 is the hole radius, R is the radius of the input pupil, $N = \pi R_0^2 n_0$ is the number of holes in the mask, $J_1(x)$ is the Bessel function of the first kind and a_0 is the parameter that removes the possibility of intersection of holes in the mask [95].

Let the number of holes be high and $r_0 \ll R_0$. Then the central maximum of function (6.1) is practically the same as without the mask and, hence, the resolving power of the optical system is retained.

The behaviour of the function $\langle I(\rho) \rangle$ at high values of ρ corresponds to diffraction on an individual hole in the mask, which creates additional noise close to the central maximum.

If the transparency of the mask is low ($T \ll 1$) the intensity distribution pattern completely corresponds to diffraction on a single hole of the mask.

In order to remove speckle in the image of an object, it is possible to use a random amplitude mask revolving at the input pupil of the optical system. To suppress noise close to the central maximum and to retain resolving power of the optical system, it is necessary to have a highly transparent mask with a large number of holes. The intensity of the diffraction distribution of the optical system, taking into account the random amplitude mask, is described by relation (6.1) after adequate averaging.

An analysis of relation (6.1) shows that using the random phase mask for the irradiator is preferable because the effect on the image of any amplitude mask for the objective may be very close to that of the phase mask but can never be better.

6.5 Formation of images using partially coherent radiation

We will consider the problem of suppressing interference fringes in the optical system for formatting the image, with the irradiator being a small-size thermal source of quasi-monochromatic radiation $\Delta \nu / \nu \ll 1$, where $\Delta \nu$ and ν are the effective frequency and the irradiator's frequency band, respectively.

According to [97] we can observe here interference fringes, provided

$$\Delta \nu \, \Delta a < \bar{\lambda}$$

where $\bar{\lambda} = c/\bar{\nu}$ is the effective wavelength, c is the velocity of light and Δa is the size of the source [98].

If a thermal source of quasi-harmonic radiation is used as irradiator, the interference fringes on the image are removed if the ratio of numerical apertures of the source and the receiver $\rho_s/\rho_0 \gg 1$ and the resulting resolution corresponds to the diffraction limit of receiver optics with non-coherent illumination.

When an object with specularly reflecting surface is observed, its image contains sparkling that suppresses the fine structure of the image. At the same time sparkling from external sources is superimposed on to the original images.

It is possible to remove this distorting noise without changes in illuminating sources, for instance, by placing in front of a flat object a nonuniform transparent scattering (refracting) plate. The possibility of removing the sparkling is determined by the scattering diagram of the plate and by the optical power of noise.

The illumination of the object in systems of direct radiovision in the millimetre wavelength band is done with coherent radiation. The image of the object is constructed in reflected radiation using special high-aperture radio objectives. An interference image is formed when layer images are added up. Furthermore, most objects have specular surface in the millimetre and sub-millimetre wavelength band. This occurs because the wavelength λ_0 at which radio vision systems work is much longer than the visible light wavelength λ_c used to visualize an object $\lambda_0/\lambda_c \approx 10^4$. Therefore the surface structure of visualized objects is smoother (more specular) with respect to microwaves than it is for light by a factor of (λ_0/λ_c); hence, an interpretation of the resulting image becomes ambiguous. For instance, the radio image of a sphere is a point.

6.5.1 Method of isotropic construction of radio images of three-dimensional objects [99]

In radio communications with short waves, interference-caused signal fading is counteracted by so-called diversity (sparsed) reception. In frequency sparsing, signals are transmitted on two frequencies and when the difference between frequencies exceeds 5 kHz, signals in the channels fade independently. Spatial sparsing is realized by using several antennas located at a distance equal to ten wavelengths [100, 101]. Reception by spaced-apart antennas is especially effective [101].

The 'spaced-apart reception' techniques can also be used to suppress interference noise in radio images.

Let us consider an object model consisting of two pointlike reflectors. With two-frequency illumination, the best suppression of interference noise is achieved if

$$\Delta\lambda/\lambda \approx 1/2m$$

where $m = d/\lambda$ and d is the distance between the reflectors.

In the case where there is only one wavelength and we have two receivers whose signals are added up, the following condition can be obtained for the angular separation Ω between two receivers:

$$\Omega = \arctan[(4m - 1)^{1/2}/(2m - 1)]. \qquad (6.2)$$

Similarly, we can derive for n-frequencies illumination and a string of n receivers the formulas

$$\Delta\lambda/\lambda = (n - 1)/mn$$

$$\Omega_k = \arctan[(2mnk - k^2)/(mn - k)], \qquad n = 2,\ldots, \quad k = 0,\ldots,(n - 1).$$

The interval $\Delta\lambda$ corresponds to the difference between the maximum and the minimum of wavelengths of n-frequencies irradiation and Ω_k corresponds to the angular separation between the first and the kth receiver of the linear string.

Consider the placement of several receivers optimized for suppressing interference noise in a finite field of view.

To suppress interference noise in a finite field of view we distribute n receivers in such a way that for each point there would be two receivers whose signals are in antiphase. We define the number of receivers as the ratio of the solid angle covering the system of receivers to the solid angle of a single receiver.

In order to maximally suppress interference effects at a point A on the axis (figure 6.8) it is necessary to place the second receiver 2 at a distance given by formula (6.2). To suppress interference effects at the edge of the field of view at points BB' it is necessary to install receivers 3 and 4 at such a distance from receiver 1 that the difference between the optical path lengths between receivers 3 and 4 and the point B equals half of the wavelength of the radiation used.

Consequently, in a string of four receivers, two pairs of them are strictly in antiphase (at the centre and at the edge of the field of view). For n receivers, the number of such locations is $n - 1$. We were looking so far at a one-dimensional case. To consider a two-dimensional case, the string of receivers has to be rotated K times around the X axis by an angle $\Delta\varphi$, so that $(K \Delta\varphi) = \pi$.

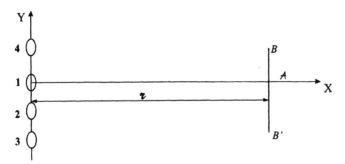

Figure 6.8. Locations of radiation receivers for suppressing interference noise in a finite field of view.

The required number of receivers can be found as

$$N = \Omega/\Delta\Omega = [2\pi(1 - \cos\theta/2)]/[(\pi/4)(D/4)^2].$$

Let us find the optimum location of two receivers for an arbitrary location of the object in the field of view. Let the object lie on the axis of the first receiver. The signals of the two receivers will be at opposite phases if

$$\frac{d(r_0 + d)}{\sqrt{(r_0 + d)^2 + a^2}} - \frac{d(r_0 + d)}{\sqrt{(r_0 + d)^2 + (a + y_0)^2}} = \frac{\lambda}{2}.$$

We conclude that for optimum suppression of interference noise the distance between the receivers must equal

$$y_0 = \left[\frac{r^2 d^2}{[(rd)/\sqrt{r^2 + a^2} - (\lambda/2)]^2} - r^2 \right]^{0.5} - a, \qquad r = r_0 + d.$$

Therefore, this method of forming radio images on the basis of an 'isotropic' receiver (source) essentially consists of implementing the principle of spatial averaging. For instance, the object may be scanned by a focused beam of electromagnetic waves, the radiation scattered by the object being received by a systems of receivers located in space on the side of the irradiator, while the signals from them are added up non-coherently and are sent to the common system of reconstruction. The receivers are placed on the surface of a hemisphere whose centre lies on the object. As a result, it becomes possible to visualize objects using the difference between their reflection coefficient and the reflection coefficient for the background signal, that is, to visualize an image of a three-dimensional object. Non-coherent adding suppresses interference effects. When integrating narrow directed signals reflected from obstacles (the background signal), the received interference signal is proportional to the coefficient of reflection from the obstacle, and this is typically much lower than the coefficient of reflection from the object. On the whole, a system of N receivers proves to be more sensitive than a single receiver by a factor of $N^{1/2}$. Furthermore, averaging considerably reduces the dynamic range of signals recorded, which makes the requirements to systems reconstructing radio images less stringent.

Another important factor must be mentioned. In systems of direct radio vision, a receiving device is as a rule large and heavy which makes using this technique very difficult (for instance, the diameter of the objective is at least 200–300λ, though even with a speed of 1 frame per second and the number of added layers $n \approx 30$, the objective that does mechanical scanning of the object over its depth must periodically move at a typical speed of about 350 m/s).

Attempts to use the conventional 'optical' approach to constructing radio images of three-dimensional objects in the millimetre wavelength band result in extremely unwieldy formulas. The resulting dimensions,

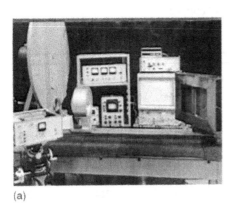 

(a) (b)

Figure 6.9. (a) System of radio vision with a two-lens radio objective. (b) Five-lens microwave objective with Nipkow-disk-type scanning system.

weight and parameters of objective radio lenses fail to satisfy today's requirements.

As an example, figure 6.9(a) shows a photograph of a quasioptical system using a two-lens radio objective lens for the millimetre wavelength band (developed at the Leningrad Institute of Precision Mechanics and Optics [202] using the optical approach). The study of the parameters of such a radio objective showed that it has narrow field of view. Attempts to improve its parameters led to creation of a five-lens radio objective. Even though the field of view and the resolving power of the new system were considerably improved, any practical utilization of such systems is highly questionable. The parameters of the five-lens radio objective designed for the 140 GHz frequency band and using spherical refractive surfaces are listed in table 6.2.

Mean-square deviations of the focal distance, aberration distortions and field curvature of the image from those prescribed for a given radio objective are 0.277 mm. The total thickness of the dielectric of the radio objective along its optical axis is 254 mm so that the radio objective transmission coefficient in the 140 GHz band is at most several percent.

Figure 6.9(b) shows a photograph of the system for plotting radio images of flat objects using the radio objective described above and a scanning system based on the so-called Nipkow disk.

The common drawbacks of the systems discussed lie in a considerable thickness of the radio objective leading to considerable losses in radiation power and in a principal impossibility of scanning an object along the third coordinate (depth) in real time.

Table 6.2.

Material of the radio objective	Polystyrene
Half-height of the object	200 mm
Front segment	−1000 mm

Number of the lens	First radius (mm)	Second radius (mm)	Thickness (mm)	Optical diameter (mm)	Gap (mm)
1	517.05	−485.49	59.01	246	33.65
2	−244.25	−541.58	27.79	264	115.93
3	626.62	−846.96	60.81	320	89.15
4	254.89	745.03	53.18	291	30.48
5	118.89	267.54	53.37	214	75.55

Layer-by-layer construction of the radio image of a three-dimensional object without mechanical scanning over depth can be implemented by using the frequency characteristics of MWDO elements in which the position of focusing area depends on the wavelength of the irradiating field. In this case the speed of the radiovision system improves considerably. It becomes possible to start with locating an object by scanning over the depth of the scene and then carefully 'scrutinizing it' in detail.

Figure 6.10(a) shows a schematic diagram of the setup for direct quasi-optical radiovision in the millimetre wave band, in which row-by-row scanning is implemented by consecutive switching of a string of generators, and frame-by-frame scanning—by mechanical scanning (using a rocking mirror). The depth scanning of the scene is implemented by using the frequency characteristics of the two-component diffractive radio lens.

If the target is moving, mechanical scanning over one of the coordinates (e.g. frame-by-frame scanning) becomes necessary. In this case the radio vision system is somewhat modified.

Figure 6.10(b) shows a pilot quasioptical radiovision system of single-user access working in the 75 GHz band. To improve the characteristics of the radiovision system, diffractive lenses on curvilinear surfaces were used in subsequent modifications.

The principle underlying the radiovision system is a multiple-angle illumination of the target object (in azimuthal and elevation angles) by a system of microwave transmitters, recording of the radio image created by the radio lens via row-by-row (column-by-column) analogue-to-digital conversion of signals of the linear array of receivers and recording of the signals into the memory of the display station 'Gamma 4.2' for visualization and processing.

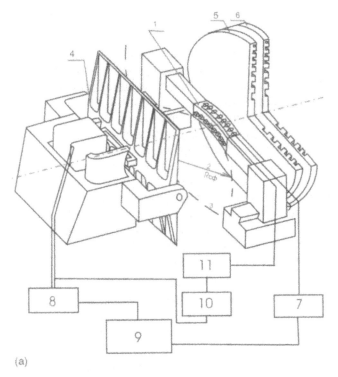

(a)

(b)

Figure 6.10. (a) Schematic diagram of a radiovision system with two-component diffractive radio objective on a flat surface [203]: 1, string of irradiators (receivers); 2, curve on which the string of irradiators is distributed (it is mirror-symmetrical to the best focusing curve); 3, 4, rocking mirror for frame-by-frame scanning; 5 and 6, two-component diffractive radio objective; 7, power supply and commutating device for a system of irradiators; 8, power supply for the rocking mirror; 9, video control device; 10, synchronizing device; 11, ADC. (b) Pilot radiovision system for millimetre wavelength band [203].

Listed below are the main specifics of designing the radiovision pilot system.

1. Row-by-row scanning for building the image of an object in real time is carried out by electronic scanning of the string of receivers.
2. Frame-by-frame scanning (column-by-column scanning) is implemented by mechanically moving the object controlled.
3. Mirror flashes in the image are removed and the images of the object in different orientations are obtained using quasi-isotropic illumination by a system of irradiators distributed in space.
4. Scanning of the controlled space is done by the electronically shifting the plane of focusing of the radiovision device (its focal length) by varying the irradiators' wavelength and using diffractive microwave optics. The use of elements of diffractive microwave optics on a non-flat surface makes it possible to extend the field of view of the radiovision device and improve the signal-to-noise ratio in the radio image.
5. Access to the object scanned is provided on one side only.

The radiovision set supports scrutinizing an object with a single frequency (scanning of a plane) or with a number of frequencies (scanning over the depth of the scene) in single-pass or continuous modes of recording the frames of radio image with an external coupling (to the motion sensor) or internal coupling to the motion of the object, and can also operate in adjustment mode to fine-tune individual components of the radiovision set.

The radiovision set consists of a system of microwave transmitters, a diffractive radio objective, a reception array with commutators, an analogue-to-digital converter, a set control module, a high-resolution display terminal 'Gamma 4.2', a video control device as a component of the display terminal, and the system that moves the object of investigation.

Let us consider the interaction between the components of the radiovision set in the process of recording one frame of radio image at one fixed frequency.

Electromagnetic field from microwave transmitters irradiates the object of monitoring during the time required to record one row. The electromagnetic field reflected from the object is focused by a diffractive radio objective on to the reception array. The reception array transforms the spatial distribution of energy flux density of the microwave field into low-frequency electric signals whose amplitude is proportional to the energy flux density in the radio image. The length of signals equals the duration of irradiation of the object by microwave transmitters. The control module provides for successive switching of outputs of the reception string to the input of the analogue-to-digital converter. Signals from the output of the ADC are recorded into the memory of the image on the display terminal and are shown on the screen of the video control device. The recording of subsequent rows in the frame proceeds in a similar manner, is synchronized by signals that couple it to the motion of the object and are sent by the system of motion of the object, or by an internal synchronizing device that couples it to the velocity of the investigated object.

Recording a frame of a radio image on several frequencies differs from the mode described above in that for each spatial position of the object, several rows of radio images are recorded on the several frequencies selected, by electronically switching the frequency of microwave transmitters.

The selection of radiovision operation modes, the size of the image frame, the number and values of scanning frequencies are controlled by the software of the 'Gamma 4.2' display terminal.

Technical parameters of the radiovision set

Working frequency	53–78 GHz
Energy flux density of the microwave field on the target object not more than	$500 \, \mu W/cm^2$
Dimensions of the target zone	
height	630 mm
depth	500 mm
Resolving power (dictated by the size of the aperture of the receiving antenna of an element/string of elements)	20 mm
Distance from radio objective to focusing plane	500–1000 mm
Number of receivers in the array	32
Size of the recorded frame of the image on video control device	32×32 to 256×256 pixels
Time of recording a reading of an element of the array	1–$5 \, \mu s$
Number of bits per reading	8–12
Number of discrete scanning frequencies of the illumination system over depth	1–256
Time of formation of a 128×128 frame on a single frequency at most	0.1 s

Figures 6.11(a–c) show radio images of a gun placed behind a dielectric screen. The images were generated by a radiovision system with optical radio objective (figure 6.11(a)), using a single illumination angle method (figure 6.11(b)) and multiple angle illumination (figure 6.11(c)) [203].

The main component of any radiovision system of quasioptical type is a radio objective that must satisfy a number of requirements [102]:

- to be multicomponent [to satisfy the requirements to field of view ($2\beta \approx 60°$) and the number of resolved pixels on it ($N^2 \approx 100 \times 100$)];
- to possess frequency characteristics adequate for inspecting three-dimensional scenes in real time;
- to have aperture ratio of at least 0.5 and lens aperture (D/λ) of at least 200 to provide high spatial resolution constrained by the diffraction limit over

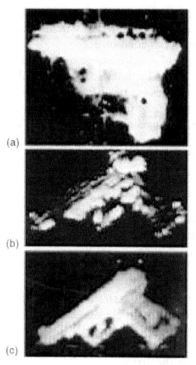

(a)

(b)

(c)

Figure 6.11. Radio images of a gun placed behind a dielectric screen, generated by a radiovision system with (a) optical radio objective, (b) using single illumination angle and (c) using multiple illumination angle reception [203].

the entire field of view;
• to be fabricated of a material possessing low absorption of microwave power, and be of minimal thickness.

Devices that satisfy these requirements are multicomponent diffractive microwave radio objectives with variable focal length, based on phase inversion zone plates [102], and diffractive radio objectives on non-flat surfaces [103]. To scan three-dimensional scenes with depth exceeding the depth of resolution of classical lenses, frequency characteristics of diffractive radio objectives are used [104, 105]. Complexity and high cost of radiovision systems based on multi-element scanning arrays prevented further progress in this field [106–108].

In the active radiovision mode, when monochromatic radiation is used to illuminate the target object, it is possible to reshape the focusing surface by controlling the frequency of the illuminating radiation and scan the space over the depth of the scene by this surface [109]. Therefore, frequency characteristics of diffractive radio objectives make it possible to remove limitations stemming from the small depth of definition of classical lenses.

Such systems of image generation are capable of scanning space over depths exceeding the depth of definition by a factor of 10 to 20 without mechanical scanners [102, 104]. The small thickness of diffractive radio objectives (on the order of radiation wavelength) allows designers to achieve high efficiency using microwave power.

6.6   Diffractive objectives in systems of holographic radiovision

Diffractive radio objectives can be used in various setups designed to visualize dielectric objects hidden behind dielectric obstacles and placed against a reflecting background, for instance, to detect weapons and plastic explosives hidden on human body [110].

To visualize phase targets it is necessary to transform changes in the phase of the electromagnetic object wave into changes in the intensity on the image of that target object. It was shown [11] that it is possible in principle to achieve this visualization by holographic methods, reconstructing the image in the visual wavelength band using a reduced negative. This procedure is very time-consuming and involves 'wet' processing.

The technique described below consists of superposing a reference wave on the focused image of the target object. In this case the so-called focused image hologram appears in the recording plane [112]; it is less sensitive to vibrations than other types of hologram and can be used as any other hologram. Using numerical methods, reconstruction can be realized layer-by-layer and on the same wavelength, so that three-dimensional images can be generated without scale distortions. However, with this method of reconstruction the usefulness of the phase method in real time becomes doubtful.

Figure 6.12 shows the experimental setup for visualizing phase objects [203]. The target object (a mannequin) is placed on a rotating stage used to

Figure 6.12. The experimental holographic setup for visualization of phase objects [203].

scan the object. The object is irradiated with an electromagnetic microwave beam focused by an objective lens incorporating a phase zone plate. The objective's diameter was 200 mm, the front segment was 220 mm, the rear segment 120 mm. The resolution achieved in this experiment was approximately 4 mm. The radiation reflected by the target object was focused by the same optical system to the same reception/transmission antenna of the phase-sensitive receiver (scatterometer). This system automatically implemented precise coupling of the focusing regions of the irradiator and receiver during scanning. In principle, the dependence of the focal distance on radiation frequency makes it possible to implement frequency scanning over depth; this was one of the features tested in this experiment. The signal separated at the output of the receiver and carrying information on the phase of the wave reflected from the current point of the object was digitized and fed to the video control device.

The object is a plastic mannequin covered with conducting paint that serves to reflect electromagnetic waves. The following objects were used, to be detected against a reflecting background: a square piece of polystyrene (refractive index of the material $n \approx 1.5$) 25 mm thick, metal rods 20 mm in diameter, a polyethylene bar 50 mm thick with an oval hole at the centre, and a phase zone plate. As dielectric shield, cotton fabric was used.

Figure 6.13 shows the result of visualization of phase objects. The image is formed of 256×128 pixels. The scanned angle was increased by setting the step between readings along the horizontal axis twice larger than that along the vertical axis. This is why all objects seem to be compressed horizontally. The target scanning angle on this image is π.

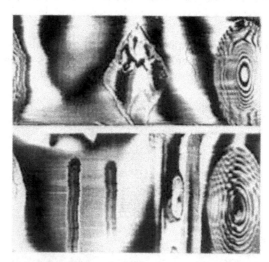

Figure 6.13. Results of visualization of phase objects in the millimetre wavelength band [203].

The figure clearly shows the contours of sharply imaged boundaries of the objects listed above. At the same time it is clear that as the shield thickness increases well-defined contours become smeared and segmented. This may stem from the defocusing action that the thick refracting medium exerts on the electromagnetic wave beam convergent at a large angle.

The phase method makes it possible to detect radio-transparent objects both against a smooth reflecting background and against a profiled background hidden behind a dielectric shield of plastic and fabric.

6.6.1   Diffractive optics in microwave object radars

It was suggested to use diffractive radio objective lenses for the millimetre wavelength band as frequency filters for harmonics or for separating signals of different frequencies [113–119].

Devices for microwave diffractive focusing elements in non-distractive detection of flaws were described in [104, 120, 121].

The feasibility of designing a simple and low-cost radio location system were investigated by Lazarus and Silvertown of Lancaster University (UK) [122]; the system was based on a Gunn diode mixing generator and polyethylene Rayleigh–Wood zone plate (figure 6.14(a)). The zone plate served to generate a plane wave. The radiation frequency used was 33 GHz, at zone plate thickness 9 mm, zone plate diameter 300 mm and focal distance 180 mm. Such antennas fabricated of plastic are sufficiently simple and can be manufactured in large batches. Figure 6.14(b) shows a millimetre

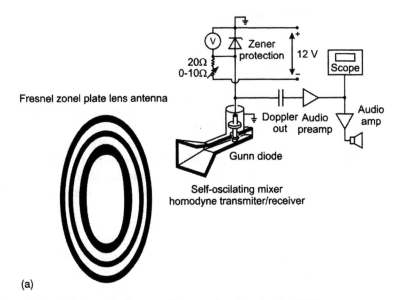

(a)

Figure 6.14. (a) Schematic diagram of a zone-plate-based radar [122].

Figure 6.14. (b) Millimetre wavelength band Doppler velocimeter using a Rayleigh–Wood zone plate.

wavelength band Doppler instrument for measuring velocity; a Rayleigh–Wood zone plate was used as a phase corrector of a horn antenna developed by the authors.

Diffractive radio objectives are used in interferometers [118] and as antennas for radio telescopes [123]. The phase inversion zone plate was first used in quasioptical transmission lines by Sobel *et al* [118]. The transmission line had insignificant losses at 210 GHz: only 2 dB over a distance of 17 m.

6.7 Microwave-band diffractive antennas

Diffractive elements are now finding wider and wider use as antennas for the microwave wavelength band [119].

Diffractive reflector elements used as antennas make it possible to reduce an antenna's effective scattering surface (ESS) owing to their frequency selectivity. Zoning the reflector surface results in considerable reduction of the antenna ESS, especially within the angle of the main lobe of the beampattern on non-working frequencies [124]. Such diffractive reflector antennas may help solve problems of electromagnetic compatibility, for instance, in radio systems in which several antenna devices for different frequency bands are used and have to be located in the immediate vicinity of one another [124].

Modified Cassegrain antennas [156]. Cassegrain antennas proved useful in some applications in the microwave band. When stability of formation of radio images of vehicles was investigated [206], a radar with Cassegrain antenna was used, with the main reflector designed as a planar reflector zone plate with modified zone structure. This led to substantial reductions in the cost of the antenna unit of the device. Another scanning Cassegrain antenna was later investigated, in which the secondary revolving parabolic mirror was replaced with a special planar diffractive antenna. This replacement decreased the antenna's gain in the 77 GHz band from 41.5 dB to 39.7 dB but made it possible to considerably level the dependence of gain on scanning angle and to widen the angular range of scanning.

6.8   Antennas for satellite communications

In view of the progress in satellite communications, flat diffractive reflector antennas, having certain advantages over parabolic antennas, were proposed as receivers [125, 126]. First of all, such antennas are technologically more convenient, they can be built flat, their weight is lower etc. Moreover, the application of antennas that focus radiation off the optical axis enables the designer to avoid blocking some of the radiation by the receiver and to reduce the noise of the antenna system [127–130].

Typically satellite television antennas require low-noise high-sensitivity amplifiers. Amplifiers of this type can be driven or saturated by short 'surges' of high-amplitude noise.

A conventional antenna simply amplifies such 'noise surges'. Diffractive antennas are less sensitive to such short noise 'surges', thus reducing the probability of noise-driving in low-noise amplifiers.

Modulated data vary slowly with time relative to the carrier wave (10–12 GHz) in TV satellite communications. The gain of the diffractive antenna is the sum of gains in each zone at the corresponding moment of time. Therefore a high-amplitude short noise 'surge' can be amplified by only a limited number of zones. Consequently, the amplification of this surge will be reduced compared with the total signal amplification, so that the low-noise amplifier cannot be driven or saturated.

Receiving antennas are now required for communications with 'Astra'-type satellites in the frequency band 11.2–11.45 GHz, with the beampattern width at most 3° [131] to ensure satellite channel reception without interference from other satellites. In a number of cases, it is expedient to use lens-type antennas. As an example, figure 6.15 shows a quarter-wavelength lens-type diffractive antenna for 10 GHz band [90].

A sufficiently serious problem of protection from antenna icing arose in a number of countries. The design of diffractive antennas working in the radiation reflection mode makes it possible to create antennas with heating that operates under conditions of snow and ice covering (figure 6.16). To achieve this, all (metal coated) radiation-reflecting zones in half-wavelength or multilevel antennas are electrically connected into a heater circuit, and electric current is run through it. Therefore, the problem of special heating devices is automatically eliminated for such antennas—their role is played by metallized Fresnel zones. Designs similar to these may also prove

Figure 6.15. Lens-type diffractive antenna for the 10 GHz band.

useful in space when it is necessary to protect a spacecraft's antenna from temperature-induced strains.

The millimetre wavelength band can be used to organize communication lines and support one or several telephone channels, TV broadcasting and transmission of data between the central computer and its terminals located in other buildings; it can work in security systems that protect territory from intruders, in technical vision systems, as elements of radar systems

Figure 6.16. Pilot model of a heated diffractive antenna.

for remote measurements and control of finished parts or monitoring of vibration in mechanisms, as stationary devices to monitor the speed on roads or railway lines, as onboard traffic security devices in coastal waters in river and sea shipping, in locking through and mooring etc. [132, 133].

For instance, a microwave system of distribution of TV programs was suggested [134]. This system would distribute up to 20 channels generated on site or received from satellites or video salons.

The programmes are transmitted in the range from 20 to 45 GHz. The transmitter is located on an elevated spot to provide maximum direct visibility range. The millimetre wavelength band limits the area serviced to 10 km^2. A subscriber's house is equipped with an antenna 15.5 cm in diameter, with a converter outside and intermediate frequency amplifier and demodulator inside the house. The antenna gain in the frequency band at the half-power level is 31 dB. The channel separation is quite small (36 MHz) which allows efficient use of cheap generators with simple dielectric resonators. Since combining 20 channels when using multiplexing entails considerable losses, each channel is given its own heterodyne and its own antenna. In view of signal attenuation due to rain (taking into account the statistics of precipitation in various regions of the UK), the maximum distance of 3.3 km can be provided with emitted power of 100 mW per channel, using a receiver with noise coefficient 12 dB for conventional TV signal. As of 1991, the British Government allocated the 40.5–42.5 GHz frequency band for local TV reception [124]. At the moment sufficiently inexpensive technology has been developed for this range to support cable-free distribution of multichannel video signals to subscribers. MWDO elements are promising as transmitting and receiving antennas for such systems.

Application of passive mirror retransmitters on line-of-sight visibility radio relay lines is discussed in [135]. A zone plate about 7.3 m in diameter, designed to work in the decimetre wavelength band, was used as a passive retransmitter.

Devices are being developed for working in the shorter millimetre wavelength band, needed for systems of automobile collision prevention [136–140]. The main purpose of the system will be to inform the driver sufficiently early about an obstacle on the route and to evaluate the danger of collision from the approach velocity or distance to the obstacle, and in extreme cases to determine the time to inevitable collision and turn on the passive safety equipment. One of the main components of this system is the antenna. A simple multibeam antenna based on lens- or reflector-type MWDO elements can be incorporated in an element of automobile design, e.g. part of the hood, without interfering with the vehicle design.

A considerable increase in the number of cell telephone networks serving mobile users and the need to create systems of personal communications demand additional communication channels, especially in cities with high population density. In view of the limitations imposed by antennas, no further increase in the number of channels can be achieved now by additional splitting of cells in the currently operating communication systems.

An alternative approach to solving this problem is to use a microcellular structure [141]. There is a plan to build an all-European 38 GHz waveband system of personal radio communications [142]. In the case of microcellular structure, line-of-site communications are implemented between mobile or portable subscribers and the microcellular-net stations mounted on posts of municipal services, for example, on street lighting posts etc., roughly every 600 m. To remove multiple scattered and reflected rays and thus suppress spurious components, directed antennas are to be used. Antennas based on MWDO elements will permit fast beampattern scanning using stationary antenna aperture or simultaneous signal reception in a sufficiently wide sector of angles.

Direct satellite TV broadcasting (DSTVB) that sends TV channels from communications satellites straight to owners of TV sets is becoming more and more widespread. An important element of receivers is the antenna whose parameters largely determine the quality of signal reception. At the moment mostly parabolic reflector antennas and their modifications are used in DSTVB systems, or their modifications—parabolic antennas with off-axis irradiator. As a rule, irradiators of parabolic reflector antennas are placed at the focal point of the reflector, that is, in the path of propagation of radio waves, which causes additional losses. Blocking of radio waves by the power supply system of the radiator, by its mounting parts, and by the frequency converter which is usually combined with the irradiator, also produces a negative effect. Parabolic reflector antennas are typically large and have low wind resistance.

Flat antennas are developed as an alternative to parabolic antennas. These are essentially stripline antenna arrays. Such flat receiver antennas can be installed on the wall of a house. Flat antenna arrays with adjustable beampattern best suit the requirements demanded of antennas of this type. However, such electronically controllable antennas are far from low-cost and are less attractive than cheap flat antennas with fixed deflection of beampattern. The advantages of flat antennas for reception of satellite broadcasting of TV channels are: compact design, low weight, easy handling and simple installation on house walls, stability under wind loads and the principal ability of adjusting the beampattern. The gain of these antennas must be on the order of 30 dB.

French TV reception networks use ground stations with antennas of 30 to 40 cm in diameter (TDF-1, TDF-2) [144]. In Germany receiving networks use antennas 60 to 90 cm in diameter (TV-Sat), in Sweden, Norway and Finland antennas diameters are 45 to 60 cm (Tele-X), and in UK antennas are 30 to 50 cm in diameter (Marco Polo, Olympus-1).

The expected occupancy of the orbital coordinates allocated on the geostationary orbit for communication and broadcasting systems will make requirements to receiving ground stations more stringent. For instance, the decision of the USA Federal Communications Commission to set the admissible angular separation between neighbour satellites at 2° precludes further reduction in the size of ground-based antennas because the main lobe of their beampattern suffers from strong effects of interference from neighbour satellites.

At the moment satellites are kept on their orbits with accuracy of at least ±0.1° in the north–south and west–east directions. This deviation results in the maximum displacement of the satellite from the prescribed position by ±0.14°. In view of the increased requirements to holding a satellite on its heliostationary orbit and to the electronic compatibility of satellites it becomes necessary to develop antenna devices that make it possible to use the same antenna for simultaneous reception of signals from several correspondents or for fast scanning by the beampattern of the antenna with the aperture remaining stationary. The main advantages of such devices are: relatively low cost of the antenna, relative simplicity of installation and maintenance (no need in the rotation mechanisms, homing system or controls), and the possibility of simultaneous reception of signals from radio broadcasting satellites within a sector of 40–50°. Such an antenna can be developed on the basis of diffractive antennas [145, 146].

A focusing device for reception of satellite TV signals simultaneously with signals of local TV stations was suggested in [147]. The focuser is a device with a glass plate having a layer of fluorine-doped stannic oxide deposited at certain areas and transparent for visual light; the layer thickness is 150–1000 nm. The shape of the coating layer corresponds to Fresnel zones. It was suggested to fabricate this focusing device, e.g. as a window in a building.

MWDO elements will find the widest applications as antennas with fixed design, meant to work in various systems of millimetre wavelength band, for instance, in transmitting data and telephone conversations, in video conferencing, in satellite TV signal reception, in TV reporting etc. [137, 139, 148]. Figures 6.17(a–c) give examples of possible types of three-dimensional diffractive antennas for the millimetre wavelength band.

The 'flat' antenna type also permits design variations. For instance, a flat 'diamond-like' diffractive antenna was discussed in [213] (figure 6.17(d)). The antenna was a conventional axisymmetric zone plate with the axial part cut out. An investigation of the properties of this antenna showed the following. With a horn irradiator that provides for field amplitude drop at the edge of the aperture by −12 to −14 dB, the side lobes were −35 dB at 14.25 GHz central frequency. The gain at this frequency was 37.8 dBi.

At 11.25 GHz the focal distance changed from 1.032 m (at 14.25 GHz) to 0.75 m. The gain simultaneously dropped to 35.4 dBi, and the gain at 10.75 GHz was already 34.9 dBi.

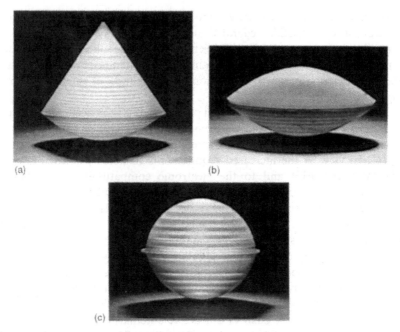

Figure 6.17. (a–c) Three-dimensional diffractive antennas of various shapes. © 2002 IEEE. Reprinted, with permission, from [137].

Diffractive antennas can find applications in most different radar level sensors. For instance, the level sensor described in [149] was designed for continuous contact-free control of level and flowrate of a liquid or dry free-flowing material in large-dimension vessels in the 37 GHz frequency band, for level measurements in the 1–20 m band with an error of ±10 cm.

The simplest diffractive optical elements are now used to demonstrate the wave theory of electromagnetic radiation [150] (figure 6.18).

MWDO elements can be widely used in most various military applications. Examples of high-priority radar systems for the shorter microwave wavelength are [151]:

- Single-pulse radars to control launches of antitank rockets; these are particularly difficult to detect because they control the power of irradiation and use optimally selected parameters and optimal polarization. Working frequencies: 94 and 140 GHz.
- 'Startle' tank radar for detection of ground targets and target display. Working frequencies: 94 and 230 GHz. Range in tank detection 3–5 km.
- Radar systems for homing of antitank third-generation rockets. Working frequency 240 GHz.
- Radiolocation systems for radiovision and detection of objects in complex meteorological conditions. Working range: 300–400 GHz.

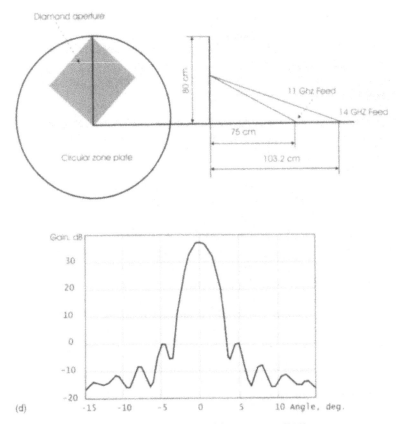

Figure 6.17. (d) Diamond-like flat antenna and beampattern [213].

Figure 6.18. Millimetre-band zone plate for demonstrations of the wave theory of electromagnetic radiation.

- 'Sadarm' radar for detection of ground targets. Working frequency 94 GHz.
- All-weather active–passive system of battle ground observation. Working frequency 220 GHz.
- Self-homing heads for 'air-to-ground' class rockets and antitank rockets. Working frequencies from 90 to 100 GHz.

Diffractive microwave focusing elements have very extensive potential, not yet implemented, and can be applied to most different fields in industry, medicine etc.

Non-medication therapy becomes more and more popular for treating various diseases. One of them is microwave resonance therapy [152–155]. The mechanism of microwave resonance therapy involves intensification of trophic processes in tissues and organs owing to a normalization and stimulation of neuro-humoral reactions and to an overall effect on the nervous system. The method is based on using electromagnetic millimetre-band waves acting on biologically active points in human body. The therapeutic action of electromagnetic millimetre waves lies in that these waves, penetrating through the skin, interact with certain frequencies at which biological object emit, at the same or near resonance frequency, which produces information signals that control the activity of physiological functions that are characteristic of a given tissue, organ or system. If a certain function of the organism is weakened, the effect of millimetre-band electromagnetic waves normalizes this function or pushes it towards the norm. Radiation used for this type of treatment is microwaves at 55 to 68 GHz, at signal power from 0.1 to 3 mW and from 10^{-14} to 10^{-13} W.

Microwave resonance therapy proved most effective with stomach and duodenum ulcers, bronchospasmodic syndrome, certain gastro-intestinal and bladder diseases, tropic ulcers, injuries of soft and bone tissue, stenocardia, high blood pressure, certain cancers and some other diseases.

Elements of MWDO can be used to focus microwave radiation with a prescribed focal curve on to human skin.

6.9 Security fence protection

Radiobeam technologies for protecting security fences have substantial advantages over other classes of similar devices:

- relatively low cost,
- simple installation and maintenance,
- high noise resistance and high probability of intrusion detection.

Using the 61.25 GHz frequency band (oxygen absorption line) provides both high electromagnetic compatibility of equipment and substantial camouflage

(a)   (b)

Figure 6.19. An example of implementation of diffractive microwave perimetre fencing protection on conical (a) and spherical (b) surfaces. © 2002 IEEE. Reprinted, with permission, from [204].

of electromagnetic radiation, plus hidden nature of operation of the equipment.

Two main factors may be quoted as principal drawbacks of these or similar systems: visually unmasking appearance of the antenna unit of the system and the shape of its directivity pattern (shape of the detection zone). The second of these drawbacks is partially removed in the 'Abris 2' modification of the security fence protection system [201] in which the detection zone is an ellipse with 1.4 m × 2 m axes. The former of the two drawbacks is practically unavoidable with classical antennas: the very existence of the protection system (its antenna) cannot be hidden, and this in itself is a considerable unmasking factor. Even though an attempt was made in the 'Abris' protection security fence to mask the antenna unit by giving it the appearance of a TV receiver, this cannot provide the solution to the problem.

However, these drawbacks can be avoided by using MWDO elements fabricated on an arbitrary three-dimensional surface [204]. In view of the basic properties of such elements, two main problems are then solved: directivity patterns of sufficiently arbitrary shape and visual masking of such antenna systems as architectural (interior decoration) elements of a building, for instance, as lighting fixtures (figure 6.19(a,b)) [156].

Diffractive antennas fabricated of optically transparent radio materials, such as optical polystyrene, can incorporate, for instance, lamps placed inside the antenna for the illumination of areas around posts. Furthermore, diffractive antennas possess additional frequency filter properties, which increases the noise protection of the system as a whole, while placing illuminating lamps inside the antenna substantially improves the working conditions of the electronic equipment at low ambient temperatures.

A two-spectrum diffractive lens antenna designed to work in two different spectral bands at the same time (infrared and millimetre bands) is shown

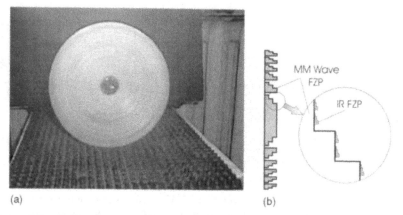

Figure 6.20. (a) Two-spectrum (IR + millimetre) diffractive lens-type antenna. © 2002 IEEE. Reprinted, with permission, from [204]. (b) Combined fabrication of phase structures of infrared and microwave diffractive optics bands.

in figure 6.20(a). This antenna has overlapping beampatterns in the two bands and can be used in protection systems designed for various applications [204].

However, replacement of a part of the central zone of a diffractive microwave-band objective with a visual- or infrared-band element of is not always optimal—if the materials of the diffractive element for two different bands are selected correctly, the diameter of the infrared band zone plate is limited by the size of the first zone of the microwave-band element. It is preferable, even though technologically more complicated, to fabricate the infrared band diffractive structure on the phase zones of the microwave band element. A schematic diagram of such an element is shown in figure 6.20(b).

6.10 Application of millimetre-wavelength-band diffractive optics in scientific research

Research into focusing properties of diffractive elements operating on radiation of different spectral composition made it possible to develop an original setup for probing plasma objects simultaneously at several discrete wavelengths within the same local zone, and to study the properties of low-temperature weakly-nonideal plasma with condensed dispersed phase (CDP).

The principle of designing the experimental setup in question was to use the spatial-selective properties of MWDO elements with an off-axis position of the focal area. These properties were used in the following manner [106, 157, 158]. The spatial position of the focal area changes in response to wavelength changes in the 'distance–angle' coordinates. Therefore, if the radiation

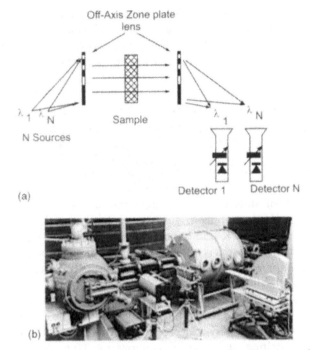

Figure 6.21. (a) Schematic diagram of a microwave setup for plasma sounding. (b) Test facility for studying phenomena caused by high-speed motion of bodies through air. © 2002 IEEE. Reprinted, with permission, from [167].

sources are placed at points f_i that correspond to a discrete set of wavelengths λ_i, the radiation leaving the diffractive element will be focused in the same local zone regardless of wavelength. The choice of the off-axial position of radiation sources is dictated by the condition of elimination of one source blocking the radiation from another. A schematic diagram of the setup is shown in figure 6.21(a).

This facility was used to experimentally study [157, 158] low-temperature weakly-nonideal plasma with a CDP generated by high-speed motion of bodies through the air; also investigated was the effect of adding flammable particles on the characteristics of a high-enthalpy turbulent non-isothermal jet. It was discovered that grain size in the material micro-structure on the surface of the body affects the electromagnetic properties of such plasma conglomerates. Figure 6.21(b) gives the appearance of an experimental test facility for studying phenomena caused when objects move through the air at high speed; the microwave section with a diffractive antenna is clearly seen.

In the past the specifics of generation and injection of CDP particles into the plasma trail (produced in the air by high-speed motion of metallic bodies) was done by high-speed filming of the motion of such bodies (recording air glow and trail spectra at visible wavelengths). Aluminium alloy, steel and magnesium were used as body materials. The bodies were shaped and then accelerated to velocities on the order of 5 km/s using cumulative devices [161].

To adjust the trajectory of hypersonic flight of a body in its motion along a fixed optical axis the effect of stabilization of a hypersonic body moving through a channel was used—for instance, in flight through a pipe. The same effect was used to correct the shape of hypersonic bodies (in motion through pipes of rectangular and elliptical cross section).

Mathematical simulation of hypersonic streamlining of bodies at such flight speeds using a software package [162] showed that when a wide-range equation of state of air that takes into account single dissociation and ionization [163] is applied, gas temperature near the surface of the body reaches $(6–8) \times 10^3$ K and the non-ideality parameter γ [164] in shock-wave-compacted air plasma is about 0.1. For aluminium bodies it was found [159] that injection of CDP particles into the plasma trail occurs by oblation of the surface, that is, by heating the surface layer of the metal to melting point and blowing this metal off by the oncoming gas flow. The amount of material blown off the surface of the body depends on the energy transferred to this body and on the physical properties of its material, such as specific heat, heat of melting and heat of vaporization [166]. We assumed that CDP particles are injected as tiny liquid metal droplets. It proved impossible to conduct similar studies with steel objects, owing to weak luminosity of the trail and insufficient sensitivity of the equipment.

Our experimental studies [167] were carried out in air at normal initial conditions on a ballistic trajectory similar to that described in [165]. The body, shaped in the process of compression by products of detonation of the cumulative lining, was moving between centred transmitting and receiving microwave antennas with an aperture ratio of about 1 and a diameter of 40λ (where $\lambda = 4$ mm is the radiation wavelength). Copper was used as lining material. The shape and velocity of bodies were controlled by a multi-positional pulsed x-ray filming along the ballistic trajectory. The time of flight to the measuring microwave section was several times longer than the time of relaxation of the local body temperature as evaluated from the thermal conduction equation,

$$t \approx c_p l^2 \Lambda^{-1} \rho$$

where ρ is density, c_p is specific heat conduction, l is the scale of heating non-uniformity in the material under the conditions of the experiment (l seemed to have the same order of magnitude as the grain size at the surface of the moving body), and Λ is the thermal conductivity of the body. The mean grain size of the microstructure of cumulative lining was monitored along the length of its generatrix at three points, using microsections. The grain

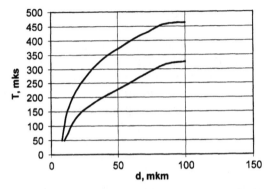

Figure 6.22. Effect of microstructure of surface layer of a moving body on the lifetime of the CDP-containing plasma. © 2002 IEEE. Reprinted, with permission, from [167].

size was controlled by adjusting the regime of thermal treatment of the lining material.

Figure 6.22 plots the lifetime of a plasma cluster with CDP as a function of grain size in the microstructure of the cumulative lining; the curves were obtained by processing the experimental data. To plot a curve, the results of three runs of experiments were used (10–15 experiments in each run). The grain size of the cumulative lining material was 7–10, 15–25 and 80–100 μm, respectively. Taking into account that the grain size on the inner surface of the cumulative lining and on the body surface are the same within an order of magnitude [168], the data indicate that the microstructural grain size of the surface material of the speeding body considerably affected the electrophysical properties of CDP-containing plasma clusters. A similar dependence was also observed for the transmission coefficient of plasma clusters.

To evaluate the effect of size of the removed metal particle on the attenuation of electromagnetic radiation, special computation of attenuation cross section by spherical particles was conducted. Computations followed the Mie theory in the single scattering approximation. Since the optical constants of metals in the millimetre wavelength band are unknown, the corresponding refractive index and coefficient of absorption for particles were calculated using metal conductivity at different temperatures.

The results of computations showed that in a wide range of variation of the optical constants of metal particles in the millimetre wavelength band the relative attenuation cross section varied insignificantly up to the particle radius ~40–50 μm.

The authors hypothesize that the discovered behaviour occurs because the heated plasma reduces the material's dynamic yield point which is an integral characteristic and determines the strength of bonding between grains, again depending on grain size, so that microcracks appear as a result of growing thermal stress [166] on the body's surface. The process of mass removal from the surface proceeds via thermomechanical erosion.

Deformation and destruction of the surface layer of the material become especially significant when the temperature region of the semi-solid state is reached. In this case surface deformation is probably possible via liquid flow of grains along boundaries. This phenomenon is observed at temperatures above $0.45T_{melt}$ [166] (where T_{melt} is the melting point of the material) at given flight velocity and specific properties of the material. The layer that lost its mechanical strength and developed thermal cracks in all directions is blown off by the oncoming airflow. This process is non-homogenous because the surface grains of metal are strained to different degrees; indeed, strain in one of the grains results in deforming its boundaries [166] and therefore deforms the adjacent grains of the polycrystalline body.

Metal particles in the CDP state injected into the plasma plume through the mechanism described above affect its electromagnetic properties through thermionic emission from their surfaces [164]; furthermore, increasing the size of microstructural grains on the surface seems to increase the concentration of charged particles in the plasma.

We conclude therefore that the results of experimental studies described above show that when CDP-containing plasma is generated in the air in the vicinity of rapidly moving bodies and when their electrophysical properties are analysed, it is necessary to take into account the microstructure of the material on the surface of the plasma-generating body.

6.11 Semiconductor zone plate

Silicon or germanium plates that change their properties under illumination by light can be used in the microwave band as converters [179]. This paper suggested creating semiconductor zone plates by generating annular plasma inhomogeneities in the semiconductor. The zone plate was fabricated from 1.5 mm thick n-type germanium plate, with resistivity 40 Ω·cm, in which plasma was initiated by illumination through a mask or by projecting an image of the zone plate on the surface of the semiconductor.

The schematic diagram of producing the semiconductor zone plate is shown in figure 6.23.

Illuminated areas correspond to regions of high-concentration electron–hole plasma and therefore to low transparency for millimetre wavelength radiation. Shaded areas in their turn correspond to regions of low concentration of electron–hole plasma, that is, to low transparency. When the light source is turned on, a Fresnel zone plate is created in the path of the electromagnetic field, whose zone radii can be varied smoothly if a variable focal distance lens is used. The values of zone radii were calculated in [179]

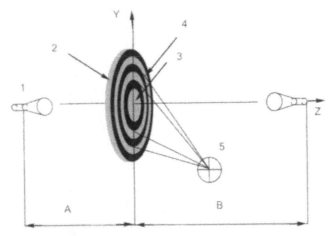

Figure 6.23. Semiconductor zone plate [179]: 1, microwave source; 2, semiconductor plate; 3, regions of high concentration of electron–hole plasma; 4, regions of low concentration of electron–hole plasma; 5, visible light source.

using an approximate formula

$$r_n = \sqrt{\frac{ab}{a+b}\,n\lambda}.$$

Measurements were conducted at wavelength $\lambda = 4$ mm for the values $a = 105$ mm, $b = 66$ mm. Plasma was generated within odd-numbered Fresnel zones. For comparison, a metal zone plate of the same size was used, made of metal foil glued to foam polystyrene.

Figure 6.24 gives experimental distribution of radiation intensity along the optical axis (on the left) and transversally to it (on the right) for different values of illumination intensity on odd-numbered zones (curves 2, 3, 4 on the left-hand plot). The position of the intensity maximum $z = 180$ mm corresponds to the nominal value $b = 66$ mm. As light intensity is raised (curves 3, 4, respectively), conduction of semiconductor zones increases and intensity of radiation at the focal point decreases. Curve 1 was obtained for a metal zone plate.

The curves on the right show the shape of the diffraction spot for the metallic (curve 1) and semiconductor (curve 2) zone plate.

Experiments showed that it is possible to change the conditions of creating nonuniform plasma in a semiconductor and the ratio of conductivities in different areas of the semiconductor and, therefore, control the position of the focal spot, its size and the intensity of radiation at this point. If the transparency can be made movable along the surface of the semiconductor, a scanning system may be created.

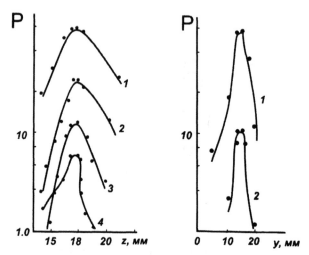

Figure 6.24. Field intensity distribution along the optical axis of a semiconductor zone plate (on the left) and transversely to it (on the right).

It is interesting to note that it took nearly twenty years for these ideas to be found useful in developing a number of new devices, for instance, optically controlled millimetre wavelength band modulators [212].

6.12   Diffractive optics in acoustics

Acoustic zone plates are used to demonstrate various physical processes [169]. Considerable difficulties are encountered, however, in creating acoustic zone plates that invert the phase of oscillations of one half of the zones. The reason is that the acoustic resistance ρu of any material is so high compared with the resistance of air that acoustic waves are nearly completely reflected. A phase zone plate can be produced using the method suggested by Kock [170] for fabrication of microwave lenses. The method consists in forcing waves to move between tilted plates; the path length then increases by a factor of $1/\cos Q$ which corresponds to the effective index of reflection $n = 1/\cos Q$ for the propagation of waves in free space.

The strip width l is found from the conventional relation $d(n-1) = \lambda/2$ or $d(n-1) = \lambda$, which in this particular case of $n = 1/\cos Q$ and strip width $l = d/\cos Q$ takes the simplest and physically transparent form:

$$l - d = \lambda/2.$$

Diffractive optical elements find interesting applications in acoustics [171–176]. Fresnel zone plates can be used in audiovision [171] and nondestructive ultrasound testing [176–178], both in reception and emission modes. In reception mode, the zone plate consists of a sequence of alternating transparent and opaque zones. This zone plate behaves as a conventional acoustic lens.

If used as a focusing emitter, the acoustic converter is fabricated as a zone plate. For instance, a binary zone plate with 10 zones made as a sequence of gold electrodes on the surface of a ceramic converter was described in [171]. The other side of the converter has a common metal coating. When used to transform electric fields, only transparent zones emit. This means that the same zone plate can be used to generate both an 'audio' picture and an image of an object.

Acoustically emitted diffractive focusing elements possess another very interesting and very promising property [170, 171]. If zones made of gold electrodes are replaced with photoconducting layers and placed between a converter and an optically transparent electrode, this device can be controlled by light. If an optical image of a diffractive element is projected on to such a converter and is moved transversally, the point of acoustical focus will also move. Therefore, two-dimensional scanning can be realized.

6.13   Zone plates for focusing shockwaves

A number of publications [35, 84, 90] reported the creation of diffractive optical elements that focus coherent monochromatic radiation into an arbitrary-shape prescribed volume of space. We have already indicated that the principle on which such devices work is based on creating specific phase delays across the wavefront, resulting in in-phase addition of different segments of the incident wavefront in a prescribed area. The mathematical problem was treated in the framework of propagation of radiation thorough vacuum, that is, as linear media and as linear mode.

Various types of reflectors with elliptic or parabolic shape of the surface [181–186] are used to focus weak shockwaves in air [180] and in liquids.

The principal possibility of controlling the shape of the shockwave front was shown in [187], for a specific example of the propagation of plane shockwave in water, as well as its focusing to a prescribed local segment of space in nonlinear mode using diffractive quasioptical elements. Note that in this particular case the term 'focusing' is defined as creation of the local area of elevated pressure in a given volume immediately behind the wavefront of the travelling shockwave, while 'shockwave' is interpreted not as an ideal discontinuity but as rapid surge of pressure to a certain value P_0 followed by slow decay.

6.13.1   Formulation of the problem

The regularities of shockwave diffraction and its focusing were studied in a numerical experiment controlled by a highly efficient computation package, 'Stereo-PC' [188, 189].

The problem was solved in two-dimensional axisymmetric formulation. Both the upper and lower bounds of the domain of computation were chosen to be 'closed'. The computation grid cell size was $128(z) \times 25(R)$ mm, with spatial resolution of 81 cells/mm^2. The time increment was selected automatically using the condition $|V_{max}| \Delta t < h$, where $|V_{max}|$ is the maximum speed of particle motion relative to the Euler grid at a given moment of time; h is the length of the side of the computational cell.

The computational grid was filled with water whose equation of state was chosen in the *D–U* form with coefficients $D = a + LU$, $\rho_0 = 1.00$ g/cm^3, $a = 1.5$ mm/μs; $L = 2.00$.

The diffractive element was placed at a distance of 20 mm from the left-hand edge of the computational domain. The shockwave in water was created with a hammer 2 mm thick acting on the entire left-hand boundary of the computational grid. The hammer's material was water with the same equation of state. The problem was solved in the streamlining mode, that is, the diffractive element was assumed to be 'absolutely rigid'.

6.13.2 Synthesis of diffractive element

To investigate the feasibility of shockwave focusing in the axisymmetric two-dimensional case we chose a diffractive element that transforms a plane incident wavefront into a ring of a prescribed radius at a fixed distance from its surface. In this case the following factors were taken into account when designing the diffractive element:

- The diffractive element is built using the laws of geometrical optics, that is, it is transplanted from the electromagnetic spectrum into the mechanics of continuous media.
- Shockwaves propagating through water are damped out relatively rapidly; hence, the focal length of a diffractive quasioptical element must be relatively short to ensure efficient functioning.
- In order to achieve acceptable quality of focusing by a diffractive optical element, the element (as shown by additional research) must have the lens aperture $D/\lambda_0 \approx 50$, where D is the diameter of the focusing element and λ_0 is the wavelength for which the element is designed. To provide for considerable gain and recalling that $G \approx N$, the maximum possible number of zones N must be placed on the element's surface.

In view of this, an element of 'intersecting' type was chosen as a diffractive optical element (rays traced from each of its zones to the focusing area intersect the optical axis before they reach the focusing plane).

The selection of the working wavelengths deserves special attention. Diffractive elements required for working with electromagnetic waves are typically designed to work with monochromatic radiation. In this particular case, it is practically impossible to form a shockwave with the frequency spectrum as a delta-function corresponding to a single wavelength. The point is that owing to the relaxation wave, the distribution of shockwave parameters (e.g. pressure) along the propagation direction is an approximately exponential decay curve. This wavefront corresponds to the frequency spectrum defined as the Fourier transform of the corresponding distribution. If, for instance, $p(t) = p_0 \exp(-t/\tau)$, $p(t) = 0$ or $t \leq 0$, the distribution function of the amplitudes of harmonic components of pressure pulse is

$$\rho(\nu) = p_0/2\pi[1/\tau^2 + \nu^2]^{-1/2}.$$

That is, for $\nu = 1/\tau$ the spectral energy density diminishes by a factor of 2 (3 dB).

On the other hand, diffractive optical elements possess specific frequency characteristics [76], that is, the position of the focusing area and its shape depend on the radiation wavelength if it differs from the nominal wavelength. However, when working with non-monochromatic non-coherent radiation, the situation becomes yet more complicated: frequency characteristics manifest themselves differently for each of the wavelength of the corresponding spectral band and, furthermore, interference effect between these sets of discrete wavelengths becomes significant.

At the same time, we can assume the width of a shockwave as a certain effective wavelength λ_s [190].

Therefore, when selecting the nominal wavelength for a diffractive optical element, the following two main factors are taken into account:

- the shockwave corresponds to wavelength λ_s;
- when the diffractive element uses a wavelength $\lambda > \lambda_0$, the position of the focusing area moves towards the diffractive element's plane.

To simplify the problem and solve it, an amplitude-type diffractive optical element was chosen. The parameters of the focusing element were: $\lambda_0 = 0.5$ mm, $R_k = 10$ mm, $f(\lambda) = 80$ mm, where R_k is the radius of the focusing ring and $f(\lambda)$ is the focal distance corresponding to the wavelength $\lambda \neq \lambda_0$. The radii of zone boundaries of the focusing element are listed in table 6.3. The thickness of the diffractive element was 5 mm (i.e. 10λ).

6.13.3   Results of the numerical experiment

The shockwave front reaches the surface of the diffractive optical element at the moment $t = 3.7\,\mu s$. The incident wavefront has the following parameters:

Table 6.3.

Zone number	1	2	3	4	5	6	7	8	9	10	11
R_k (mm)		1.7	3.3	4.6	5.9	7.1	8.2	9.2	10.2	11.1	12.1	12.9

Table 6.4.

Pressure level	0.9	0.7	0.5	0.3	0.1
Δx_p (mm)		1.6	8.3	14.2	22.1	34.3

pressure at the shockwave front reaches 12.5 GPa and its velocity is 2.1 mm/μs. The width of the pressure pulse at the appropriate levels characterizes the shape of the shockwave front.

The shockwave front is 'spread' over 0.8 mm. The axial component of velocity decreases linearly over the path of 33.3 mm down to $V_z \approx$ 1.25 mm/μs (see table 6.4).

Nonuniformity of parameter distribution along the shockwave front:

- pressure $\Delta p/p_{max} = 4.4\%$
- axial velocity component ~2%
- radial velocity component ~2.5%.
- energy conservation law is satisfied within an error of ~3%.

Some $\Delta t = 0.3$ μs later the shockwave incident on the diffractive optical element is reflected back and about 50% (area-wise) is transmitted. At this moment the front of the transmitted shockwave is located mid-way through the thickness of the focusing element. The wavefront of the reflected shockwave propagating through the already compressed matter has a complicated structure. As we know from Huygens' theory of diffraction, each segment of the propagating wavefront can be treated as a pointlike source of a spherical wave. A similar pattern is observed in the case in hand: each zone of the diffractive optical element that transmits the shockwave is a source of toroidal waves. Consequently, elevated values of pressure and velocity are observed in the regions of interference of two adjacent toroidal waves. The pressure distribution along the wavefront surface parallel to the plane of the focusing element oscillates quasiperiodically. The extremums of oscillations fall on the positions of transparent zones of the diffractive optical element and the number of peaks coincides with the number of transparent zones.

For instance, the distribution of the axial velocity component along the reflected shockwave front varies at a given moment of time (figure 6.25(a)) as a cosine curve, with the maximum in the region of the middle zone of the diffractive element ($V_z = 1.24$ mm/μs) and falling off at the edges ($V_z = 0.35$ mm/μs at the centre of the focusing element and 0.2 mm/μs at the edges of the computational grid). Note that this distribution is modulated as pointed out earlier, by a rapidly oscillating function with the modulation depth of about 0.2 mm/μs. A similar distribution is also observed for the radial velocity component but the envelope of the oscillating function is

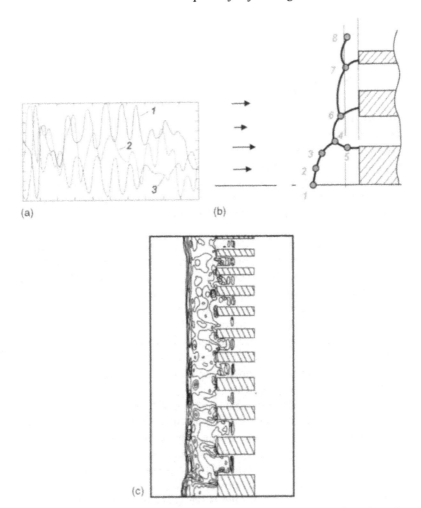

Figure 6.25. Flow pattern at a time moment $t = 4\,\mu s$. (a) Distribution along the reflected shockwave: 1, axial velocity component; 2, radial velocity component; 3, pressure; (b) configuration of pressure waves; (c) pressure contours.

very nearly linear. The maximum value of this component is $V \approx 0.18\,\text{mm}/\mu s$ and is observed in the region of the interference of the waves from the central and first zones of the diffractive element; the maximum negative value is observed at the edge of the computational grid and reaches about $0.17\,\text{mm}/\mu s$.

Another characteristic feature of this flow is that the distribution of the axial and radial velocity components are 'quasi-synchronized', that is, the positions of extremums of these functions nearly coincide and are displaced

Table 6.5.

Point	1*	2	3	4*	5	6*	7*	8*
P (GPa)	38.1	21.3	18.3	25.5	21.7	34.3	30.1	25.6

relative to one another by $\Delta\varphi$, a small constant fraction 0.5–1.0 of grid period, or $(0.1–0.2)\lambda_0$.

In contrast to this, pressure distribution along the wavefront surface is in 'antiphase' to the distribution of the parameters mentioned above, since owing to the interference phenomena this distribution does not possess such a sharply defined quasiperiodical structure, even though oscillations do survive. As an example, figure 6.25(b) shows the configuration of reflected shockwaves.

Table 6.5 lists pressure at a number of points in the vicinity of the first three zones of the diffractive optical element. Asterisks mark points where pressure is higher than the pressure of shockwave reflected from a rigid wall (doubled pressure at the front of the incident shockwave is $P_{refl} = 25$ GPa). Shockwave configurations in this figure were identified and analysed using three techniques allowed by 'Stereo-PC' software package [189–191]: using a differential analyser, from the maxima of artificial viscosity (a quantity proportional to pressure gradient), and from pressure maxima in a given cross section of the computational grid. The first two methods are essentially equivalent.

Note [189] that the differential analyser technique makes it possible to identify zones of rapid change of the quantities being calculated, which are thus associated with shockwave fronts. The width of such zones was not more than 3–5 grid cells, and drawing the line of the maximum rate of change of a quantity localizes shockwave fronts within a single cell.

Wavefronts are found from pressure maxima on the assumption that a transition from incident to reflected wave is accompanied with a deviation from the position of the maxima along the normal to the OZ axis in the direction opposite to the front of the incident shockwave. The analysis makes it possible to identify the position of coupling points of two shockwaves to within one cell. The computed value of the shockwave reflection (or incidence) angle is then found from the average slope of the reflected (incident) wavefront over the length of almost five cells. Figure 6.25(c) shows pressure contour lines for the same moment of time (the number of contour lines is 7, the initial value is 15 GPa, the step between the two neighbouring contours is 2.5 GPa).

A Mach configuration is therefore formed as a result of a specific nature of flow—interference of compression waves that produces non-regular interaction between waves from different segments of neighbouring zones. For instance, by the moment $t \approx 4.2$ µs a Mach disk is formed on the symmetry

Table 6.6.

Number of a transparent zone	V_z^c	V_R^c	V_z^M	P^M	V_R^M
1	2.63	−0.08	1.94	0.96	−1.05
10	3.11	−0.15	1.56	4.29	0

axis, about 0.92 mm in diameter (about 36.6 GPa), with incidence angle $\psi \approx 45°$. It should be emphasized that this particular configuration has not been deliberately monitored in a process of computations and therefore the values of parameters given above are not the upper or lower bounds.

In $\Delta t = 0.3\,\mu s$ the shockwave front passes through the zones of the focusing element and moves through the not yet perturbed medium. By analogy to the situation described above, each transparent zone on the element is a source of a toroidal shockwave. The main features of the flow are retained: the axial and radial velocity components along the wavefront are quasiperiodical. The 'phase' shift between them is also maintained. However, the envelope of these distributions is not a cosine curve but an exponential decreasing towards the periphery of the diffractive element.

Some specifics of the flow caused by unequal widths of the zones transparent for shockwaves are listed in Table 6.6.

A clear representation of the flow pattern is obtained from pressure contours given in figure 6.26(a). The number of contours is six, the initial

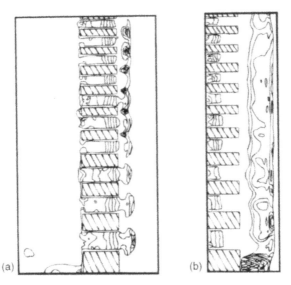

Figure 6.26. Pressure contours at a moment t: (a) 4.51 μs; (b) 4.92 μs.

value of pressure is 4.5 GPa, the step between neighbouring contours is 1.5 GPa. It is obvious, e.g. in figure 6.26(a), that the fronts of the transmitted shockwaves have in the region above the fourth zone of the diffractive optical element already interacted and formed local zones of elevated pressure.

After 4.9 μs following the start of the process the fronts of all shock-waves transmitted through diffractive elements have already interacted and the process of formation of a local annular region of elevated pressure has started, even though until this moment the axial and radial velocity component distributions along the wavefront surface of the transmitted shockwave are very clearly quasiperiodical, maintaining the 'phase' shift between them.

By this time a well-pronounced configuration of non-regular shockwave interaction is observed on the symmetry axis. The size of the Mach disk is about 0.62 mm or 0.36 of the radius of the first zone of the focusing element. Pressure is $P \approx 31.5$ GPa, the axial velocity component $V_z \approx 2.83$ mm/μs, and the shockwave incidence angle is approximately 70°.

Figure 6.26(b) plots pressure contours that characterize this moment of time (11 contour lines, initial pressure 4.3 GPa, the step between contours corresponds to 0.7 GPa). The figure shows that a well-pronounced annular region of elevated pressure starts to form, with a boundary near 5 GPa.

Approximately $\Delta t = 1$ μs later ($t = 5.91$ μs) the annular zone of elevated pressure has completely formed right behind the front of the shockwave that traversed the diffractive optical element. The corresponding pressure contours are shown in figure 6.27(a) where the initial pressure corresponds to 4.5 GPa, the step between adjacent contours is 0.25 GPa and the number of contours is 9. The annular high pressure zone is delineated by the value

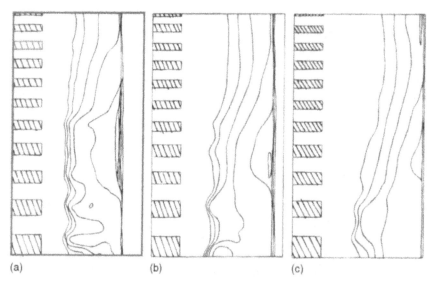

(a)   (b)   (c)

Figure 6.27. Pressure contours at a time t: (a) 5.91 μs; (b) 6.16 μs; (c) 6.42 μs.

$P \approx 5.5\,\text{GPa}$. The pressure and the velocity component distributions along the shockwave front are not oscillating any more; they are smoothly varying functions with a local maximum on the symmetry axis and with the maximum reached within the focusing ring. The maximum pressure is $P \approx 6.2\,\text{GPa}$, and the maximum axial velocity component $V_z \approx 1.39\,\text{mm}/\mu\text{s}$.

From the data given in figure 6.27(a) it was possible to evaluate the efficiency of the diffractive optical element as the ratio of the force applied by pressure of the shockwave on to the entire area of the focusing element (P_{max} at the front of the incident shockwave times the surface area) to the force applied by pressure within the ring delineated by the contour $P \approx 4.5\,\text{GPa}$. The obtained value of diffraction efficiency was $\eta \approx (7\text{--}8)\%$ which agrees sufficiently well with a similar value for a binary amplitude-type diffractive optical element designed for working in a range of electromagnetic wavelengths [30].

Note that focusing of compression waves in a given local area of space occurs in the nonlinear mode because by the order of magnitude the value of pressure P^* in the focusing are is

$$P^* \approx P_0 = \rho_0 c_0^2/(\gamma - 1)$$

so the acoustic approximation is not valid.

The further behaviour of flow consists of the local annular pressure maximum 'melting away' and converging on the symmetry axis of the diffractive element. We will only point to the main features. For instance, by the time $t = 6.16\,\mu\text{s}$ the area of the annular region of high pressure decreases by a factor of 1.5 (figure 6.27(b); the contour line parameters are retained here by analogy to the preceding figure). However, another local high-pressure zone has formed—the one on the symmetry axis of the diffractive optical element. In the electromagnetic spectral range this local maximum has been well studied [90]: it is formed by the in-phase summation of boundary toroidal waves (formed by boundaries of each zone) on the optical axis of the focusing element. The emergence of a local axial pressure maximum in the present problem is also caused by diffraction phenomena, and it is located where the beam traced from the middle of the diffractive optical element through the centre of the annular focusing area intersects the symmetry axis. The maximum pressure in the focusing region at this moment of time is $P \approx 5.73\,\text{GPa}$, and the axial velocity component is $V_z \approx 1.42\,\text{mm}/\mu\text{s}$.

Note that until this moment the area to the left of the diffractive optical element continues to pump energy to the shockwave transmitted through the focusing element. Table 6.7 list the values of the axial velocity components in the flow for each transparent zone in the cross section of the diffractive optical element. We see that the corresponding velocity distribution is of an oscillatory nature, even though on average the velocity increases with increasing zone width.

Table 6.7.

Zone number	1	2	3	4	5	6	7	8	9
V_z (mm/µs)	2.89	2.92	2.67	2.91	2.70	2.85	2.88	3.06	3.03

After an increment of $\Delta t = 0.3\,\mu s$ ($t = 6.42\,\mu s$) the area of the high-pressure annular zone greatly decreases but still survives (see figure 6.27(c)).

Ultimately, at $t = 7.42\,\mu s$ the focusing area has 'melted away' and contracted towards the symmetry axis (figure 6.28(a)). The pressure and axial velocity component distributions along the shockwave front resemble in appearance the shape of the Airy function although they continue to have certain specific properties. Thus the corresponding distributions close to the symmetry axes have a locally constant value (a 'plateau'). The height at such a plateau is 5.25 GPa, the axial velocity component $V_z \approx 1.29\,\text{mm/µs}$, the length of the plateau is about 3.7 mm or 2.12 of the radius of the first zone of the diffractive element. Note that these zones of

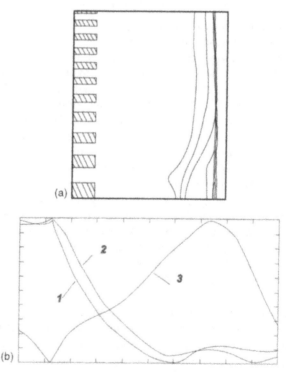

Figure 6.28. (a) Pressure contour lines and (b) distribution of parameters along the wavefront at a moment $t = 7.42\,\mu s$; 1, pressure; 2, axial velocity component; 3, radial velocity component.

constant pressure and velocity continue to slowly collapse on to the symmetry axis, as indicated by the distribution of the radial velocity component (figure 628(b)).

Figures 6.29(a–g) present the results of a numerical experiment that studied the interaction between shockwaves and diffractive optical element in a liquid [187].

We shall evaluate the effective wavelength λ_s on which the diffractive element actually operates. We know that the resolving power of a diffractive optical element that focuses radiation to a spot of diffractionally limited size is found from the Lagrange–Helmholtz invariant

$$\Delta x \approx 1.2 f(\lambda_s) \lambda_s / D.$$

The values of Δx and $f(\lambda_s)$ will be found from figure 6.29(b). Knowing the diameter D of the diffractive element, we find an estimate of the effective wavelength: $\lambda_s \approx D \Delta x / (1.2 f(\lambda_s)) \approx 23$ mm. This value of the effective wavelength is in good agreement with the width of the shockwave pressure pulse at the 0.3–0.2 level.

Let us now evaluate to what extent λ_s corresponds to the maximum wavelength λ_{max} possible for a given type of diffractive optical element. This wavelength is evaluated as the one required for the focal plane (focusing plane) to coincide with the plane of the diffractive optical element. Following [192] we can derive from the condition $f(\lambda) \to 0$ in the electrodynamic case of the geometrical analysis that

$$\lambda_{max} \approx 2(R_k + 2R_N)/N.$$

In this particular case $\lambda_{max} \approx 0.27$ mm.

Therefore, the value of λ_s found above is reasonable, not exceeding the maximum possible value of λ_{max}.

The main conclusions drawn from the behaviour of flow discussed above are as follows:

- The diffractive element borrowed from the field of electromagnetic waves can be efficiently used to focus shockwaves in essentially nonlinear modes (pressure in the focusing area is comparable with pressure of single longitudinal compression of the medium ($P_0 = \rho_0 c_0^2 / (\gamma - 1)$)).
- The diffractive optical element maintains its focusing and frequency characteristics, that is, we observe the dependence of the position in space and of size of focusing area on the current value of the effective wavelength.
- The efficiency of focusing by the diffractive optical element (diffraction efficiency) is comparable with a similar quantity for diffractive optical elements in the electromagnetic wavelength band.
- The results of studying the properties of diffractive elements in the electromagnetic field can be transferred (obviously, taking into account certain

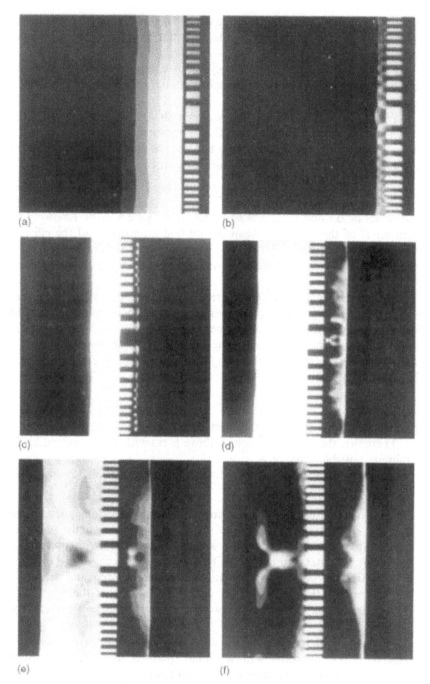

Figure 6.29. (a–f) The initial stage of interaction between a shockwave and the diffractive optical element in a liquid.

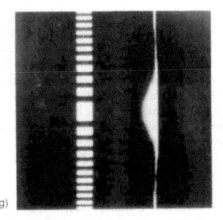

(g)

Figure 6.29. (g) Formation of the focusing region when a shockwave interacts with the diffractive optical element in a liquid.

specifics, like wave attenuation, non-stationarity of the process, its non-linearity etc.) to the mechanics of continuous media.

To illustrate the complete analogy of shockwave propagation problems in nonlinear mode and the functioning of diffractive elements in the electro-magnetic band, figure 6.30 shows a hardcopy of the screen at the last stage of interaction between the planar shockwave in water and an analogue of the zone plate. The configuration of shockwaves is shown in the upper left

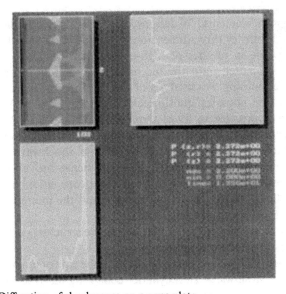

Figure 6.30. Diffraction of shockwaves on a zone plate.

corner, pressure distribution in the horizontal line is shown below, and the corresponding distribution along the vertical line is on the right. It is clear that pressure distribution transversally to the symmetry axis of the diffractive optical element qualitatively coincides with the Airy function, and that along the axis coincides with a corresponding field intensity distribution created by a microwave-band diffractive element. Pressure at the focal point exceeds that at the front of the incident shockwave by a factor of approximately four.

The results given in [187] make it possible to discuss the possibility of creating an entirely new class of diffractive converters of shockwave fronts and possibly even of detonation waves. There is hope that devices may be developed some day that would focus shockwaves into a prescribed arbitrary local spatial configuration with a fixed spatial distribution, in demand for most different purposes. Furthermore, it appears possible to create 'phased' diffractive devices with much higher diffractive efficiency.

Another approach to developing such diffractive converters may be the development of special generators of Mach shockwaves, also in a previously prescribed region of space. Moreover, it is possible to generate not one but several spatially separated Mach configurations. It is not inconceivable, for example, that devices cancelling out shockwaves may be developed.

There is no doubt that all these aspects need further theoretical analysis and experimental study. The results given in [187] only point to the feasibility-in-principle of focusing shockwaves.

6.14   Suppression of shockwaves using diffraction gratings

Diffractive optical elements [76] designed for focusing electromagnetic radiation on to an arbitrary three-dimensional configuration were first shown in [193] to be usable in physics of high-energy densities [194]; for instance, they can be 'implanted' into the mechanics of continuous media such as for shockwave focusing. An analysis of the dynamics of focusing of shockwaves incident on complex quasioptical structures [194–196] showed that in principle it is possible to focus shockwaves into arbitrary three-dimensional configuration in an essentially nonlinear mode. The study was conducted as a numerical experiment [197–199] on shockwaves propagating through water, with the diffractive element being an 'absolutely rigid' body; this means that in this treatment the propagation of shockwaves through the 'body' of the diffractive element and the interaction of shockwaves within this element were ignored.

In what follows, we will show that diffractive structures may in principle offer solutions to all these problems.

The geometry of the problem as applied to the case of magnetic discharge welding looks as follows. A steel cylinder collapsing on to its symmetry axis with a radial velocity typical of magnetic discharge welding and explosion

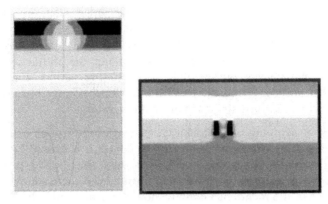

Figure 6.31. 'Suppression' of shockwaves by a diffraction grating [203].

welding, about 200 m/s, interacts with a cylindrical steel target whose outer surface is a periodic (in this particular case) diffraction grating. The goal is to damp out the high-intensity shockwave in the material of the target. The study was conducted as a numerical experiment, following the methods described in [198, 199] and using the software package also described in these references.

In order to clarify the specifics of interaction between a cylindrical shockwave converging on the symmetry axis and an element ('tooth') of the diffraction grating, a number of model problems were solved; the simplest of them is shown in figure 6.31. The lower part of figure 6.31 shows pressure distribution in the material of the target along the line shown by the dot-dash curve for the moment of time corresponding to the shockwave having covered the distance equal to the height of one tooth of the diffraction grating. We see that pressure in the toroidal shockwave that traversed a single grating element is less by an order of magnitude than in the case of using a monolithic target. This effect is caused by the specifics of interaction (interference) of shockwaves and depression waves in the region of contact of the hammer with a tooth of the diffraction grating on the surface of the target.

An analysis of pressure distribution inside the material of the hammer showed that the shockwave reflected by the diffraction grating is rapidly dissipated.

The investigation of the characteristics of dynamics of pressure parameter variation within the hammer in the geometry given in figure 6.31 demonstrated that the pressure distribution along the line parallel to the symmetry axis of the problem over the diffraction grating area changes with time to an anti-phase curve, with the pressure amplitude rapidly decreasing [203].

Doubling the grating period (simultaneously with proportionally increasing the width of the grating's teeth) produces two qualitative changes:

- It correspondingly increases the period of pressure oscillations, which is an obvious result.
- It nearly triples oscillation amplitude, which means that the oscillation amplitude in the pressure distribution along the diffraction grating depends nonlinearly on the diffraction grating period. Furthermore, increased grating period increases the 'lifetime' of pressure oscillations within a fixed volume above the diffraction grating.

The effect of damping of a shockwave propagating through a target with diffraction grating can be conveniently explained: it is caused by the inter-action of toroidal shockwaves and depression shockwaves propagating from the lateral surfaces of diffraction grating teeth, because owing to the interference of the mentioned compression waves and unloading waves the shockwave propagating through the teeth of the target is rapidly suppressed. Therefore, diffractive structures offer a way of damping out shockwaves propagating through solids.

6.15 Zone plates for use in deep space

The feasibility of manufacturing diffractive antennas on arbitrary three-dimensional surfaces opens vast new fields for their application. In fact, scientists' and engineers' fantasy is the only constraint on possible applications. A few typical examples are described below.

One of the concepts of interstellar probe now being developed at NASA design centres includes a solar engine—a kind of ultra-thin (gossamer) sail that provides the probe with a thrust coming from the solar wind acting on it. Evaluations show that in order to provide the necessary thrust, the typical overall dimensions of the thin solar sail will be on the order of 0.5 km [207].

James Early of Lawrence Livermore National Laboratory discusses in [208] the possibility of combining a solar sail of a fly-by mission probe in deep interstellar space with a space telescope antenna made as a zone plate. In principle, this 'gossamer optics' antenna can be used for laser telecommunications and as a solar light concentrator. An antenna-cum-zone plate will be made of the material of the solar sail.

In the same paper Timothy R Knowles of Energy Science Laboratories Inc points out that a zone plate-based antenna has such advantages as low weight, high efficiency, relatively low cost, the possibility of designing a furled deployable antenna, and possible 'extensibility' of the antenna's size by adding more outer zones at relatively low cost.

The possibility of combining a solar sail and a Fresnel zone plate is mentioned in [209] in connection with maintaining extra-remote communications with a probe in deep space in the microwave wavelength band. In this connection the results of studies on diffractive antennas on an arbitrary curvilinear surface with arbitrary (non-circular) zone profile appear to be

Figure 6.32. The concept of a solar sail combined with a diffractive element on a curvilinear surface.

important. The idea of one possible version of such a solar sail antenna shown in figure 6.32 takes into account what was mentioned above and the results outlined in chapter 3. Furthermore, diffractive antennas may make it possible to realize communications on several not very different wavelengths, thus making use of the inherent frequency characteristics.

We wish to note that a solar sail can also be incorporated in the concept of interstellar probe with ionic engine [214]. In this version of a solar sail combined with thin-film solar power converters [214], the converters can be arranged as a diffractive element, thereby making them function simultaneously as a microwave communications antenna.

To extend the frequency band of a zone plate-based antenna, it was suggested to use its frequency characteristics [210]. For instance, if the zone plate was designed to work with the parameters: focal length $F = 10$ m at 8.4–9.25 GHz, antenna diameter $D = 10$ m and the number of Fresnel zones $N = 33$, then the focal length would change to 8.239 m at 7.1675 GHz. The authors of this publication used the following relation to determine the position of the focal point at the new frequency:

$$F_{new}^2 + \left(\frac{D}{2}\right)^2 = (F_{new} + \Delta D)^2, \qquad \Delta D = N\lambda_{new}.$$

Rewriting these expressions in a more familiar form, we obtain

$$F_{new} = \frac{D^2}{8N\lambda_{new}} - \frac{N\lambda_{new}}{2}.$$

In fact, novel applications of zone plate-based antennas stimulate re-investigation of their frequency characteristics. At the same time, the results described in previous chapters show that the position of the focusing region does not coincide for short-focal-length zone plates with the co-ordinates calculated using the laws of geometrical optics, and Fresnel zones themselves do not coincide with the classical ones. Therefore, the wave theory should be used to optimize the relevant parameters.

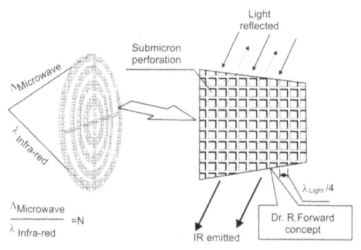

Figure 6.33. Antenna + solar sail combination.

Landis [211] also indicated that one of the possible engine versions for interstellar flight space probe is a high-power laser whose emission is focused by a lens about 1000 km in diameter. This lens is a phase-inversion zone plate, or the so-called O'Meara lens.

We wish to emphasize that two essential feature are recognizable in projects of this type: the possibility of using the effect of focusing of electromagnetic waves by diffractive elements simultaneously at several multiple wavelengths and/or harmonics, and the possibility of placing these elements on arbitrary three-dimensional surfaces. For instance, it is possible—in principle—to create an antenna that works simultaneously in the microwave and infrared bands (figure 6.33). Low efficiency of such an antenna in the infrared band (the frequency harmonic of order ~100 is used) is safely counterbalanced by the size of its aperture. Furthermore, it is important that Fresnel zone boundaries in antennas with high D/F ratio do not coincide with classical ones.

In many techniques of radio monitoring of objects in space it is necessary to form very special beampatterns of antennas to separate and tilt the radio axes of several irradiators located on the same antenna [200]. Technological complications do not allow such operations on large-size antennas using counter-reflectors, even though it is preferable to have precisely such high-sensitivity antennas for space monitoring jobs.

It was suggested in [200] to use for such tasks a Fresnel zone plate inserted into the last segment of the focusing path of radio telescopes. This should provide an additional opportunity of focusing an antenna beam, reducing the noise contribution via narrowing the aperture without losses in the signal, and reducing the effect of side lobes of the beampattern. The tilting of the radio axis is implemented by displacing the zone plate in the focal plane. The transformation of the circular beampattern into an elliptical one is realized by tilting the Fresnel zone plate towards the radio axis. Experiments conducted on the RT-70 radio telescope [200] showed that application of zone plates to solve such problems is quite efficient.

6.16 Segmented apertures

Achieving acceptable quality (resolving power) in radio image construction of objects in the millimetre wavelength band is constrained by the antenna (lens) diameter. This limitation can be substantially alleviated in a number

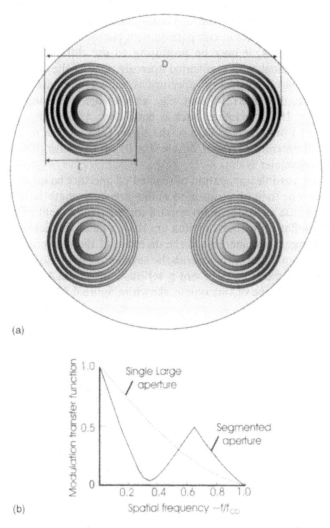

Figure 6.34. The concept of segmented aperture and antenna's transfer function.

of practical applications by using an antenna with segmented aperture. However, the usefulness of parabolic reflector antennas for this purpose is limited by the following practically important factors:

- the high cost of fabrication of parabolic mirrors;
- the need for massive mounting devices.

Comparisons of the focusing properties and efficiencies of parabolic and diffractive antennas (based on zone plates), as conducted in a number of publications and summarized in [215], showed that the efficiency of zone plates reached 70% of that of parabolic antennas. Therefore, this shortcoming is easily compensated for by a small increase in the diameter of the diffractive antenna. On the other hand, diffractive antennas have much better technological and cost parameters in comparison with parabolic antennas.

In view of this, it may be promising to use diffractive antennas as elements of antennas with segmented aperture. One of the concepts of such a zone plate-based segmented aperture is shown in figure 6.34(a). It is fairly simple to evaluate [216] that, e.g., an aperture consisting of four segments $D_1/\lambda \approx 10^3$ in diameter each is equivalent, taking into account the techniques of Fourier processing of the signal [216], to the resolution of a conventional antenna of about $D/\lambda \approx 7 \times 10^3$ in diameter (figure 6.34(b)). The gain in diameter is thus ~ 7.

Another possible application of segmented aperture based on diffractive antennas may be their application to millimetre-wavelength-band telescopes [217]. In this case again the main criteria for the replacement of a parabolic reflector by a flat diffractive antenna are the cost and simpler technological demands. Insignificant increase in the diameter of the segmented diffractive antenna in comparison with a parabolic segment that would be required to make their efficiencies equal is not a serious constraint. A schematic view of a possible telescope of this type is shown in figure 6.35.

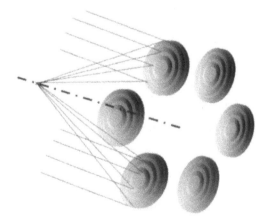

Figure 6.35. Principal diagram of a telescope with segmented aperture.

Bibliography

[1] Aristov V V and Erko A I 1991 *X-Ray Optics* (Moscow: Nauka) p 150 (in Russian)

[2] McGee J F, Hesser D R and Milton J W 1969 X-*Ray Reflection Optics* (*Recent Developments*) (Berlin: Springer)

[3] Aristova V V (ed) 1987 *Optics of Soft X-Ray Radiation* (Moscow: MTsNTI) p 70 (in Russian)

[4] Shabel'nikov P G 1987 'Microelectronics technology in the fabrication of focusing elements' *Zarubezhnaya Elektronnaya Tehnika* **8**(315) 24–52

[5] Schmalt G and Rudolph D (eds) 1984 *X-Ray Microscopy* (Berlin: Springer) Springer Series in Optical Science, vol 43

[6] Erko A I 1990 'Synthethized Fresnel optics of nanometer wavelength band' in *Collected Papers. Holographic Techniques in Science and Technology* (Leningrad: FTI im. A F Ioffe) pp 171–205

[7] Aristov V V, Erko A I and Martynov V V 1988 *Revue Phys. Appl.*, **23**(10) 1623–1630

[8] Cornbleet S 1976 *Microwave Optics* (New York: Academic Press)

[9] Aristov V V, Gaponov S V, Genkin V M *et al* 1986 'Focusing properties of profiled multilayer x-ray mirrors' *Pis'ma v ZhETF* **44**(4) 202–209

[10] Martynov V V 1988 'Diffractive principles of high-resolution nanometer wavelength band optics' PhD thesis, Chernogolovka Institute of Problems in the Technology of Microelectronics and Extrapure Materials

[11] Tatchyn R, Csonka P L and Lindau I 1984 'The constant-thickness zone plate as a variational solution' *Opt. Acta* **31**(7) 729–733

[12] Tatchyn R, Csonka P L and Lindau I 1984 'A unified approach to the theory and design of optimum transmission diffraction system in soft X-ray range' *SPIE* **503** 168–180

[13] Tatchyn R, Csonka P L and Lindau I 1985 'Symmetry restricted profiles for maximizing the output diffraction efficiencies of plates in the soft X-ray range' *J. Opt. Soc. Amer.* **B2**(8) 1287–1293

[14] Yun W B, Viccaro P J, Lai B and Chrzas J 1992 'Coherent hard X-ray focusing optics and applications' *Rev. Instrum.* **63**(1) 582–585

[15] Davydov A V, Erko A I, Panchenko L A *et al* 1987 'Single-crystal Fresnel optics for nanometer wavelength range' *Pis'ma v ZhTF* **13**(6) 1017–1020

[16] Aristov V V, Basov Yu A and Snegirev A A 1987 'Experimental discovery of focusing of x-rays in Bragg diffraction by Fresnel-zone-structured perfect crystals' *Pis'ma v ZhTF* **13**(6) 114–118

[17] Aristov V V, Basov Yu A, Snigirev A A and Yunkin V A 1992 'Bragg–Fresnel lens with the slope zone structure' *Rev. Sci. Instrum.* **63**(1) 586–587

[18] Bonse V, Rieker C and Snigirev A A 1992 'Kirkpatrik–Balz microprobe on the bases of two linear single crystal Bragg–Fresnel lenses' *Rev. Sci. Instrum.* **63**(1) 622–624

[19] Aristov V V, Erco A I and Martynov V V 1988 'Principles of Bragg–Fresnel multilayer optics' *Rev. Phys. Appl.* **23**(10) 1623–1630

[20] Aristov V V and Shekman V Sh 1971 'Properties of three-dimensional holograms' *UFN* **104**(1) 51–56

[21] Unger P, Bogli V, Beneking H *et al* 1988 'Imaging zone plates for X-ray microscopy fabricated by electron beam lithography' *J. Vac. Sci. Technol.* **B6**(1) 323–327

[22] Aristov V V, Babin S V, Davydov A V *et al* 1987 'Precise lithography for component integral optics of nanometer range' in *Microelectronic Engineering* (Amsterdam: North-Holland) vol 6, pp 129–134

[23] Aristov V V, Bashkina G A, Dorozhkina L V *et al* 1983 'Projective x-ray lithography of periodic structures' *Surface, Physics, Chemistry, Mechanics* **12** 113–118 (in Russian)

[24] Thime J 1984 *Construction of Condenser Zone Plates for a Scanning X-Ray Microscope* (ed G Schmalh and G Rudolph) (Berlin: Springer) Springer Series on Optical Sciences, vol 43, pp 91–96

[25] Agafonov Yu A, Bryunetkin B A, Erko A I *et al* 1992 'Generation of image of 'hot point' of a fast pinch discharge using a linear Bragg–Fresnel lens' *Pis'ma v ZhTF* **18**(15) 56–59

[26] Bonse U, Riekel C and Snigirev A A 1992 'Kirkpatrick–Baez microprobe on the basis of two linear single crystal Bragg–Fresnel lenses' *Rev. Sci. Instrum.* **63**(1) 622–624

[27] Kirz J, Abe H, Anderson E and Attwood D *et al* 1990 'X-ray microscopy with the NSLS soft X ray undulator' *Phys. Scr.* **31** 12–17

[28] *New Scientist*, 19.01.91, vol 129, No. 1752, p 32

[29] Cowley J M 1995 *Diffraction Physics*, 3rd edn (Amsterdam: North-Holland)

[30] Bobrov S T and Greisuh G I 1985 'High-resolution diffractive-optics-based projective lenses' *Avtometriya* **6** 3–7

[31] G J Swanson and W B Veldhamp 1985 'Binary lenses for use at 10.6 micrometers' *Optical Engineering* **24**(5) 701–795

[32] Svechnikov G S 1988 *Integrated Optics* (Kiev: Nauk. Dumka) (in Russian)

[33] Semenov A S, Smirnov V L and Shman'ko A V 1990 *Integrated Optics for Systems of Data Transfer and Data Processing* (Moscow: Radio i Svyaz')

[34] Anikin V I and Shokol S F 1984 'Focusing elements of integrated optics' *Zarubezhnaya Radioelectronika* **5** 67–77

[35] Engel A, Steffen J and Herziger G 1974 'Laser machining with modulated zone plates' *Appl. Opt.* **13**(12) 269–273

[36] Ageshin V I, Azarov A A, Popov V V *et al* 1988 'Application of focusators to laser processing of materials' *Kompyuternaya Optika* **3** 91–93

[37] Sisakyan I N, Shorin V P, Soifer V A *et al* 1988 'Technological promise in application of focusators to laser processing of materials' *Kompyuternaya Optika* **3** 94–97

[38] Goto K, Mori K, Hatakoshi G and Takahashi S 1987 'Spherical grating objective lenses for optical disk pickups' *Jap. J. Appl. Phys.* **26** Suppl. 26-4, p 135

[39] Kimura Y, Sugama S and Ono Y 1988 'Compact optical head using a holographic optical element for CD players' *Appl. Opt.* **27**(4) 668

[49] Leger J R 1992 'Design of diffractive optics for concentration of light from diode laser arrays' in *Diffractive Optics: Design, Fabrication, and Applications. Technical Digest* (Washington, DC: Optical Society of America), vol 9, pp TuA2-1/55–TuA2-3/57

[41] Streible N 1992 'Diffractive optical elements for optoelectronics' in *Diffractive Optics: Design, Fabrication and Applications. Technical Digest* (Washington, DC: Optical Society of America), vol 9, pp 162/WC1-1–164/WC1-3

[42] Freeman M H 1988 'Improving vision' *Phys. Bull.* **2**

[43] Jurgen Jahns 1992 'System design for planar optics' in *Diffractive Optics: Design, Fabrication and Applications. Technical Digest* (Washington, DC: Optical Society of America) vol 9, pp 26/MC1-1–Mc1-2/27

[44] Ono Y and Nishida N 1982 'Holographic laser scanning using generalized zone plates' *Appl. Opt.* **21**(24) 4542

[45] Stewens R F 1988 'Zone plate interferometers' *J. Modern Opt.* **35**(1) 75
[46] Collected papers 1990 *Holographic Techniques in Science and Technology* (Leningrad: FTI im. A F Ioffe) p 237
[47] Feldman M R and Guest C C 1987 'Computer-generated holographic optical elements for optical interconnection of very large-scale integrated circuits' *Appl. Opt.* **26**(20) 4377
[48] Morrison R L, Walker S L, Cloonan T J *et al* 1992 'Diffractive optics in a free-space digital optic system' in *Diffractive Optics: Design, Fabrication and Applications. Technical Digest* (Washington, DC: Optical Society of America) vol 9, pp 26/ MC1-1–MC1-2/27
[49] Lindvold L R 2001 'Commercial aspects of diffractive optics' *DOPS-NYT* **2** 62–65
[50] Spaulding K E and Morris G M 1991 'Achromatic waveguide lenses' *Appl. Optics* **30**(18) 2558–2569
[51] Stone T and George N 1988 'Hybrid diffractive–refractive lenses and achromats' *Appl. Optics* **27**(14) 2960–2971
[52] Koronkevich V P, Kiriyanov V P, Kokoulin F I *et al* 1984 'Fabrication of kinoform optical elements' *Optick* **67**(3) S259–266
[53] Koronkevich V P and Lenkova G A 1984 *Diffraction Interferometer* **3** 61–67
[54] Akopyan V S, Danilenko Yu K, Danilov V A *et al* 1985 'Planar non-axially symmetric focusators in laser-assisted ophthalmology' *Kvantovaya Electronika* **12**(2) 401–402
[55] Petrov N I, Sisakyan I N and Sysoev V S 1988 'Elements of computer optics in the diagnostics of disperse systems' *Kompyuternaya Optika* **3** 97–98
[56] Walsh A 1952 'Echelette zone plate for use in far infrared spectroscopy' *J. Opt. Soc. Amer.* **42**(3) 213
[57] Firth K 1985 'Recent developments in diffractive optics' *GEC J. Res.* **3**(1) 1–10
[58] Greisuh G I and Stepanov S A 1987 'Synthesized diffractive elements for reading data off optical disks' *Kompyuternaya Optika* **1** 173–177
[59] Shiono T and Setsune K 1989 'Wavelength independent focus sensor a reflection twin micro Fresnel lens' *Appl. Opt.* **28**(3) 5115–5121
[60] Bobrov S T and Turkevich Yu G 1990 'Objective lens with diffractive corrector for laser CD player' *Kompyuternaya Optika* **7** 26–32
[61] Lenkova G A and Churin E G 1991 'High-aperture micro-objective lens for kinoform corrector' *Avtometriya* **6** 70–76
[62] Komma Y, Kudowaki S, Hori Y and Kato M 1990 'Holographic optical element for an optical disk head with spot—size detection servo optics' *Appl. Opt.* **29**(34) 5127–5130
[63] Stewens R F 1988 'Zone plate in interferometers' *J. Modern Opt.* **35**(1) 75
[64] Mustafin K S 1990 'Holographic optics. Developments and applications' in *Collected Papers. Holographic Techniques in Science and Technology* (Leningrad: FTI im. A F Ioffe) pp 217–225
[65] Haruna M, Takahashi M, Wakahayashi K and Nishinara H 1990 'Laser beam lithographed micro-Fresnel lenses' *Appl. Optics* **29**(34) 5120–5126
[66] Sutera S P, Metzger H W and Erway D D 1957 *Concentrating Solar Collectors. An Analysis of Solar Energy Utilization*, vol II, part III, Sec. 2, WADC Tech. report 59-17, February
[67] Truppendienst 1982 **5** S512
[68] Hewish M 1982 'British Developments in wide-angle head-up display' *Int. Defence Review* **15**(8) 1033–1036

[69] Warwick G 1986 'Head up holograms show the way flight' *Electronic News* 129(4002) 26–28
[70] Vorontsov M A, Koryabin A V and Shmal'gauzen V I 1988 *Controlled Optical Systems* (Moscow: Nauka) p 272
[71] Kakichashvili Sh D and Vardosanidze Z V 1989 'Anisotropic profile zone plate' *Pis'ma v ZhTF* 15(17) 41–44
[72] Minin V and Minin O V 1990 Inventor's certificate No. 1596417 USSR, Í01Q 15/12. 'Zone plate I' Bull. No. 36
[73] Voevodin G G, Dianov E M, Kuznetsov A A and Nefedov S M 1989 'Polychromatism in optical computers' *Doklady Akademii Nauk SSSR* 308(2) 370–374
[74] *New Scientist* (UK) 116(1580), 1 October 1987, pp 45–48
[75] Abraham E, Seaton C T and Smith D 1983 'The optical computer' *Scientific American* 4 15–25
[76] Minin I V and Minin O V 1992 *Diffractive Optics* (Moscow: NPO InformTEI)
[77] Minin I V and Minin O V 1989 'Elements of optical computer based on diffractive quasioptics' in *Abstracts of the USSR seminar 'Processing of Two-Dimensional Signals: Methodology and Techniques'*, Part II (Moscow: TsNIINTIKPK) 38–39
[78] Barzhin V Ya and Zvarskii V I 1981 Inventor's certificate No. 881650 USSR, G 02 F 3-00. 'Optical gate' Bull. No.42
[79] Minin O V and Minin I V 1989 Inventor's certificate No. 1679458 USSR, G 02 F 3-00. 'Optical gate'
[80] Bobrov S T, Greisuh G I and Turkevich Yu G 1986 *Optics of Diffractive Elements and Systems* (Leningrad: Mashinostroenie) p 223
[81] H Andersson, Ekberg M, Hard S *et al* 1990 'Single photomask, multilevel kinoforms in quarts and photoresist manufacture and evaluation' *Appl. Optics* 28(29) 4259–4267
[82] Poleschuk A G 1992 'Fabrication of profiled phase structures with continuous and multilevel profile for diffractive optics' *Avtometriya* 1 66–79
[83] Korol'kov V P, Koronkevich V P, Polischuk A G *et al* 1989 'Kinoforms: technologies, novel elements and optical systems' *Avtometriya* 4 47–64
[84] Koronkevich V P, Lenkova G A, Mihal'tsova I A *et al* 1985 'Kinoform optical elements: methods of computation, fabrication technology, practical applications' *Avtometriya* 1 4–25
[85] Spektor B I, Trubetskoi A V and Scherbachenko A V 1989 'Laser plotter of kinoform optical elements' *Kompyuternaya Optika* 4 53–59
[86] Zvereva V A and Stepanova I N (eds) 1979 *Experimental Optics* (Moscow: Nauka) p 256
[87] Zelkin E T and Petrova R A 1974 *Lens Antennas* (Moscow: Soviet Radio) p 279
[88] Schukin I I 1975 'Generation of radio images by phase inversion zone plates' *Radiotehnika i Elektronika* 20(2) 405–406
[89] Minin I V and Minin O V 1988 'Information properties of zone plates' *Kompyuternaya Optika* 3 15–22
[90] Minin I V and Minin O V 1989 'Diffractive radio optics systems: achievements and prospects' in *Abstracts of the USSR Conference 'Optical, Radiowave and Thermal Methods of Nondestructive Monitoring'*, Mogilev, pp 204–205
[91] German Patent 1762, 406, 2301800, 2655525
[92] Kozma A and Kristensen Ch 1978 'Effect of speckle structure on the resolving power of optical systems' *Avtometriya* 2 93–108

[93] Pearson J E 1976 'Speckle phenomena in optics, microwaves and acoustics' in OSA Topical Meeting, 24–26 February; 1976 *Appl. Opt.* **15**(6) 1362

[94] Yu F T S and Wang E Y 1973 'Speckle reduction in holography by means of random spatial sampling' *Appl. Opt.* **12**(7) 1653

[95] Gal'pern A D *et al* 1976 'On intensity distribution in the pattern of diffraction over a random amplitude mask' *Optika i Spektroskopiya* **41**(5) 870–876

[96] Bowman M 1968 'Two methods of improving optical image quality' *Appl. Opt.* **7**(11) 2280

[97] Perina J 1972 *Coherence of Light* (London: Van Nostrand)
Tarlykov V A and Magurin V G 2002 *Foundations of Coherent and Statistical Optics* (St Petersburg: SPbGITMO) (in Russian)

[98] Kaliteevskii N I 1971 *Wave Optics* (Moscow: Nauka) (in Russian)

[99] Minin I V and Minin O V 1998 'Generation of radio images of 3D objects in the mm wavelength range' in *Proc. NVOKU, Novosibirsk*, No. 4, pp 40–46

[100] Sifonov V I 1974 *Radio Receivers* (Moscow: Soviet Radio) (in Russian)

[101] Doluhanov M P 1960 *Propagation of Radiowaves* (Moscow: Svyaz' i Radio) (in Russian)

[102] Minin I V and Minin O V 1986 'Wide-angle multicomponent diffractive microwave lens' *Radiotehnika i Electronika* **31**(4) 800–806

[103] Minin O V and Minin I V 1988 'Diffractive lenses on parabolic surfaces' *Kompyuternaya Optika* **3** 8–15

[104] Baibulatov F H, Minin I V and Minin O V 1985 'Focusing properties of Fresnel zone plate' *Radiotehnika i Elektronika* **30**(9) 1681–1688

[105] Bazarskii O V 1984 'On the feasibility of generation of images by frequency scanning over distance' *Radiotehnika i Electronika* **29**(3) 597

[106] Lyubchenko V E and Borisov V I 1996 'Development of methodology and instruments for detection of metal objects in opaque media' in *Physics of Microwaves* (Nizhnii Novgorod) (in Russian) pp 118–121

[107] Krivoruchko V I, Kulikov A V and Kupriyanov P V 1996 'Development of coherent multi-element radiometric receiving matrix for the mm wavelength band' in *Physics of Microwaves* (Nizhnii Novgorod) (in Russian) pp 257–262

[108] Pirogov Yu A *et al* 'High-resolution radiovision systems based on phase arrays' in *Physics of Microwaves* (Nizhnii Novgorod) (in Russian) pp 248–252

[109] Minin I V, Minin O V, Skarbo B A *et al* 1985 'Application of microwave holographic radio lenses in nondestructive testing and plasma diagnostics' *Abstracts of 5th USSR Conf. on Holography, Riga*, pp 233–234 (in Russian)

[110] Goldsmith P F, Hsich C T, Huguenin G R *et al* 1993 'Focal plane imaging systems for millimeter wavelengths' *IEEE Trans. Microwave Theory and Techn.* **41**(10) 1664–1675

[111] Bazarskii O V, Kotonosov N V and Hlyavich Ya L 1972 'Analysis of microwave holographic techniques for generation of visual images of phase objects' *Radiotehnika i Elektronika* **17**(8) 1733–1734

[112] Ginzburg V M and Stepanov B M 1981 *Holographic Measurements* (Moscow: Radio i Svyaz') (in Russian)

[113] Wiltse J C 1985 'The Fresnel zone-plate lens' *Proc. Soc. Photo-Opt. Instrum. Eng.* **544** 41–47

[114] Wiltse J C 1985 'The phase correcting zone plate' *10th Conf. Infrared and Millimeter Waves, Lake Buena Vista, FL*, 9–13 December, No. 4, pp 345–347

[115] Minin I V and Minin O V 1989 Inventor's Certificate No. 1762651 USSR, G02B5/32
[116] Markin A S and Studenov V B 1987 'Spatial filtering of harmonics of coherent radiation' *Kompyuternaya Optika* **1** 147–151
[117] Markin A S, Studenov V B and Ioltuhovskii A 1980 'A filtering of harmonics of coherent emitter using zone plates' *ZhTF* **50** 2482
[118] Sobel F, Wentworth E L and Wiltse J C 'Quasi optical surface waveguide and other components for the 100 to 300 Hz region' *IRE Trans. Microwave Theory and Techniques* 1961 **MMT-9**(6) 512–518
[119] Minin I V and Minin O V 1994 'Elements of diffractive quasioptics. Part II. Primary application' *Avtometria* **4** 66–74
[120] Minin I V and Minin O V 1987 'Radio-optical diffractive systems in the microwave band' in *Abstracts of the 6th USSR Conf. on 'HF and Microwave Instruments and Techniques for Measuring Electromagnetic Characteristics of Materials'*, *Novosibirsk*, pp 169–170
[121] Schukin I I and Obteranskii Yu S 1975 'Applications of Fresnel lenses to introscopy' *Radiotehnika i Electronika* **20**(4) 1073–1074
[122] Lazarus M J and Silvertown A 1979 'Fresnel-zone plate aids low-cost Doppler design' *Microwaves* 78–80
[123] Buskirk L F and Hendrix C E 1961 'The zone plate as a radiofrequency focusing element' *IRE Trans. Antennas and Propagation* **AP-9**(3) 319–320
[124] Danilov Yu N and Fedorova L A 1989 'Scattering properties of reflector antennas' *Izv. Vyssh. Ucheb. Zaved. Ser. Radioelectronika* **32**(2) 61–65
[125] Lambley R 1989 'Fresnel antenna' *Electronics and Wireless World* **95**(1642) 792
[126] German Patent 3536348, H 01 Q 15/14. Fresnel'sche Zonenplatte zum Fokussieren von langes einer Achse einfallender Mikrowellenstrahlung
[127] German Patent 3801301, H 01 Q 15/23. Fresnel'sche Zonenplatte des Reflektor für eine Mikrowellen-Sende/Empfangsantenne
[128] German Patent 3728976, H 01 Q 19/19 Cassegrain-Antenne für den Mikrowellen-bereich
[129] US Patent 3189907, H 01 Q 15/16. Zone plate radio transmission system
[130] US Patent 4825223, H 01 Q 15/16. Microwave reflector assembly
[131] *New Scientist* (UK) vol 127, No. 1733, 8 September 1990, p 42
[132] Bolomey J C 1989 'Recent European developments in active microwave imaging for industrial, scientific and medical applications' *IEEE Trans. Microwave Theory and Techniques* **37**(12) 2109–2117
[133] Bierman H 1990 'Non-military microwave applications' *Microwave J.* **4**(4) 190–202
[134] Pilgrim M, Carver R D and Barnes B C 1990 'M3VDS—40 GHz multichannel TV to the home' in *20th European Microwave Conf., Budapest*, 10–13 September; 1990 *Conf. Proc.* vol 1, Tunbridge Wells, pp 299–304
[135] Kalmaev R R, Orozobakov T, Geide E K and Dalbaev K 1983 'Passive reflector retransmitters used in line-of-sight relay lines' *Elektrosvyaz'* **3** 19–21
[136] Amiryan R A 1992 'Millimeter wavelength band devices for national economy' *Vestn. Mosk. un-ta, ser. 3. Fizika, Astronomiya* **33**(3) 111–117
[137] Minin I V and Minin O V 2001 'Curvilinear diffractional lenses and antennas for mm-wave communications and wireless' in *Proc. Int. Conf. on Electromagnetics and Communications ICECOM-2001, Dubrovnik, Croatia*, 1–3 October, pp 155–158
[138] Schneiderman R 1991 'Millimeter-waves find commercial markets' *Microwaves and Radio Frequencies* **30**(7) 35–36, 38, 39

[139] Kitazume S and Kondo H 1991 'Advances in millimeter-wave subsystems in Japan' *IEEE Trans. Microwave Theory and Techniques* **39**(5) 775–781

[140] Rohling H, Meinecke M M, Klotz M and Mende R 1998 'Experiences with an experimental car controlled by a 77 GHz radar sensor' in *Int. Radar Symp. Munich, Germany*, 15–17 September

[141] Bystrov R P, Samoilov S I and Sokolov A V 1999 'Millimeter band waves in communication systems' *Zarubezhnaya Radioelectronika* 3 60–71

[142] Bystrov R P, Petrov A V and Sokolov A V 2000 'Millimeter band waves in communication systems' *Elektronika* **5**

[143] *Radioelectronika za Rubezhom* 1989 (Moscow: NIIEIR), 7(39) 1–15 (in Russian)

[144] Sokolov V V, Moguchev V I, Pyl'tsov V A and Fomin A N 1999 'Evaluation of potentials of satellite communication systems with various types of satellite orbit' *Zarubezhnaya Radioelectronika* 2 64–68 (in Russian)

[145] Minin I V and Minin O V 1991 'Millimeter wavelength band antenna systems based on elements of diffractive quasioptics' in *Collected Papers 'Radiotechnical Systems in mm and Sub-mm Wavelength Bands'* (Har'kov: Institute Radiofiziki i Elektroniki AN Ukrainy) pp 120–127

[146] Minin I V and Minin O V 1991 'Elements of diffractive quasioptics and mm-band systems based on them' in *Collected Papers 'Radiotechnical Systems in mm and Sub-mm Wavelength Bands'* (Har'kov: Institute Radiofiziki i Elektroniki AN Ukrainy) pp 102–109

[147] 'Focusing device for a microwave antenna'. International publication number WO 90/07199, 28 June 1990

[148] Button K J (ed) 1983 *Infrared and Millimeter Waves*, vol 9, part I (New York, London: Academic Press)

[149] Potapov A V, Parilov V A, Kuznetsov Yu N *et al* 1991 'Radar level sensor' *Elektronnaya Tehnika SVCh* **8**(442) 51–53

[150] Perkal'kis B Sh 1971 *State-of-the-Art Research Tools Used for Physics Demonstrations* (Moscow: Nauka) p 207

[151] Bastrov R P, Potapov A A, Sokolov A V *et al* 1997 'Problems in propagation and application of millimeter band waves in radars' *Zarubezhnaya Radioelectronika* **1** 4–19

[152] Devyatkov N D, Golant M B and Betskii O V 1991 *Millimeter Waves and Their Role in Vital Functions of Organism* (Moscow: Radio i Svyaz') p 168

[153] Zaitsev A E, Himenko L P, Nud'ga A N *et al* 1989 'Practical application of microwave resonance therapy in a multidisciplinary emergency hospital. Millimeter and sub-mm bands of radio waves: applications' in *Har'kov AN USSR IRE*, pp 20–23

[154] Nud'ga A N, Himenko L P, Zaitsev A E *et al* 1989 'Electromagnetic mm-band resonance therapy in the treatment of bronchospasmodic syndrome. Millimeter and sub-mm bands of radio waves: applications' in *Har'kov AN USSR IRE*, pp 23–26

[155] Kuz'menko A P, Lobachev V E, Solov'ev I E *et al* 1989 'Microwave resonance therapy at 10.5–14.0 W power level. Millimeter and sub-mm bands of radio waves: applications' in *Har'kov AN USSR IRE*, pp 26–30

[156] Minin I V and Minin O V 2001 'New horizons of diffractive quasioptics' *Kompyuternaya Optika* **22** 98–102

[157] Minin I V and Minin O V 1991 'Application of diffractive quasioptics elements for microwave plasma diagnostics' in *Abstract of Reports at the Ukrainian Symposium*

on 'Physics and technology in the mm and sub-mm Wavelength Bands', Har'kov, part 2, pp 11–12

[158] Minin I V and Minin O V 1988 Inventor's certificate No. 134769 USSR, G 01N 22/00. Device for measuring parameters of plasmas. Bull. No. 7

[159] White W C, Rinehart J S and Allen W A 1952 'Phenomena associated with the flight of ultra-speed pellets. Part II. Spectral character of luminosity' *J. Appl. Phys.* **23**(2) 198–201

[160] Rinehart I S, Alleen W A and White W C 1952 'Phenomena associated with the flight of ultra-speed pellets. Part III. General features of luminosity' *J. Appl. Phys.* **23**(3) 297–300

[161] Thiel M and Levation J 1980 'Jet formation experiments and computations with a Lagrange code' *J. Appl. Phys.* **51**(12) 6107–6113

[162] Dushin V R, Minin I V, Minin O V *et al* 1989 'Investigation of three-dimensional supersonic flow of real gases around axisymmetric bodies' *Vestnik MGU Ser. I* **4** 41–49

[163] Kuznetsov M M 1965 *Thermodynamic Functions and Shock Adiabats in Air at High Temperatures* (Moscow: Mashinostoenie) p 433

[164] Fortov V E and Yakubov I T 1984 *Physics of Non-Ideal Plasmas* (Chernogolovka: IVT AN SSSR) p 264

[165] Chernyi G G and Chernyavsky Yu C (eds) 1979 *Theoretical and Experimental Hypersonic Flow Studies in Flow Around Bodies and in Trails* (Moscow: MGU) p 140

[166] Urvantsev L A 1966 *Erosion and Anti-Erosion Protection of Metals* (Moscow: Mashinostroenie) p 236

[167] Minin I V and Minin O V 2002 'The possibility of impulse plasma antenna creation' in *6th Russian–Korean Int. Symp. on Science and Technology, Novosibirsk, Russia*, 24–30 June, vol 2, pp 289–292

[168] Hayes G A 1984 'Linear shaped-charge collapse model' *J. Materials Sci.* **19** 3049–3058

[169] Kirilov V A, Tverdohlebov V I and Homenko V I 1964 'Demonstration experiment with acoustic zone plate' *UFN* **70**(1) 166

[170] Kock W 1965 *Sound Waves and Light Waves* (New York: Plenum Press)

[171] Greguss P 1980 *Ultrasonic Imaging* (London, New York: Focal Press)

[172] Shattuck D P and Nowhi J 1988 'Focusing on optoacoustic transducer' *IEEE Trans. Ultrason., Ferroelec. Freq. Contr.* 35(4) 445–449

[173] Metherell A F (ed) 1969 *Acoustical Holography* vol 1 (New York: Plenum Press)

[174] Wu J-y, Wang C-h and He O-g 1988 'Focusing and scanning properties of acoustic beam in solid using a Fresnel array' *Acta Phys. Sinica* **37**(10) 1575–1584

[175] Cao J-b, Zhang W-c and Zhao H J 1988 'Focusing properties of acoustic Fresnel lenses' *Acta Acoust.* **13**(5) 369–375

[176] Stamnes J J, Cravelsxter J and Bentsen O 1982 'Image quality and diffraction efficiency of a holographic lens for sound waves' in *Acoustical Imaging* (ed P Alais and A F Tetherell (New York: Plenum Press) vol 10, pp 587–606

[177] Ermolaev I N, Kanevskii I N, Kofolev V D *et al* 1980 'Focusing zoned finder for ultrasound non-destructive testing' *Defektoskopiya* **1** 94–96

[178] Chernoverskii M P 1988 'Focusing zoned converter with low electric capacitance' *Defektoskopiya* **2** 94

[179] Dolmatova E A 1978 'Semiconductor Fresnel zone plate' *Vestn. Har'kovskogo Universiteta* **7**(163) 71–72

[180] Gustafsson G 1987 'Experimental on shock-wave in an elliptical cavity' *J. Appl. Phys.* **61**(11) 5193–5195

[181] Kitagama O, Ise H, Saito T and Takayama K 1988 'Non-invasive gallstone disintegration by an underwater shock focusing' in *Shock Tubes and Waves* (Weinheim: VCH) pp 897–903

[182] Matsuo H 1983 'Cylindrically converging shock and detonation waves' *Phys. Fluid.* **26** 1755–1762

[183] Nishida M and Kishige H 1988 'Numerical simulation of focusing process of reflected shock waves' in *Shock Tubes and Waves* (Weinheim: VCH) pp 551–557

[184] Sturtervant B and Kulnarny V A 1976 'The focusing of weak shock wave' *J. Fluid Mech.* **73** 651–671

[185] Cramer S 1981 'The focusing of weak waves at an axisymmetric area' *J. Fluid Mech.* **110** 249

[186] Cramer S 1983 'Three-dimensional effects on the focusing of weak waves' *J. Sound and Vibration* **90** 25

[187] Minin I V and Minin O V 1991 'Possibility of shockwave focusing by diffractive quasioptics elements' in *Research in Materials' Properties Under Extreme Conditions* (Moscow: IVTAN SSSR) pp 225–230

[188] Minin V F, Agureikin V A, Kryukov B P *et al* 1986 'Integrated software–hardware complex for mathematical simulation of nonstationary processes in mechanics of continuous media' in *Abstracts of Reports to the USSR Conf. on 'Problems in automation of design and development work'* (Moscow: VIMI) pp 31–33

[189] Minin V F, Landin A A and Kryukov B P 1988 ' 'Stereo-PC' software package applied to nonstationary problems in dynamics of compressible media' in *Abstracts of Talks at Conference-School on 'Current problems in fluid mechanics'*, *Irkutsk*, pp 227–229

[190] Askar'yan G A and Klebanov L D 1988 'Focusing and cumulative phenomena in reflection of opto-thermoacoustic shock pulses from concave surface heated by laser flash' *Kvantovaya Electronika* **15**(11) 2167–2168

[191] Bushman A V, Zharkov A P, Minin V F *et al* 1989 'Numerical simulation of non-regular reflection of shockwaves in condensed media', Chernogolovka, Preprint RISO OIHF AN SSSR, p 71

[192] Minin I V and Minin O V 1988 'Information characteristics of zone plates' *Kompyuternaya Optika* **3** 15–22

[193] Minin I V and Minin O V 1990 'On possible focusing of shock waves by diffractive elements' in *Waves and Diffraction 90* (Moscow: MFTI) pp 187–189

[194] Minin I V and Minin O V 1990 'Diffractive optics in high energy density physics' in *Waves and Diffraction 90* (Moscow: MFTI) pp 194–196

[195] Minin V F, Minin I V and Minin O V 1991 'The dynamics of shock wave focusing with the elements of diffraction quasioptics' in *18th Int. Symp. on Shock Waves*, *Sendai*, 21–26 July, pp 39–40

[196] Minin I V and Minin O V 1992 'The numerical experiments on dynamics of shock wave interaction with regular and quasi-regular diffraction gratings' in *Proc. Int. Symp. on Intense Dynamic Loading and its Effects, Chengdu, China*, 9–12 June, pp 434–437

[197] Kobayashi S, Adachi T and Suzuki T 1991 'Unsteady behaviour of Mach reflection over particulate layer' in *18th Int. Symp. on Shock Waves, Sendai*, 21–26 June, pp 55–56

[198] Minin V F, Minin I V and Minin O V 1992 'The calculation experiment technology' in *Proc. Int. Symp. on Intense Dynamic Loading and its Effects, Chengdu, China*, 9–12 June, pp 431–433

[199] Minin V F, Minin I V and Minin O V 1992 'Numerical experiment techniques' *Mathematical Simulation* 4(12) 65–67

[200] *http://www.snezhinsk.ru/asteroids*

[201] Larin A I 2000 'Use of the 60 Ghz frequency band for radio beams two-position security equipment for perimeter fence protection' *Spetsial'naya tehnika* 1 20–24

[202] Krylov K I, L'vova N A and Smirnov S A 1980 Report on the 'Raduga' research project, LITMO

[203] Minin I V and Minin O V 2000 'The system of microwave radiovision of three-dimensional objects in real time' *Proc. SPIE 'Subsurface Sensing Technologies and Applications. II*, ed. Cam Nguyen, vol 4129, pp 616–619

[204] Minin I V and Minin O V 2002 'Diffractional lenses and mirror antennas for mm-waves wireless net' in *Proc. 6th Russian–Korean Int. Symp. on Science and Technology, Novosibirsk*, 24–30 June, vol 2, pp 347–350

[205] Legkii V N, Minin I V and Minin O V 2002 *Methods and Tools for Detection of Sabotage and Terrorist Devices* (Novosibirsk: NGTU) p 132

[206] Minin I V and Minin O V 1999 'Non-stability of mm-wave radar imaging of the car in dynamics' *Computer Optics* 19 151–153

[207] *http://antwrpgsfc.nasAgov/apod/ap000526.html*

[208] *http://research.hq.nasAgov/code_s/nra/current/NRA-00-OSS-06/winners.html*

[209] Khayatin B, Rahmat-Samii Y and Pogorzelski R 2001 'An antenna concept integrated with future solar sails' *Antennas and Propagation Soc., IEEE Int. Symp. 2001*, vol 2, pp 742–745

[210] Khayatin B and Rahmat-Samii Y A 2002 'Dual-band dual-feed Fresnel zone antenna concept: application in solar sails missions' *Antennas and Propagation Soc., IEEE Int. Symp. 2002* pp 638–641

[211] 'Small laser-propelled interstellar probe' Presented at the 46th International Astronautical Congress, October 1995, Oslo, Norway. Reprint in: *http://www.qedcorpcom/pcr/pcr/starflt.html*

[212] Tateishi A and Kikuchi K 1998 'Application of Fresnel zone plate to millimeter wave modulator' in *Int. Conf. on Microwave and Millimeter Wave Technology Proceedings ICMMT '98*, pp 642–645

[213] Wilcockson P C 1994 'Evolution of Flat Reflectors for Space and Ground Antenna Applications' Report, European Space Agency. Order 142600, 22 December

[214] Landis G A 1989 'Optics and materials considerations for a laser-propelled lightsail' Presented as paper IAA-89-664 at the 40th International Astronautical Federation Congress, Málaga, Spain, 7–12 October. Revised December 1989

[215] Reprint in: *http://www.sff.net/people/Geoffrey.Landis/lightsail/Lightsail89.html*

[216] Hristov H D 2000 *Fresnel Zones in Wireless Links, Zone Plate Lenses and Antennas* (Boston, London: Artech House)

[217] Goodman J W 1968 *Introduction to Fourier Optics* (McGraw-Hill)

[218] Kuhn J R, Moretto G, Racine R, Roddier F and Coulter R 2001 'Concepts for a large-aperture, high dynamic range telescope' *Publications Astronom. Soc. Pacific* **113** 1486–1510

Additional bibliography

Brandt A, Pashin Yu and Petelin V 1958 'Study of focusing properties of zone antenna in centimeter wavelength band' *Sci. Reports of Higher Schools* **6** 201–207 (in Russian)

C F Ye and S Y Tan 2000 'A reflective half-cylindrical Fresnel zone plate antenna with low backward radiation for wireless LAN' *Microwave and Optical Technology Lett.* August

Cosma I, Ristoiu T and Nicoara S 1998 'Physical principles and calculus of the zone grating plate antennas for receiving satellite TV-signals' in *9th Mediterranean Electrotechnical Conference MELECON '98* vol 1, 1998, pp 270–273

Delmas J-J, Toutain S, Landrac G and Cousin P 1993 'TDF antenna for multisatellite reception using 3D Fresnel principle and multilayer structure' in *Antennas and Propagation Society International Symposium, AP-S Digest*, vol 3, pp 1647–1650

Fan Z, Lu Z, Liao J, Zhang J and Chen F 1997 'Design of mm-wave Fresnel zone plate reflector antenna with continuous phase structure' in *Asia–Pacific Microwave Conference Proceedings, APMC '97*, vol 1, 1997, pp 449–451

Gouker M A and Smith G S 1991 'Measurements of substrate-mounted millimeter-wave integrated-circuit antennas at 230 GHz' in *Antennas and Propagation Society International Symposium, 1991, AP-S Digest*, vol 2, pp 991–994

Hoashi T, Onodera T, Kimura E and Hagio F 1993 'TV receiving systems of broadcasting satellite using fresnel zone plate lens' in *4th Int. Symp. on Recent Advances in Microwave Technology, New Delhi, India*, pp 450–453

Izadian J S 2001 'Considering antenna options for LMDS' *Microwaves and RF* 65–74, 107

Jacobsson S, Lundgren A and Johansson J 1990 'Computer generated phase holograms (kinoforms) for millimeter and submillimeter wavelength' *Int. J. Infrared and Millimeter Waves* **11**(11) 2151–2161

Minin I V and Minin O V 'Antennas of mm-band based on the quasioptical diffraction elements for the commmication systems' in *50th Vehicular Technology Conference VTC 1999—Fall, Amsterdam, The Netherlands*, 19–22 September, pp 3040–3046

Minin I V and Minin O V 1999 'Diffraction elements of the mm-waves planar integral optics' in *USNC/URSI National Radio Science Meeting, Orlando, FL*, 11–16 June, p 287

Minin I V and Minin O V 1999 'New conception of the low cost multibeam antennas in the mm-wave regime for radar systems of transport means' in *50th Vehicular Technology Conference VTC 1999—Fall, Amsterdam, The Netherlands*, 19–22 September, pp 2053–2056

Shuter W Lh *et al* 1984 'A metal plate Fresnel lens for 4 GHz satellite TV reception' *IEEE Trans. AP* 306–307

Webb G W, Vernon W, Sanchez M S, Rose S C and Angello S 1999 'Optically controlled millimeter wave antenna' in *Int. Topical Meeting on Microwave Photonics, MWP '99*, vol 1, pp 275–278

Wiltse J C 1999 'History and evolution of Fresnel zone plate antennas for microwaves and millimeter waves' *IEEE International Symposium* (Antennas and Propagation Society), vol 2, pp 722–725

Wright T M B 1989 'Large reflecting zone plates for satellite signal reception' in *IEE Colloquium on Mechanical Aspects of Antenna Design 1989*, pp 12/1–12/3

Wright T M B and Wilcockson P C 1995 'Experimental verification of the multifocal properties of reflecting zone plate antennas' *9th Int. Conf. on Antennas and Propagation 1995* (Conf. Publ. No. 407), vol 1, pp 291–294

Chapter 7

Diffractive elements using man-made dielectrics

7.1 Optical constants of materials in sub-millimetre and millimetre spectral bands

Let us consider a number of weakly absorbent materials that can be used to fabricate diffractive elements for the shortwave part of the microwave band. Note that depending on the problem needing solution, materials with different values of diffractive index, isotropic and anisotropic, and working in response to various external stimuli may be required.

A few words about terminology are in order. In the millimetre and sub-millimetre bands, a spectrum is normally defined as the spectral dependence of either the complex refractive index $N = n - ik$ of the material, or of the complex dielectric permittivity $\varepsilon = \varepsilon_1 - i\varepsilon_2$, or of the dielectric permittivity ε_1 and loss angle tangent $\tan \delta$. Here n is the refractive index of the material, k is the absorption coefficient related to the absorption coefficient α (cm^{-1}) at a given wavelength λ by the expression:

$$\alpha(\text{cm}^{-1}) = 4\pi k / \lambda(\text{cm}).$$

ε_1 and ε_2 are the real and imaginary parts of the complex dielectric permittivity, and $\tan \delta = \varepsilon_2/\varepsilon_1$. The familiar relation $\varepsilon = N^2$ leads to the following relations between the real and imaginary parts of the complex dielectric permittivity and the complex absorption coefficient:

$$\varepsilon_1 = n^2 - k^2, \qquad \varepsilon_2 = 2nk$$

and conversely

$$n = [0.5\varepsilon_1 + (0.25\varepsilon_1^2 + 0.25\varepsilon_2^2)^{1/2}]^{1/2}$$

$$k = [-0.5\varepsilon_1 + (0.25\varepsilon_1^2 + 0.25\varepsilon_2^2)^{1/2}]^{1/2}$$

$$\tan \delta = 2nk/(n^2 - k^2).$$

As a rule, the absorption coefficient k in the millimetre and sub-millimetre wavelength bands increases with decreasing wavelength of radiation, and its dispersion is small.

The following materials possess the lowest absorption in the millimetre and sub-millimetre bands: polyethylene, polypropylene, Teflon-4, polytetramethylpentane, polystyrene, Rexolite. Composite materials (glassfibre plastics and mixtures of various materials) make it possible to span a certain range of values of refractive coefficient. For instance, quartz- and silica-based glassfibre fabrics have $n = 1.68$–1.85 with $\tan \delta = (1.5$–$9) \times 10^{-3}$ at wavelengths $\lambda = 1.5$–2.5 mm [1]. Composites based on Teflon-4 cover the range of n values from 1.05 to 1.40 and $\tan \delta = (3$–$5) \times 10^{-4}$ for $\lambda = 1$–2 mm. Composites with fillers of Al_2O_3 or TiO_2 in a Teflon-4 matrix may have n up to 2. However, $\tan \delta$, at wavelengths shorter than 1.5 mm does not exceed 10^{-2}.

The refractive index as a function of density of polyolefins (polyethylene, polypropylene) is well described by the formula [1]

$$n = [2.27 + 2.01(\rho - 0.92)]^{1/2}$$

where ρ is density (g/mm^3).

Table 7.1.

Material	n	k	λ (mm)	Source
Polystyrene	1.59–1.60	$\sim 10^{-3}$	0.58–4.2	[1–7]
	1.60	7×10^{-4}	30	
Polyethylene	1.51–1.52	3×10^{-4}	0.3–4	[1, 2, 8, 9]
	1.503	5×10^{-4}	30	[7]
Polypropylene	1.510	5×10^{-4}	0.5–2	[1, 10, 8]
Poly-4-methylpentane (TRH)	1.46	6×10^{-4}	0.2–2	[1, 8, 9, 10]
Teflon-4	1.43–1.48	5×10^{-4}	0.58–4.2	[1, 3,6, 11, 12]
	1.44		30	[7]
Teflon-44B	1.42	10^{-3}	0.63	[24]
Ebonite	1.64–1.67	3×10^{-2}	0.58–4.2	[5, 12]
	1.63	6×10^{-3}	30	[7]
Plexiglas	1.59–1.60	10^{-2}	0.58–5	[3, 5, 6, 12, 13]
	1.616	8×10^{-3}	30	[7]
Fused quartz	1.95–2.00	5×10^{-4}	0.3–2.5	[6, 11, 13]
	1.950	1.7×10^{-4}	30	[7]
Sapphire n_0	3.07	1.6×10^{-3}	0.33	[24]
Sapphire n_1	3.415	3.0×10^{-3}		
Germanium	3.92–4.0	7×10^{-3}	0.3–2.5	[1, 14, 24]
Silicon	3.41	10^{-3}	0.3–2.5	[1, 14, 24]
Gallium arsenide	3.6	3×10^{-4}	0.7–3	[24]
ZnS	2.896	2×10^{-3}	0.88–5	[21]
ZnSe	3.01–3.02	3×10^{-3}	0.88–5	[21]
Paraffin	1.48–1.51	10^{-3}	1.3–4.2	[3, 6]
Lavsan	1.79	10^{-2}	1.8	[15]

Table 7.2. Solidified epoxy binding agents.

Material	n	k	λ (mm)	Source
Epoxy resin	1.7	10^{-2}	2.2–7	[16]
EDT-10	1.69–1.73	10^{-2}	3–30	[17]
SPE-14	1.67–1.73	10^{-2}	3–30	[17]
SPE-16/4	1.74–1.79	10^{-2}	3–30	[17]
SPE-15	1.68–1.70	10^{-2}	3–30	[17]
ES-1	1.60–1.72	10^{-2}	3–30	[17]

For Teflon-4, with $\rho = 0.4$–$2.7\,\text{g/mm}^3$, the following expression holds [1]:

$$N = 1 + 0.196\rho.$$

For beryllium ceramics (BeO) at density $\rho = 2.8$–$3.0\,\text{g/cm}^3$ the refractive index in the wavelength band 3.5–15 mm is a linear function of density [2]:

$$n = 0.66678\rho + 0.66865.$$

The optical parameters of a number of materials are listed in tables 7.1–7.5.

Ferrites, ferroelectrics and semiconductors are among the continuous media with controlled refractive index that are used in the microwave band.

The refractive index of ferrites can be changed by applied magnetic field. Longitudinal magnetization of such media produces the Faraday effect (rotation of polarization plane) so that ferrite lenses with longitudinal magnetization (along the optical axis of the lens) operate on circularly polarized waves. If magnetization is transversal relative to the direction of propagation of electromagnetic waves, refractive index can also be controlled.

Note that the lowest losses in ferrites are found in medium-frequency yttrium ferrites, lithium ferrites (LT) and nickel–zinc ferrites (NZ).

Table 7.3. Foam plastics.

Material	n	k	λ (mm)	Source
PS-1, $\rho = 0.1\,\text{g/cm}^3$	1.05	10^{-3}	30	[18]
PS-1, $\rho = 0.2\,\text{g/cm}^3$	1.13	2×10^{-3}	30	[18]
PS-2, $\rho = 0.2\,\text{g/m}^3$	1.17	10^{-4}	30	[18]
FF, $\rho = 0.2\,\text{g/m}^3$	1.145	10^{-2}	30	[18]
FK-20-ST	1.374	10^{-2}	30	[18]
K-40, $\rho = 0.4\,\text{g/cm}^3$	1.225	3×10^{-3}	30	[18]
PVH-1, $\rho = 0.17\,\text{g/cm}^3$	1.26	3×10^{-2}	30	[19]
VPG-2L, $\rho = 0.2\,\text{g/cm}^3$	1.64	3×10^{-2}	30	[19]

Table 7.4. Ceramics.

Material	n	k	λ (mm)	Source
TSM-303	3.06	10^{-4}	0.9–30	[20]
Al_2O_3	3.031	4×10^{-3}	1	[22]
$SiO_2 + Cr_2O_3$	1.867	8×10^{-4}	1	[22]
Mg_2F_2	2.162	1.4×10^{-3}	1	[22]
$MgTiO_3$	4.65	4×10^{-4}	6	[24]
ST-47	2.93		2	[23]
TsM-4	2.17		2	[23]
BeO	2.583	2×10^{-3}	0.88–5	[21]
	2.583	10^{-3}	2.5	[24]
Talcum-based	3.87	4×10^{-3}	1.3–1.9	[6]
	2.28	3×10^{-3}	30	[7]
$MgAl_2O_4$	2.89	10^{-3}	0.83–2.5	[24]
Polikor	3.099	3×10^{-4}	1.5	[24]
	3.01–3.15	10^{-3}	3.0	[24]
22XC	3.072	3×10^{-3}	1.93	[24]
Boron nitride, $\rho = 1.47\,g/cm^3$	1.73	10^{-3}	0.65	[24]
T-150	12.3	0.13	2.71	[28]
	12.4	0.17	1.75	[28]
	12.4	0.31	0.552	[28]
	12.4	0.41	0.372	[28]
T-250	14.75	0.21	2.53	[28]
	14.78	0.22	1.775	[28]
	14.80	0.51	0.604	[28]
SN-750	9.50	0.05	2.757	[28]
	9.50	0.06	1.66	[28]
	9.53	0.10	0.789	[28]
	9.55	0.15	0.421	[28]
	9.54	0.21	0.316	[28]
TL-75	6.42	0.02	1.989	[28]
	6.44	0.03	0.68	[28]
	6.47	0.09	0.35	[28]
	6.56	0.096	0.322	[28]
TL-0	6.35	0.01	2.54	[28]
	6.34	0.014	1.813	[28]
	6.57	0.04	0.605	[28]
	6.26	0.07	0.338	[28]
TBNS	8.96	0.06	2.64	[28]
	8.96	0.07	1.758	[28]
	8.53	0.11	0.982	[28]
	8.56	0.16	0.657	[28]
	8.60	0.18	0.565	[28]
TTB	5.60	0.01	2.135	[28]
	5.60	0.04	0.35	[28]
	5.70	0.05	0.315	[28]
LLTK	6.34	0.004	2.439	[28]
	6.32	0.03	0.437	[28]
MgF_2	2.165	6×10^{-4}	1.2	[24]
TB-8	6.15	10^{-3}	0.95	[24]

Table 7.5. Ferrites.

Material	n	k	λ (mm)	Source
1S44	3.62–3.65	1.5×10^{-3}	1.3–2.2	[24]
10SCh46	3.88–3.97	7.5×10^{-4}	2	[4]
LT-9	3.95	1.2×10^{-3}	2.16	[24]
NTs-11	3.54	10^{-3}	2	[24]
10SCh46B	3.82	7×10^{-3}	0.85	[29]
30SCh3B	3.80	2.8×10^{-3}	0.63	[29]
30SCh9B	3.77	2×10^{-2}	0.85	[29]
90SCh	3.78	1.4×10^{-2}	0.85	[29]
NTs-7	3.65	0.016	0.8	[29]
11-V	3.66	0.005	0.85	[29]
11-G	3.95	0.005	0.85	[29]
16BA-190	4.30	0.21	2.18	[29]
Li-33	3.74	0.05	0.47	[29]
LT-8	3.94	0.01	2.18	[29]
LT-6	3.85	0.01	2.18	[29]
	3.64	0.038	0.87	[29]
6SChA-1	4.15	0.04	0.85	[29]
6SChA-16	3.98	0.12	0.85	[29]
7SChA	3.69	0.06	0.85	[29]
8SChA	3.91	0.07	0.85	[29]
9SChA	3.62	0.07	0.85	[29]
7SChA-14	3.70	0.03	0.85	[29]

The effective value of ferroelectrics' refractive index changes in electrostatic field. The main shortcomings of such devices are the strong temperature dependence of their characteristics, technological difficulties in achieving reproducibility of properties, and the need to use high control voltages.

7.2 Man-made dielectrics

One of the techniques for reducing the weight and losses in the microwave diffractive elements is to use the so-called man-made dielectrics [25].

Metal–dielectric media are one of the forms of artificially created dielectrics. These media incorporate grids of rods and ribbons, or can be fabricated as a three-dimensional lattice of metal spheres, disks, rods etc. that imitate the spatial molecular structure of conventional dielectrics.

The incident wave induces surface currents in metal elements of the structure. These currents become sources of secondary emission of radiation. The resulting field in the structure is a superposition of the primary and secondary fields; phase relations between them dictate the phase velocity of the resulting field. The structure is a retarding one if the secondary emission is delayed in phase, or advancing if the secondary emission is advancing on the primary one.

Small-size metal bodies placed in a volume and isolated by air gaps play the same role in the metal dielectric that polarizing molecules play in a conventional dielectric. Therefore in the static approximation the calculation of parameters of an artificial dielectric is based on the equivalence of conventional and metal dielectrics. The polarization vector of a metal dielectric containing N metal elements per unit volume is

$$P = Np = N\alpha\varepsilon_0 E \tag{7.1}$$

where α is the mean polarizability of particles.

The displacement vector is

$$D = (\varepsilon_1 + N\alpha)\varepsilon_0 E$$

where ε_1 is the dielectric permittivity of the medium into which the elements of the artificial dielectric are embedded.

Therefore the effective dielectric permittivity of a metal–dielectric is

$$\varepsilon = \varepsilon_1 + N\alpha.$$

Similarly, the effective magnetic permeability of a metal–dielectric is

$$\mu = \mu_1 + N\beta.$$

This treatment is approximate because it ignores the difference between the true field E_0 acting on a dipole and the mean field E in the dielectric.

The values of α and β of a number of elements are listed in table 7.6.

More accurate calculations [25] show that equation (7.1) remains valid if we replace $\alpha = 4\pi\varepsilon_1 R^3$ in the case of spheres with

$$\alpha_0 = \alpha/(1 - 4\pi N\lambda/3).$$

Table 7.6.

Shape of elements	α	β
Sphere radius R	$4\pi\varepsilon_1 R^3$	$-2\pi\mu_1 R^3$
Ellipsoid of revolution with semiaxes $a > b$ in field E parallel to the larger semiaxis	$4\pi\varepsilon_1 ab^2$	—
Thin disk of radius R in field E parallel to the plane of the disc	$5.33\varepsilon_1 R^3$	—
Thin a wide ribbon of length $b > a$ in field E parallel to the narrower side	$0.25\pi\varepsilon_1 a^2$	—

Then $\varepsilon = (1 + 2V)\varepsilon_1/(1 - V)$, where $V = N\alpha/3\varepsilon_1$, which for spheres gives the volume of metal elements per unit volume.

If the geometrical shape of elements is different, these expressions can be used for calculating the approximate correction factor using the formula

$$\eta = (1 + 2V)/(1 + 2V - 3V^2).$$

This equation is not quite accurate either because it does not take into account quadrupole, octupole and other higher-order electric moments induced on elements owing to mutual interactions at sufficiently high volumetric fraction of inclusions.

Greater precision is achieved by using the equivalence of a three-dimensional dielectric to a linear arrangement of these elements in a waveguide (as far as field structure and phase velocity are concerned).

One example is an arrangement of metal ribbons placed transversally to the electromagnetic wave propagation direction. The ribbons are w wide, a long and are placed at a distance l from one another in the direction of wave propagation and spaced by b in the transversal direction. Such ribbon structures can be modelled by a structure of parallel waveguide channels with a periodic structure formed of capacitance diaphragms.

The effective dielectric permittivity can be found from the expression [25]

$$\frac{\varepsilon^2}{\varepsilon_1^2} = \frac{\lambda}{2\pi l}\arccos[\cos 2\pi l/\lambda - B\sin 2\pi l/\lambda] \qquad (7.2)$$

where $B = 2b\ln(\csc(\pi b_0/2b))/\lambda$ and $\varepsilon^{0.5} = n$ is the refractive index of the dielectric without metal elements.

For a metal–dielectric made of a system of long rods of radius r each, arranged at a distance a in the transversal direction, the equivalent shunting inductance has the conductivity

$$B = -\lambda/2a\ln(l/2\pi r).$$

The effective dielectric permittivity and magnetic permeability of a composite medium with spherical inclusions can be calculated in the quasistatic approximation [30]. It is assumed in this case that the size of the spherical particles is infinitely small and that the distances separating particles are also small in comparison with the wavelength of the electromagnetic field.

The composite medium chosen is a cubic lattice of spherical particles within the embedding medium. If S is the lattice period, a is the particle radius ($a < S/2$), ε_1, ε_2 are the dielectric permittivities and μ_1, μ_2 are the magnetic permeabilities of this medium and of particles, respectively, $k_1 = \omega(\varepsilon_1\mu_1)^{1/2}$, $k_2 = \omega(\varepsilon_2\mu_2)^{1/2}$ are the wave numbers in the corresponding media, and ω is the circular frequency of the electromagnetic field, the effective parameters of the composite medium are described by the expressions

$$\varepsilon_{\text{eff}} = \varepsilon_1\left(1 + \frac{3fL}{1 - FL}\right), \qquad \mu_{\text{eff}} = \mu_1\left(1 + \frac{3fM}{1 - FM}\right)$$

where

$$f = 4\pi(a/S)^3/3$$

$$f = f(1 - kk_1^2 S^2)$$

$$L = \frac{\varepsilon_p - \varepsilon_1}{\varepsilon_p + 2\varepsilon_1 - (\varepsilon_p - 2\varepsilon_1)k_1^2 a_2/2}$$

$$M = \frac{\mu_2 - \mu_1}{\mu_p + 2\mu_1 - (\mu_p - 2\mu_1)k_1^2 a_2/2}$$

$$\frac{\varepsilon_p}{\varepsilon_2} = \frac{\mu_p}{\mu_2} = \frac{2(\sin k_2 a - k_2 a \cos k_2 a)}{(k_2^2 a^2 - 1)\sin k_2 a + k_2 a \cos k_2 a}$$

$$k = 9[\ln(2 + 3^{1/2}) + \arcsin(0.5(2 - 3^{1/2}) - 1/3]/4\pi \approx 0.892.$$

The actual physical design of metal–dielectric diffractive elements may be very diverse. For example, a diffractive element may be made of metal balls that are fixed in space by insulator rods or are embedded layer by layer into foam polystyrene. Fixing of metal foil disks in a diffractive element is carried out by relatively thin sheets of foam polystyrene or polyethylene film. Platelet elements are formed of square films of conducting paint deposited through a stencil on to plates of foam polystyrene. Printed circuit board technology can be used to fabricate artificial metal dielectrics.

7.3 Perforated artificial dielectrics

Man-made metal dielectrics discussed above have a significant drawback connected with the difficulty of fixing metal elements of a structure in their proper places. Using artificial dielectrics with holes allows the designer to eliminate this drawback.

A dielectric made of metal plates with periodically aligned holes is simple in both design and fabrication. The relative dielectric permittivity and refractive index of such a dielectric can be calculated on the basis of the parameters of a metal–dielectric. With small hole diameters (a fraction of wavelength), the static approximation is valid so that the dielectric permittivity can be found using equation (7.1) and the values of disk polarizability listed in table 7.6.

If hole diameter is comparable with wavelength, the dielectric permittivity must be calculated using equation (7.2). The shunting conductivity of the equivalent circuit in the waveguide is

$$Y_{\text{equiv}} = j2B_{\text{equiv}} = -2\Gamma/(1+\Gamma)$$

and can be found either experimentally or by calculations using the reflection coefficient Γ.

A dielectric plate with cavities of various shape is another type of man-made dielectric in which the effective dielectric permittivity is reduced in comparison with a solid dielectric because the fraction of volume that the dielectric occupies is reduced. The value of permittivity can be obtained in the first approximation from equation (7.1) but polarizability must then be calculated for a cavity inside the dielectric. For instance, for a spherical cavity we have

$$\alpha = 4\pi\varepsilon_1 R^3(1-\varepsilon_1)/(2\varepsilon_1 = 1).$$

An approximate calculation of the effective value of permittivity and refractive index can be reduced to calculating the capacitance of a flat capacitor between whose planes the element of the dielectric's structure is placed. For instance, for a dielectric containing cavities in the shape of parallelepipeds, we have

$$\frac{\varepsilon}{\varepsilon_1} = 1 + \frac{S_2}{S}\left(1 + \frac{1}{1+d_{20}(\varepsilon_1-1)/d}\right)$$

where $S_2 = S - S_1$ is the area of the top face of the cavity of height d_{20}, S is the area of the dielectric plate with cavities, whose height is the side edge d, S_1 is the area of the capacitor plate, and ε_1 is the dielectric permittivity of the dielectric.

Petosa and Ittipiboon [35] published their results of studying a four-level phase inversion plate fabricated of perforated dielectric.

A zone plate with flat surfaces in which the refractive index of the material varied continuously along the radius within each zone seems to have been suggested first in [36]. To ensure the required value of the refractive index this paper proposed to use a mixture of two or more dielectrics with different values of the refractive index, in proper ratios. Later Wiltse suggested in [37] a zone plate with flat surfaces in which each zone would be fabricated out of a different dielectric and the refractive index within each zone would not be radius-independent. However, Dr Aldo Petosa and Prof Apisak Ittipiboon were very likely pioneers of using such zone plates.

The principle of constructing an artificial dielectric used in [35] is shown in figure 7.1(a).

For a triangular perforation cell, the effective dielectric permittivity is calculated using the expression

$$\varepsilon_{\text{eff}} \approx \varepsilon_r(1-\alpha) + \alpha$$

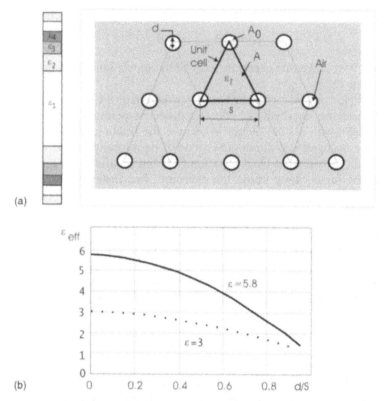

(a)

(b)

Figure 7.1. (a) Principle of constructing an artificial dielectric (Reproduced with permission of Dr Aldo Petosa). (b) Effective dielectric permittivity of the material as a function of d/s.

where the coefficient α for the geometry in question is the ratio of the area of one circular perforation to the area of the triangular cell:

$$\alpha = \frac{\pi d^2/4}{2(\sqrt{3}/4)s^2}.$$

Figure 7.1(b) plots the effective dielectric permittivity of the material as a function of d/s for two values of dielectric permittivity of the initial dielectric. Obviously, higher dielectric permittivity of the original dielectric allows a wider range of possible values of the effective dielectric permittivity of the artificial dielectric, although losses to reflection inevitably grow.

The authors of [35] chose the initial dielectric with dielectric permittivity 5.8. The antenna diameter was 15 cm, the nominal frequency was 30 GHz, with aperture ratio $D/F = 1$. On the whole the dielectric contained about 8800 holes, and the number of Fresnel zones was seven. As the irradiator, a horn was used that provided for field amplitude of 14.5 dBi at the edge

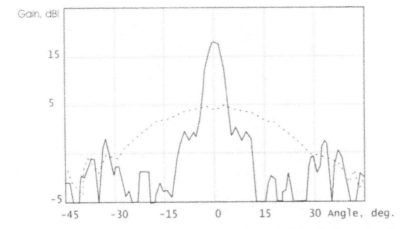

Figure 7.2. Beampattern of an antenna made of perforated dielectric (Reproduced with permission of Dr Aldo Petosa). The dotted curve traces the directivity pattern of the irradiating horn.

of the antenna's aperture. The results of measuring the beampattern are shown in figure 7.2.

The antenna gain was 28 dBi, the transmission band at 1 dB level was 14%, the aperture utilization coefficient was 35%, the beampattern width in the E and H planes was almost the same, about 4° at 30 dB cross polarization level.

It must be mentioned that using perforation for creating an artificial dielectric reduces the weight of the antenna in proportion to the mass of the removed dielectric.

Artificial dielectrics with perforated structures can find applications as coupling media for planar diffractive antennas. As we know, this reduces losses to reflection and also the level of side lobes. One example of such a configuration is shown in figure 7.3.

In this connection it is of interest to discuss the publication [38]. It gives the results of an experimental investigation of a perforated dielectric in the resonance region. The artificial dielectric is schematically shown in figure 7.4, which shows the characteristics of a dielectric prepared as a three-dimensional lattice. The parameters of the dielectric were $\varepsilon_1 = 1$, $\varepsilon_2 = 2.08$, $p = 5\,\text{mm}$, $2a = 2.5\,\text{mm}$. The lattice was made of Teflon-4 and contained

Figure 7.3. Diffractive antenna fabricated using printed circuit board technology, with a coupling layer.

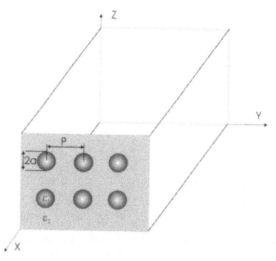

Figure 7.4. Schematic diagram of a dispersive man-made dielectric.

four rows of holes. An E-polarized plane wave (vector E pointing along the generatrix of a cylinder of radius a) is incident on this three-dimensional lattice (the wave vector points along the axis Z). Measurements were conducted in the wavelength band from 1.8 to 8 mm. The results of measurements are plotted in figure 7.5.

The quantity n_{eff} is defined as $n_{\text{eff}} = \sqrt{\varepsilon_{\text{eff}}\mu_{\text{eff}}}$, the effective loss angle tangent is defined as $\tan \delta_{\text{eff}} = (2n_{\text{eff}}\chi)/(n_{\text{eff}}^2 - \chi^2)$, $\chi = \alpha\lambda/4\pi$ and α is the wave attenuation in the medium.

Figure 7.5(a) shows that resonance dispersion n_{eff} is observed in the region of the narrowest resonance dip at $a/\lambda = 0.24$. The next area of noticeable dispersion is observed at $a/\lambda = 0.19$. When $a/\lambda > 0.5$, the value of n_{eff} varies very little. Furthermore, the results shown here imply that if a/λ is small, then n_{eff} is described quite adequately by the quasioptical formula [38],

$$n_{\text{eff}}^2 = Q_1\varepsilon_1 + (1 - Q_1)\varepsilon_2$$
$$\tan \delta_{\text{eff}} = \tan \delta_1 + \mu(\tan \delta_2 - \tan \delta_1)(\eta + \mu)$$
$$\mu = (1 - Q_1)/Q_1 \tag{7.3}$$
$$\eta = \varepsilon_1/\varepsilon_2$$

where Q_1 is the volume fraction of the component with dielectric permittivity ε_1. In the version of the artificial dielectric described above, equation (7.3) gives $n_{\text{eff}} = 1.36$.

The results presented indicate that a perforated dielectric has four typical frequency ranges:

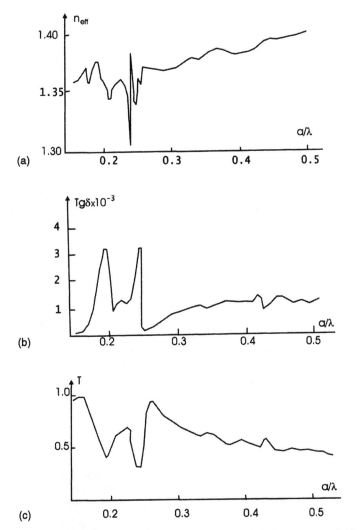

Figure 7.5. (a) Effective refractive index, (b) effective loss angle tangent and (c) transmission coefficient of the artificial dielectric as functions of parameter a/λ [38].

1. $a/\lambda < 0.16$. Here the artificial dielectric has practically constant value of n_{eff} and low loss angle tangent. This is a quasistatic region in which equation (7.3) works quite well.
2. In the region $0.16 < a/\lambda < 0.26$ we see typical peaks of $\tan \delta$ at $a/\lambda = 0.19$ and $a/\lambda = 0.24$ that correspond to resonance dispersion $n_{\text{eff}}(a/\lambda)$.
3. In the region $0.26 < a/\lambda < 0.55$ both parameters (n_{eff} and $\tan \delta$) increase monotonously.
4. In the region $a/\lambda > 0.55$ both these parameters are practically constant.

Note that segments with negative slope of $n_{eff}(a/\lambda)$ in range 2 can be used to broaden the coupling band by using a quarter-wavelength layer of artificial dielectric: the condition of quarter-wavelength $l = \lambda_0/4n_{eff}$ is satisfied at constant l in a wider range of frequencies if n_{eff} decreases with decreasing λ_0.

By varying the ratios $\varepsilon_1/\varepsilon_2$ and a/p of the artificial dielectric in this structure, it is possible to shift ranges 1–4 to change the values of n_{eff} and the slope of their frequency characteristics.

7.4 Metal–dielectrics

A man-made dielectric incorporating a metal ribbon structure was used to build microwave diffractive elements (figure 7.6) [26]. The equivalent refractive index depends on the distance between ribbons:

$$n = [1 - (\lambda/2a)^2]^{1/2}.$$

For $a = 18\,\text{mm}$ and $\lambda = 3.2\,\text{cm}$, the refractive index $n = 0.46$.

This principle of creating artificial dielectrics was applied in [27] to designing a metal-plates Fresnel lens for receiving TV signals from satellites. The antenna size was $3.66\,\text{m} \times 3.66\,\text{m}$. The irradiator was a rectangular pyramidal horn with the following parameters: aperture 2.78×1.95 wavelengths, focal distance of the antenna 4.4 m, distance between 8.1 mm thick aluminium plates 4.69 cm, wavelength 7.5 cm, antenna mass 173 kg. In order to receive signals from the Westar V satellite, the axis of the antenna was pointed at this satellite; if the irradiator was deflected by 9°, the antenna was capable of receiving vertical polarization signals from the Satcom F3 satellite. The beam width was about 1.6°. The efficiency of

Figure 7.6. Microwave lens made of artificial dielectric [26].

Figure 7.7. Microwave lens made of artificial dielectric [28].

using the surface of the Fresnel multiple-plate lens in the 500 MHz waveband was 0.3–0.5. The stability under wind loads was mentioned as one of the advantages of this antenna; to minimize icing damage, a protecting plastic lid was suggested.

A so-called focusing pseudo-lens with variable gradient of refractive index was discussed in [28] (see figure 7.7). In this case the artificial metal–dielectric was a waveguide structure.

The following relation gives the equivalent refractive index of the waveguide structure as a function of the radius r of the waveguide cross section:

$$n = c/v = [1 - (\lambda/3.42R)^2]^{1/2}.$$

The refractive index of the waveguide structure in the converging pseudo-lens changes in the radial direction from 0.95 for $R = 30$ mm (radius of the central waveguide cross section) to 0.37 for $R = 10$ mm. At the wavelength $\lambda = 3.2$ cm the thickness of the pseudo-lens was 5.3 cm.

Artificial dielectrics based on liquids in which finely dispersed metal particles were suspended are described in [32–34]. For the millimetre wavelength band, nonane, hexane, heptane etc. can be used as suspending liquids. The shape of the suspended metal particles can be described as elongated ellipsoids. If no external stimuli were present, the particles were randomly distributed and the effective dielectric permittivity of this dielectric was very nearly 2. The required alignment of metal particles is achieved by applying an external electric field. If most of the particles are aligned with their greater axis in the direction of the electric vector of the microwave field, the effective dielectric permittivity increases. The dielectric permittivity of suspensions can also be changed by ultrasonic irradiation.

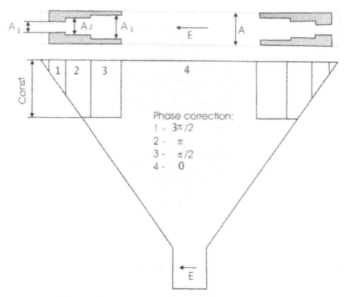

Figure 7.8. Diffractive lens in sectoral horn (reproduced with permission of Dr Donald Black).

Two types of diffractive elements incorporating metal–dielectrics were discussed by Black and Nguyen [39]. The design of the corresponding diffractive elements was based on the dependence of the equivalent refractive index on the distance between metal ribbons. Below we give two versions of waveguide diffractive element design.

Figure 7.8 shows a zoned lens in sectoral horn [39]. The aperture of the sectoral horn was $9.13\lambda_0$ at $0.852\lambda_0$, where λ_0 is the wavelength in free space. The focal distance was $F = 6.84\lambda_0$, that is, $D/F = 1.33$. Overall dimensions of phase-correcting zones are listed in table 7.7.

Figure 7.9 shows the structure of the metal zone plate for the sectoral horn [39] and the measured directivity pattern [40].

The transmission band at $-3\,\text{dB}$ level was about 9.6%, with the reflection coefficient $-15\,\text{dB}$; the beam width at the $-3\,\text{dB}$ level was approximately $7°$.

Table 7.7.

Phase shift (degrees)	Width (λ_0)	Height (λ_0)
90	0.884	0.192
180	0.708	0.276
270	0.615	0.319

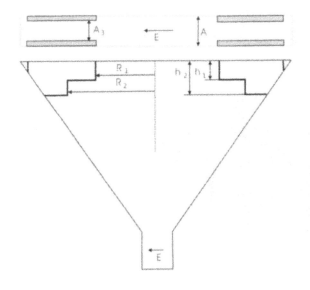

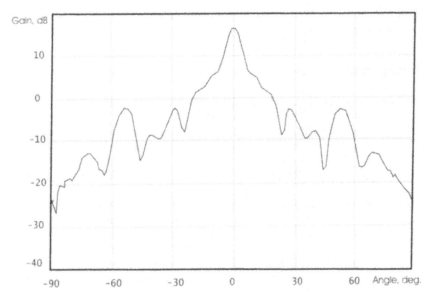

Figure 7.9. Zone plate for sectoral horn (reproduced with permission of Dr Donald Black).

As a comparison, figure 7.10 gives the beampattern of a sectoral horn with a dielectric lens [40]. The deterioration of directivity characteristics in this type of antenna may be caused by the focal point of the zone plate not coinciding with the phase centre of the horn, but also by the multimode composition of radiation in the cross section of the zone plate.

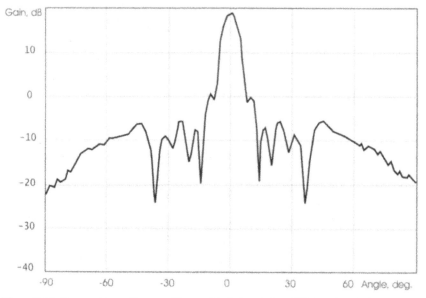

Figure 7.10. Beampattern of sectoral horn with dielectric lens [40] (reprinted with permission of Dr Donald Black).

Bibliography

[1] Valitov R A and Makarenko B I (eds) 1984 *Measurements in Millimeter and Sub-millimeter Waves* (Moscow: Soviet Radio)

[2] Sattler P and Simonis G J 1983 'Dielectric properties of beryllia in the near-millimeter wavelength range' in *8th Int. Conf. Infrared and Millimeter Waves, Miami Beach, FL*, 12–17 December; 1983 *Conf. Digest*, New York, M6.9/1

[3] Sarafanov V I 1955 'Measurements in dielectrics with millimeter waves' *UFN* **55**(1) 127–131

[4] Vinogradova E A, Dianov E M and Irisova N A A 1965 'Michelson interferometer for measuring refractive index of dielectric materials in the 2-mm wavelength band' *Radiotehnika i Electronika* **10**(10) 1804–1808

[5] Al'tshuler Yu G, Kats L I and Revzin R M 1970 'Measurement of refractive index of dielectrics in sub-millimeter wavelength band' *PTE* **6** 145–146

[6] Lubyako L V 1971 'Investigation of electric and magnetic properties of materials in the millimeter and sub-millimeter wavelength ranges' *Izvestiya Vuzov, Radiofizika* **14**(1) 133–137

[7] Zhuk M S and Molochkov Yu B 1973 *Designing Wide-Range Scanning Lens Antenna and Feeder Systems* (Moscow: Energiya)

[8] Meriakri V V and Chigryai E E 1976 'Measurements of absorption index in the sub-millimeter wavelength band' *PTE* **1** 210–212

[9] Afsar M N, Chaberlen J and Chantry G W 1976 'High-precision dielectric measurements on liquids and solids at millimeter and submillimeter wavelengths' *IEEE Trans. Instrumentation and Measurement* **25**(4) 291–294

[10] G W Chantry and J W Fleming 1971 'Far infrared and millimetre wave absorption spectra of some low-loss polymers' *Chem. Phys. Lett.* **10**(4) 473–477

[11] Igoshin F F, Kir'yanov A P *et al* 1974 'Measurements of refractive index of some dielectrics in the millimeter wavelength band' *Izv. Vuzov, Radiofizika* **17**(2) 291–293

[12] Kats L I and Traitel'man L A 1963 'Bridge interferometer used for dielectrics measurements in the mm wavelength band' *Izv. Vuzov, Radiotehnika* **6**(2) 143–147

[13] Mon K K and Sievers A J 1975 'Plexiglas: a transmission filter for the FIR spectral region' *Appl. Opt.* **14**(5) 1054–1055

[14] Igoshin F F, Kir'yanov A P, Mozhaev V V *et al* 1973 'Interference method of measuring optical constants of some semiconductors and dielectrics in far infrared spectral band' *Trudy Mosk, fiz-tehn. in-ta, ser. Obschaya i molek. fiz., Moscow* pp 23–29

[15] Dryagin Yu A and Chuhvichev A N 1969 'Resonance measurements of parameters of solid dielectrics in the shortwave part of the millimeter spectral band' *Izv. Vuzov, Radiofizika* **12**(8) 1245–1248

[16] Konev V A, Lyubetskii N V and Tihonovich S A 1989 *Radiowave Ellipsometry of Dielectric Structures* (Minsk: Nauka i Tehnika)

[17] Gurtovik I G and Sportsmen V N 1987 *Glassfiber Plastics of Radiowave Applications* (Moscow: Himiya)

[18] A A Moiseeva (ed) 1960 *Styrophomes* (Moscow: Oborongiz)

[19] Bazarova F F 1974 *Organicheskie i Neogranicheskie Polimery v Konstruktsiyah REA* (Moscow: Soviet Radio)

[20] Anisimova T I, Dem'yanov V V, Ushatkin E F *et al* 1987 'Frequency absorption spectra of some materials in the microwave and sub-mm wavelength ranges' in *Scientific Foundations of Materials Science* (Moscow: Nauka) pp 173–177 (in Russian)

[21] 'Measuring dielectric characteristics in the mm wavelength band' 1985 *TIIER* **73**(1)

[22] Meriakri V V *et al* 1976 'Weakly absorbing materials for sub-mm wavelength band applications' in *USSR Symposium on Instruments, Techniques and Propagation of mm and Sub-mm Electromagnetic Waves in the Atmosphere* (IRE AN SSSR) pp 149–151

[23] Pasechnik V F and Skurlov V M 1974 'Dielectric permittivity measurements of sheet dielectrics in the mm wavelength band' *Radiotehnika* **19** 157–160

[24] Meriakri V V 1988 'Weakly absorbing materials for mm and sub-mm wavelength ranges' in *Proc. Har'kov IRE AN USSR* pp 144–151

[25] Konkina P A (ed) 1969 *Foundations of Electrophysics Engineering, Part 1. Foundations of Technical Electrodynamics* (Moscow: Vysshaya Shkola)

[26] Molotkov N Ya 1988 'Study of focusing in the university course of optics' Deposition No. 4701-V88, Tomsk; Editorial Board of 'Izvestiya vuzov MV i SSO SSSR', seriya 'Fizika'

[27] Shuter W L H, Chan C P, Li E W P and Yeung A K C A 1984 'Metal plate Fresnel lens for 4 GHz satellite TV reception' *IEEE Trans. Antennas and Propagation* **32**(3) 306–307

[28] Molotkov N Ya 1987 'Study of focusing and scattering systems with variable refractive index in the university course of optics' Deposition No. 114-B87, Tomsk; Board of 'Izvestiya vuzov MV i SSO SSSR', seriya 'Fizika'

[29] Goncharov Yu G, Kozlov G V, Lugovskii V V *et al* 1988 'Study of ceramic materials with high dielectric permittivity in the millimeter wavelength band' *Elektronnaya Tehnika Ser. Elektronika SVCh* **9**(423) 68–69

[30] Meriakri V V and Murmuzhev B A 1990 'Ferrites for the millimeter and sub-millimeter wavelength band' in *Magnetoelectric Materials* (Moscow: Nauka) (ed Yu N Venevtsev and Â N Lyubimov, pp 143–151 (in Russian)

[31] Levin L 1981 *Theory of Waveguides* (Moscow: Radio i Svyaz') (in Russian)

[32] Buscher H T 1979 'Electrically controllable liquid artificial dielectric media' *IEEE Trans.* **MMT-27**(5) 540–545

[33] US Patent No. 3805197

[34] US Patent No. 3631501

[35] Petosa A and Ittipiboon A 2002 'A Fresnel lens designed using a perforated dielectric' in *Proc. 'Symposium on Antenna Technology and Applied Electromagnetics, ANTEM 2002', Canada, Montreal*, 31 July–2 August, pp 399–402

[36] Skellett A M 1951 'Dielectric lens' US Patent 2,547,416, 3 April

[37] Wiltse J C 1985 'The Fresnel zone-plate lens' *Proc. SPIE Symp., Arlington, VA*, vol 544, pp 41–47, 9–10 April

[38] Meriakri V V and Nikitin I P 1989 'Man-made dielectric with dispersion in the millimeter wavelength band. Quasioptical devices in millimeter and sub-millimeter wavelength ranges' in *Proceedings of IRE, Har'kov*, pp 65–70

[39] Black D and Nguyen C 1999 'Metallic zone plates for sectoral horns' *IEEE Int. Antennas and Propagation Symp., Orlando, FL, Symp. Digest*, vol 2, 11–16 July, pp 730–733

[40] Donald Black, Private communication, 2002

Additional bibliography

Colburn J S and Rahmat-Samii Y 1999 'Patch antennas on externally perforated high dielectric constant substrates' *IEEE Trans. Antennas and Propagation* **47**(12) 1785–1794

Du L J and Scheer D J 1976 'Microwave lens design for a conical horn antenna' *Microwave J.* September 99–102

Gauthier G P, Courtay A and Rebeiz G M 1997 'Microstrip antennas on synthesized low dielectric-constant substrates' *IEEE Trans. Antennas and Propagation* **45**(8) 1310–1314

Griffiths H D, Vernon A M and Milne K 1989 'Planar phase-shifting structures for steerable DBS antennas' in *ICAP 89, Conf. Proc., University of Warwick, UK*, 4–7 April, Part 1, pp 45–49

Lee C K, R J Langley and E A Parker 1985 'Single-layer multi-band frequency-selective surface' *IEEE Proc. H* **132**(6) 411–412

Minin I V and Minin O V 1999 'Antennas of mm-range based on the quasioptical diffraction elements for the commumication systems' in *50th Vehicular Technology Conference VTC 1999—Fall, Amsterdam, The Netherlands*, 19–22 September, pp 3040–3046

Muldavin J N and Rebeiz G M 1999 'Millimeter-wave tapered-slot antennas on synthesized low permittivity substrates' *IEEE Trans. Antennas and Propagation* **47**(8) 1276–1280

Noll J and Edenhofer P 1991 'Focusing effects of quasiperiodically distributed frequency selective surfaces' in *7th Int. Conf. on Antennas and Propagation, University of York, UK*, 5–18 April, pp 632–635

Parker E A and Vardaxoglou J C 1985 'Plane-wave illumination of concentric ring frequency selective surface' *IEE Proc. H* **132**(3) 176–180

Vaughan M J, Hur K Y and Compton R C 1994 'Improvement of microstrip patch antenna radiation patterns' *IEEE Trans. Antennas and Propagation* **42**(6) 882–885

Appendix

The electric field strength in the Kirchhoff–Kottler approximation is found using the expression

$$
\mathbf{E}(\mathbf{R}) = \frac{1}{4\pi} \oint_L G[\mathbf{E}_0, \boldsymbol{\tau}] \, dL + \frac{1}{4\pi} \iint_{S_0} \left(G \frac{\partial \mathbf{E}_0}{\partial n} - \mathbf{E}_0 \frac{\partial G}{\partial n} \right) d\mathbf{S}
$$

$$
+ \frac{1}{4\pi k} \oint_L (\mathbf{H}, \boldsymbol{\tau}) \operatorname{grad} G \, dL \tag{A1.1}
$$

where $\mathbf{R}$ is the radius vector of the observation point, $\mathbf{n}$ is the outside normal to the screen S_0, $G(r)$ is a Green's function, $\mathbf{H}$ is the magnetic field strength, $\mathbf{E}_0$ and $\partial \mathbf{E}_0/\partial n$ correspond to the unperturbed incident electromagnetic wave, and $\boldsymbol{\tau}$ is the unit vector tangent to the contour L of the stop opening. The first and third integrals are taken over the aperture contour and the second is taken over the area of the aperture opening.

A stop with a round hole is a particular case of a flat screen for which we know the corresponding Green's functions,

$$
G_1 = \frac{\exp(ikr)}{r} - \frac{\exp(ikr_1)}{r_1}, \qquad G_2 = \frac{\exp(ikr)}{r} + \frac{\exp(ikr_1)}{r_1} \tag{A1.2}
$$

where $[\exp(ikr_1)]/r_1$ is a function of a pointlike source placed at a point which is a mirror image of the observation point in the plane of the stop. Using (A1.2) in (A1.1), we obtain

$$
\mathbf{E}_1(\mathbf{R}) = \frac{1}{4\pi} \iint_{S_0} \mathbf{E}_0 \frac{\partial G_2}{\partial n} \, dS + \frac{i}{4\pi} \oint_L (\mathbf{H}, \boldsymbol{\tau}) \operatorname{grad} G_2 \, dL
$$

$$
\mathbf{E}_2(\mathbf{R}) = \frac{1}{4\pi} \oint_L G_2[\mathbf{E}_0, \boldsymbol{\tau}] \, dL + \frac{1}{4\pi} \iint_{S_0} G_2 \frac{\partial \mathbf{E}_0}{\partial n} \, dS \tag{A1.3}
$$

$$
+ \frac{i}{4\pi k} \oint_L (\mathbf{H}, \boldsymbol{\tau}) \operatorname{grad} G_2 \, dL.
$$

Expression (A1.3) makes possible finding a solution of the diffraction problem for arbitrarily large diffraction angles and arbitrary distances from the stop, and works also in the near field. The Maggi–Rubinowicz transformation [10, 12] is used for transforming the integrals over the area of the aperture to integrals over the contour of the hole. In the case of normal incidence of plane wave $\exp(ikr)$ on to a round aperture, we have the following expression for the complex wave amplitude in cylindrical coordinates,

$$E(z, r) = \chi \exp(ikz) - \frac{a}{4\pi} \int_0^{2\pi} \frac{(a - r \cos \varphi) \exp(ik\rho)}{\rho(\rho - z)} \, d\varphi \qquad (A1.4)$$

where $\delta = 0$ in the region of geometric shadow, $\delta = 1$ in the geometrically 'illuminated' region, and ρ is a distance from the point on the aperture contour to the point (z, r), and a is the aperture radius. The first term in (A1.4) corresponds to the non-perturbed incident field and is formally generated by removing the singularity at the shadow border. If we use G_1 and G_2, the corresponding formulas are obtained by rewriting (A1.4) for a point that is a mirror image of point (z, r) in the shadow border, by adding up or subtracting these two expressions for $\delta = 0$:

$$E_1(z, r) = \delta \exp(ikz) - \frac{az}{4\pi} \int_0^{2\pi} \frac{\exp(ik\rho)(a - r \cos \varphi)}{\rho(\rho - z)} \, d\varphi$$

$$E_2(z, r) = \delta \exp(ikz) - \frac{a}{4\pi} \int_0^{2\pi} \frac{\exp(ik\rho)(a - r \cos \varphi)}{\rho^2 - z^2} \, d\varphi \qquad (A1.5)$$

$$\rho = (z^2 + a^2)^{1/2}.$$

Pyatakhin and Suchkov [10] obtained the complex field amplitude for the diffraction of a converging spherical wave on a round aperture, using G_1:

$$E_{1c}(z, R) = \pm \delta \frac{\exp(\mp ikr_1)}{r_1} + \frac{a \exp(-ikr_0)}{4\pi r_0} \int_0^{2\pi} \frac{\exp(ikr)}{r}$$

$$\times \left\{ \frac{a(z - f) + R \cos \varphi}{r_0 r - a^2 + aR \cos \varphi - zf} + \frac{a(z + f) - fR \cos \varphi}{r_0 r - a^2 + aR \cos \varphi + zf} \right\} d\varphi.$$

$$(A1.6)$$

where the upper sign in the first term is taken before, and the lower one after, the passage through the focal point; r_0 and r are the distances from the point on the aperture contour to the focal point and to the observation point, respectively, r_1 is the distance from the focal point to the observation point, and f is the distance from the stop to the focal point. The expression for $E_{2c}(z, R)$ obtained by using G_2 differs from $E_{1c}(z, R)$ in that $\text{sgn}(+)$ is replaced by $\text{sgn}(-)$ within the expression in braces. The field on the axis of the aperture is (A1.6):

$$E_{1c}(z,0) = u_0 + \frac{a^2 \exp[ik((a^2 + z^2)^{1/2} - (a^2 + f^2)^{1/2})]}{2[(a^2 + z^2)(a^2 + f^2)]^{1/2}}$$

$$\times \left\{ \frac{z - f}{\{[(a^2 + z^2)(a^2 + f^2)]^{1/2} - a^2 - zf\}^{1/2}} \right.$$

$$\left. + \frac{z + f}{\{[(a^2 + z^2)(a^2 + f^2)]^{1/2} - a^2 - zf\}^{1/2}} \right\}. \qquad \text{(A1.7)}$$

For $z = 0$ the expression (A1.7) gives a correct result: the non-perturbed field incident on the stop. If the focusing and diffraction angles are small, $z^2 \gg a^2$, $f^2 \gg a^2$, we obtain from (A1.7) the expression $E_{1c}(z,0) = u_0[1 - \exp(1/2ika^2(1/z - 1/f))]$, which coincides with the Fresnel approximation.

Pyatakhin and Suchkov [10] obtained expressions for the projections of the field on to the coordinate axes in the Kirchhoff–Kottler approximation. They singled out the dependence on the azimuthal observation angle Φ (the incident field is assumed polarized along the axis x and the stop lies in the plane $z = 0$). In cylindrical coordinates they obtained for a Green's function G_1. Scalar approximation results for the vector Kirchhoff approximation depend on the choice of Green's function. The scalar Kirchhoff approximation does not allow us to calculate the wave depolarization due to its diffraction on a stop. The vector Kirchhoff approximation using G_2 for small diffraction angles, including the field on the aperture axis, is in good agreement with the results of the physical theory of diffraction at arbitrary distances from the aperture and for arbitrary observation angles Φ.

In all the enumerated cases with small diffraction angles and for different distances from the plane of the stop, E_x exceeds E_y and E_z by approximately two orders of magnitude. The components E_y and E_z are in a complicated oscillating dependence on R/D (R is the distance from the stop and D is its characteristic size); the amplitude of the E_z for the observation angle $\Phi = 0°$ is approximately twice as large as for E_y for $\Phi = 45°$ and correspondingly almost by a factor of 1.5 greater than when both are for the angle $\Phi = 45°$. The angular dependences of these two components look almost identical in the vector Kirchhoff approximation with G_2 and in the physical theory of diffraction: $E \approx \sin 2\Phi$, $E \approx \cos \Phi$, respectively. As R/D increases, E_x at $\Phi = 45°$ approaches E_y and E_z and becomes of the same order with them at diffraction angles of greater than $45°$, which means that stray depolarization in this region is considerable.

$E_{x1}(z, R, \Phi) = E_1$, which is identical to the solution of the scalar problem

$$E_{y1}(z, R, \Phi) = 0$$

$$E_{z1}(z, R, \Phi) = \frac{az}{\pi i k} \cos \Phi \int_0^\pi \frac{\cos \varphi}{r^2} \exp(ikr)(1/r - ik) \, d\varphi \qquad \text{(A1.8)}$$

where $r = (z^2 + R^2 + a^2 - 2Ra \cos \varphi)^{1/2}$. We see that the transverse character of the field is perturbed and the cylindrical symmetry disappears: the field component along the axis Z depends on the observation angle Φ. The field gains no cross-polarization component E_y. For the second Green's function G_2 [10, 11] we have

$$E_{x2}(z, R, \Phi) = E_2 + \frac{a}{2\pi i k} \int_0^\pi (R \cos \varphi - a) \frac{\exp(ikr)}{r^2} (1/r - ik) \, d\varphi$$

$$+ \frac{a}{2\pi i k} \cos 2\Phi \int_0^\pi (R \cos \varphi - a \cos 2\varphi) \frac{\exp(ikr)}{r^2} (1/r - ik) \, d\varphi$$

$$\text{(A1.9)}$$

$$E_{y2}(z, R, \Phi) = \frac{a}{2\pi i k} \sin 2\Phi \int_0^\pi (R \cos \varphi - a \cos 2\varphi) \frac{\exp(ikr)}{r^2} (1/r - ik) \, d\varphi$$

$$E_{z2}(z, R, \Phi) = \frac{-a}{\pi} \cos \Phi \int_0^\pi \cos \varphi \frac{\exp(ikr)}{r^2} \, d\rho.$$

Contrary to the case when G_1 is used, here we find a nonzero field component along the axis y. Now each of the three projections of the field of the diffracted waves onto the coordinate axes depends on the angle Φ. Note a curious fact that in the Kirchhoff approximation with G_2 the field on the axes is different in the scalar and the vector cases:

$$E_{z1}(z, 0, \Phi) = E_{y2}(z, 0, \Phi) = E_{z2}(z, 0, \Phi) = 0$$

$$|\mathbf{E}_2(z, 0, \Phi)| = \exp(ikz) - \exp(ik\rho) + \frac{1}{2}\left(\frac{a}{\rho}\right)^2 \exp(ik\rho)$$

$$- \frac{1}{2}\left(\frac{a}{\rho}\right)^2 - \frac{1}{2}\left(\frac{a}{\rho}\right)^2 \frac{\exp(ik\rho)}{ik\rho}.$$

The results of calculations in the scalar Kirchhoff approximations at small distances from the stop greatly depend on the Green's function used. The Kirchhoff approximation with $G = \exp(ikr)/r$ is more reliable for calculating the field on the axis near the stop opening; the best results in the polarization plane of the original wave $\Phi = 0°$ are found with the function G_1, and for $\Phi = 90°$ with G_2. A similar correlation is observed at large distances from the stop. An analysis of results in the vector case demonstrates that if the stop is ideally conducting, than the Kirchhoff–Kottler approximation using G_2 provides sufficient accuracy of determining the field component

E_{x1} in the original polarization plane at arbitrary distances from the plane of the stop. It is especially important that the approximation also works well for arbitrary observation angles, correctly reproducing the angular dependence of the transmitted field in the near-field zone. The vector Kirchhoff approximation makes it possible to determine the component E_z of the diffracted field with sufficient accuracy and reliability in a wide range of diffraction angles and distances from the stop, in the cases of both G_2 and G_1.

Index

Aberration 118
Achromatic diffraction element 117
Aperture diaphragm 257
Application diffraction elements 252,
    262, 282, 287, 294, 302, 319, 328,
    336, 352
Artificial dielectric 376
Axicon 166, 167, 168

Bandwidth 61

Classification of zone plates 32
Conformal antenna 188
Conical diffractive element 110, 296
Convex diffractive element 37
Cylindrical zone plate 268

Diffraction 1, 10, 13
Diffraction antenna 23, 87, 192, 203, 205,
    216, 230, 268, 320
Diffraction element 20, 89, 123, 142, 146,
    172, 176, 257
Diffraction lens 64, 103
Diffraction objective lens 65, 107
Diffraction radome 218, 244, 248
Diffraction reflector 223, 274

Efficiency 68, 183

Fast Fourier transform 18
Femtosecond pulses 195
Field of view 53, 89, 104, 115, 130, 132,
    135, 138, 150
Flat antennas 207
Focusing shock waves 337, 350

Fraunhofer zone 14
Fresnel–Kirchhoff diffraction integral 4,
    8, 9, 11, 124, 160
Fresnel zone 6, 87
Fresnel zone lens 33, 44
Fresnel zone plate 29, 37

Harmonics 92, 96
Huygens–Fresnel principle 2, 4
Huygens–Kirchhoff integral 14

Information properties 50, 148
Invariant properties 108

Kirchhoff–Kotler approximation 390

Lens-type diffraction antenna 216
Logarithmic axicon 166

Metal-dielectric 382
Microwave antenna 193, 203
Microwave lens 382, 383
Multifocal 98
Multilayer flat antenna 203

Non-flat surface 100, 118, 223

Off-axis zone plate 76, 124
Omnidirectional antennas 263, 270, 272
Optical constants 369
Optic gate 300
Optimization 22, 179, 181, 185

Phase profile 39, 43, 83, 96, 113
Polarization diffractive element 121

Q-factor 63

Radiovision systems 302, 317
Rayleigh resolving 29, 50

Segmented apertures 355
Semiconductor zone plate 334
Soret zone plate 30
Spectrum range 59
Square zone plate 128
Synthesis 82, 159, 162

Theory of diffraction 1
Three-dimensional diffractive antennas
    325

Two-order diffractive elements
    172
Two-spectrum diffractive lens
    329

Wood zone plates 34

X-ray diffractive optics 282

Zone plate on parabolic surface 101, 104,
    107, 218, 241
Zone plate on a spherical surface 22, 101,
    240, 249
Zone plate on a conical surface 242

Milton Keynes UK
Ingram Content Group UK Ltd.
UKHW021827071024
449327UK00021B/1458